George L. Barron, Ph.D.

Associate Professor
Department of Botany
Ontario Agricultural College
University of Guelph
Ontario, Canada

# of Hyphomycetes from Soil

ROBERT E. KRIEGER PUBLISHING COMPANY
MALABAR, FLORIDA

Originally published, 1968
Reprint 1971, 1977, 1983



Printed and Published by
ROBERT E. KRIEGER PUBLISHING CO., INC.
Krieger Drive
Malabar, Florida 32950



*Reprinted by arrangement*

Library of Congress Catalog Card Number: 68-14275

*Printed in U.S.A. by*
NOBLE OFFSET PRINTERS, INC.
NEW YORK 3, N. Y.

# The Genera of Hyphomycetes from Soil

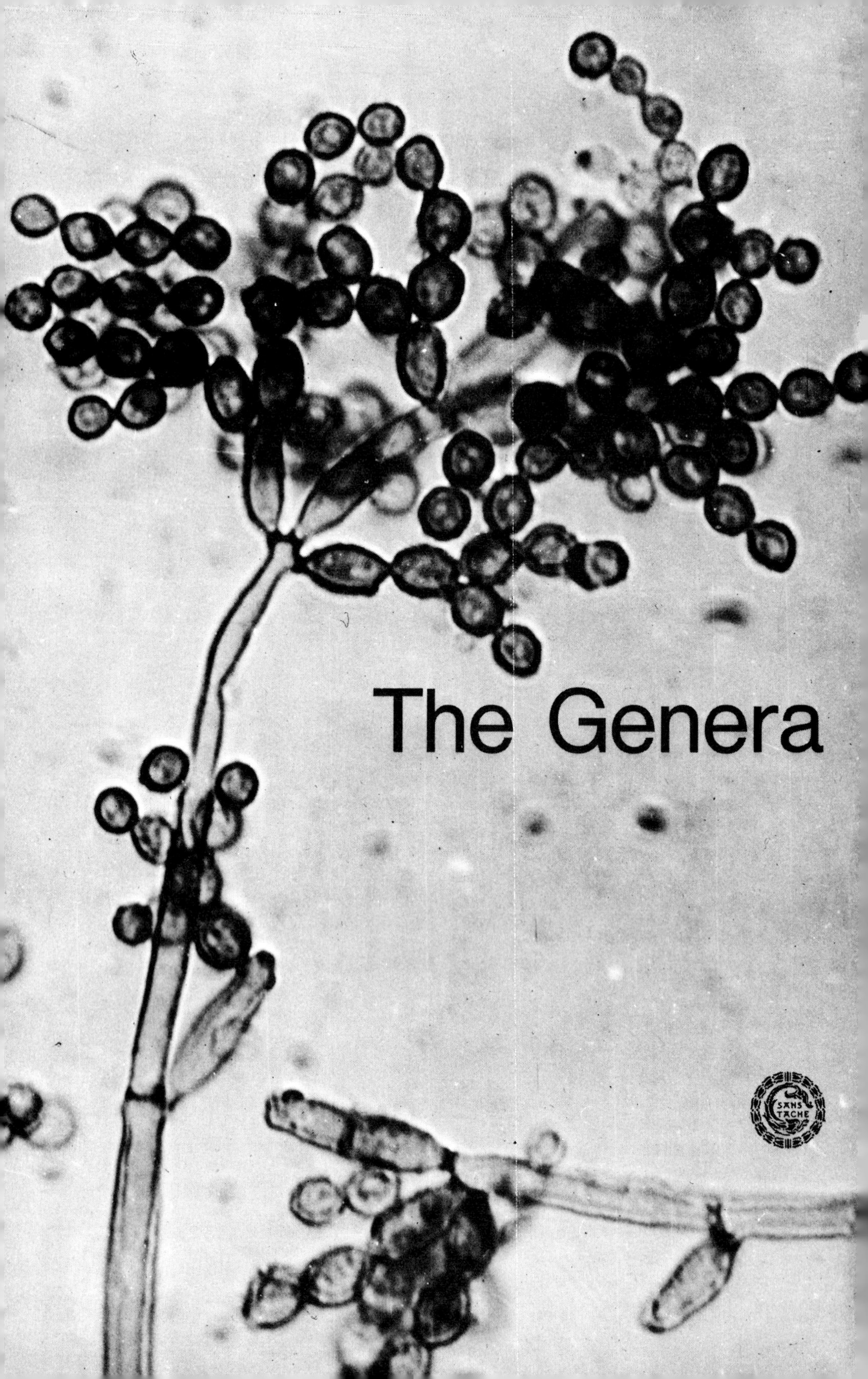

# The Genera

DEDICATED TO THE MEMORY
OF
JOSEPH C. GILMAN

# Preface

This text deals with Hyphomycetes from soil. No pretence is made, however, that the fungi recorded herein inhabit the soil as their natural ecological habitat. Some are unquestionably soil fungi in the most restricted sense and play an important role in the breakdown of organic debris. Others are equally obviously transients, transported by wind, water or other agent to an essentially foreign or casual habitat. Some are possibly facultative parasites, rain-washed from their hosts or released by slow decay, and may grow and thrive or linger for a while, to disappear in competition with the more vigorous indigenous saprophytes. Others are perhaps litter fungi or predators, lying in wait to snare the unsuspecting nematode or rotifer, or other minute soil animals. About many, we have no knowledge.

Members of the Melanconiales which have been recovered from soil have also been included in this discussion. There are two reasons for this. First, many fungi which produce an acervulus on their natural substratum fail to do so in culture. Second, it is very difficult for the beginner, or even the expert, to make the distinction between an acervulus (Melanconiales) and a sporodochium (Moniliales) in culture.

A number of taxonomists have expressed dissatisfaction with the classical concepts of Saccardo in the classification of the Hyphomycetes and in recent years there have been several attempts to present modern arrangements of this group based on more fundamental taxonomic criteria related particularly to the nature of the conidiophore or sporogenous cell and the precise method by which spores are produced. These classifications, presented by Hughes, Subramanian, Tubaki, Luttrell and others, are essentially experimental in nature and to date no comprehensive treatment of the Hyphomycetes is available.

The outstanding contribution in this area is that of Hughes (1953), who divided the Hyphomycetes into eight major sections, one of which involved two subsections. The classification adopted in this text is essentially that of Hughes, with minor modifications which involve the elevation of Hughes' subsections to secton rank and the splitting of one section into two. Excluding the *Mycelia Sterilia*, therefore, I have recognized ten sections within the Hyphomycetes recovered from soil. Reference to these sections by Roman numerals as proposed by Hughes, while perfectly adequate for an experimental classification, is not satisfactory for a general treatment at the applied level. I have therefore followed Tubaki and named the sections on the basis of the spore type, *e.g.* Blastosporae, Phialosporae, etc.

From my own experience, I am extremely sympathetic with the difficulties which students of the new system will encounter. In the first instance, it is much more difficult to apply the modern taxonomic concepts in Hyphomycetes to the practical problems of identification of unknown fungi. In the long run, however, the modern system is infinitely more rewarding. Its use is mandatory for serious students of the group and, from the teacher's standpoint, a student obliged to evaluate critically even a single member of each of Hughes' sections will have a far greater understanding of the Hyphomycetes as a group than if he had examined several times as many fungi by using the superficial criteria of Saccardo.

The organization of the text is relatively straightforward, with the bulk of the material consisting of taxonomic appraisals of genera which are arranged in alphabetical order. A descriptive paragraph is given in which the salient features for each genus are pointed out. The discerning reader will quickly be aware that these descriptions are not always the diagnoses of the original authors of the various genera. In many cases, genera have not been redefined since their original descriptions in the classical literature. Such descriptions are frequently very brief, consisting of one or a few lines, and do not adequately describe the genus in the light of either our revised criteria for diagnosis or our present understanding of the limits of the particular genus concerned. I have presented, therefore, more or less standardized descriptions circumscribing what might be regarded as the current concept of each genus. There is the probability that some of these concepts will prove to be erroneous when we have a chance to examine the type specimens on which these genera were based.

Quite often generic descriptions are lengthy and contain a wealth of information which is supportive rather than critical. I have therefore taken from each generic description one or two characters which are considered the most important for identification and stressed these under the heading "Diagnostic Features." It must be appreciated that these abbreviated diagnoses should be used with discretion.

Descriptions are accompanied by photographs or line drawings illustrating a typical or common species of the genus. More information and detail can be given with line drawings; nevertheless, I have favored photographs wherever possible, as they give a degree of naturalness and authenticity that a drawing cannot match. With few exceptions, all illustrations were prepared by myself and acknowledgment is given for borrowed material. Following the descriptions and illustrations of each genus are notes which include pertinent information on the genus, particularly with respect to further information on identification at the species level. Unfortunately, monographic treatments (or indeed any treatments) are not available for a number of the genera considered. This is compensated to some extent by the fact that a number of the genera are monotypic.

The purpose of this book is to aid the nonspecialist in identifying genera of Hyphomycetes recovered from soil. It is hoped that the text will be of value, not only to soil microbiologists and plant pathologists concerned with the soil environment, but also to anyone interested in fungi and the decay of organic products, in which the Hyphomycetes play a significant role.

# Acknowledgments

An important part of the text is the illustrations of the various genera. In the preparation of these illustrations, it was essential that I study good culture material, herbarium specimens or slides. This would not have been possible without the generous donations of cultures and specimens from many friends and colleagues. I would like to thank the following individuals for their cooperation and the time and trouble they took on my behalf.

Dr. H. L. Barnett, University of West Virginia, Morgantown, West Virginia; Mr. G. C. Bhatt, University of Waterloo, Ontario; Dr. C. Booth, Commonwealth Mycological Institute, Kew, England; Dr. G. E. Bunschoten, Centraalbureau voor Schimmelcultures, Baarn, Holland; Dr. J. W. Carmichael, University of Alberta Mould Herbarium, Edmonton, Alberta; Dr. W. Bridge Cooke, Robert A. Taft Sanitary Engineering Centre, Cinncinnati, Ohio; Dr. Martha Christensen, University of Wyoming, Laramie, Wyoming; Dr. C. Dickinson, University of Dublin, Ireland; Dr. C. L. Fergus, Pennsylvania State University, University Park, Pennsylvania; Dr. J. C. Frankland, Nature Conservancy, Merlewood Research Station, Grange-over-Sands, Lancs, England; Dr. R. D. Goos, American Type Culture Collection, Rockville, Maryland; Dr. R. H. Haskins, Prairie Regional Laboratory, Saskatoon, Saskatchewan; Dr. L. F. Johnson, University of Tennessee, Knoxville, Tennessee; Dr. L. R. Kneebone, Pennsylvania State University, University Park, Pennsylvania; Dr. Aedine Mangan, Department of Pathology, Bristol Royal Hospital, England; Dr. E. F. Morris, Western Illinois University, Macomb, Illinois; Dr. G. F. Orr, Dugway Proving Ground, Utah; Dr. P. A. Orpurt, Manchester College, North Manchester, Indiana; Dr. J. A. Parmelee, Plant Research Institute, C.D.A., Ottawa, Canada; Dr. G. J. F. Pugh, University of Nottingham, England; Professor R. Smith, University of Guelph, Ontario; Dr. M. I.

Timonin, C.D.A., Forest Pathology Laboratory, Saskatoon, Saskatchewan.

I am especially grateful to John Elphick of the Commonwealth Mycological Institute who freely made available to me the culture collection in his charge.

The National Research Council of Canada and the Department of University Affairs of the Province of Ontario have for many years supported the research on soil fungi which provided the background and stimulation to write this book. I am pleased to acknowledge their help.

In 1965 I had the good fortune to spend several months at the Commonwealth Mycological Institute, Kew, England. I would like to thank the Nuffield Foundation, Regents Park, London, for their generosity in granting me the fellowship which supported this trip. I would also like to thank Dr. J. C. Ainsworth, the Director of the CMI, and Dr. M. B. Ellis, senior mycologist, for making laboratory, herbarium and other facilities available to me during my stay. Regarding the staff of the Institute, I would like to take this opportunity to thank all of them, not only for their help and encouragement, but for the many little things which went to make my visit a memorable and enjoyable experience.

It is a pleasure to acknowledge the help of Dr. Bryce Kendrick, University of Waterloo, and Dr. Roger Goos of the American Type Culture Collection. I am indebted to them for critically reviewing the manuscript and making many helpful comments and suggestions.

# Contents

CHAPTER

# I

# Classification of the Hyphomycetes

## INTRODUCTION

The classification of the *Fungi Imperfecti* and keys to their identification are based principally on the system presented by Saccardo (1886) in his *Sylloge Fungorum*, or later modifications of this system by Lindau (1900), Saccardo (1906) and others.

In the current interpretation of this system (Ainsworth, 1963), the *Fungi Imperfecti* are divided into four orders according to the type of fructification produced. If the conidia and conidiophores are produced within a pycnidium, then they are included in the Sphaeropsidales and if on an acervulus then in the Melanconiales. Those forms in which the conidiophores are neither in a pycnidium nor on an acervulus are classified in the Moniliales (=Mucedineae of Saccardo) or if no spore form is produced, in the *Mycelia Sterilia*. Saccardo equated his Mucedineae with the Hyphomycetes, other authors (Ainsworth, 1963) regard the Hyphomycetes as including both Moniliales and *Mycelia Sterilia*.

Saccardo divided the Hyphomycetes into four families viz. Mucedinaceae (=Moniliaceae), Dematiaceae, Tuberculariaceae and Stilbaceae (=Stilbellaceae). In the Tuberculariaceae, the conidiophores are generally short and borne in a more or less tightly-packed, parallel series arising from a cushion-shaped, stromatic mass referred to as a sporodochium. In the Stilbaceae, the conidiophores are generally long and more or less fused along their length to form erect fructifications referred to as synnemata. In the Moniliaceae and Dematiaceae, the conidiophores are solitary, sometimes in groups or fascicles, but are never in synnematal or sporodochial arrangements. The Moniliaceae is characterized by hyaline or lightly pigmented conidia and conidiophores and in the Dematiaceae the conidia or conidiophores are more or less darkly pigmented.

The various families were divided further by Saccardo into sections based primarily on conidium morphology, with emphasis on the shape and septation of spores as follows: Amerosporae, conidia nonseptate; Didymosporae, conidia one-septate; Phragmosporae, conidia multiseptate; Dictyosporae, conidia longitudinally and transversely septate; Scolecosporae, conidia filiform; Staurosporae, conidia stellate, radiate or trifurcate; Helicosporae, conidia convolutely spiral.

Saccardo (1906) modified the section names with the prefixes hyalo- or phaeo-, depending on whether the conidia were hyaline or pigmented, *e.g.* Hyalophragmae, Phaeophragmae.

Over the years there has been increasing dissatisfaction with the classification of Saccardo, particularly as it applied to Hyphomycetes. There have been many critics, notably Vuillemin (1910, 1911) and Mason (1933, 1937) in the earlier literature and Hughes (1953), Goos (1956), Tubaki (1958, 1963), Subramanian (1962) and Luttrell (1963, 1964) in the more recent literature. The reliability of the criteria used for delimitation of families and genera has been seriously questioned by these and other students of the group. It has been pointed out that application of the suggested criteria frequently results in the separation of genera which are closely related and, at the same time, brings together forms which have little in common. Most workers have now lost confidence in the use of color, septation, presence or absence of synnemata or sporodochia and the like, as primary criteria for separating the Hyphomycetes into families and genera. The findings and opinions of the critics are commented on in the following section.

## COMMENTS ON CRITERIA USED FOR CLASSIFICATION

### Color of Conidia and Conidiophores

Great stress is placed in the Saccardo system on the pigmentation of conidia and conidiophores. Degree of pigmentation is the basis for separation of the two families, Moniliaceae and Dematiaceae, which contain the majority of the genera of Hyphomycetes. At first thought, color seems a rather easy and objective character to use. This, unfortunately, has not proved to be the case. The main difficulty is based on the fact that the family Moniliaceae includes those forms in which the conidia and conidiophores are either hyaline *or* lightly pigmented. Immediately, therefore, we become subjective. For example, in *Botryotrichum piluliferum* (Fig. 36) the young aleuriospores are golden-brown, and this genus is therefore frequently classified in the Moniliaceae, as for example by Barnett (1960). On the other hand, in older, mature colonies the conidia may be dark-brown and could easily be considered as Dematiaceae, as by Clements and Shear (1931). Similarly, in *Cephaliophora tropica* (Fig. 7*C*) the spores are relatively lightly pigmented and pale orange in mass to the unaided eye. *Cephaliophora*, however, is placed in the Dematiaceae by some authorities (Barnett, 1960) and in the Moniliaceae by others (Clements and Shear, 1931). If the authorities have problems, beginning or casual students of the group have even greater difficulty in applying the color criterion. Pigmentation in a fungus may be influenced not only by age and environment but also by the intensity of the illumination source by which the spores and conidiophores are examined. It is not unusual for beginning students to classify spores (viewed under the microscope) as hyaline when they are black in mass to the unaided eye.

With the continuing discoveries of new species of Hyphomycetes there is an increasing number of instances in which we find a group of species which are, in all respects, congeneric, except that some are hyaline and others strongly pigmented. If the species group is large, we usually find all gradations from hyaline to darkly pigmented within the complex. In the *Aspergillus niger* series, conidia are darkly pigmented as they are in *Penicillium megasporum* (Fig. 156) and the *nigricans* series of the genus *Penicillium*. There has been no trend, however, to segregate these as "new" genera in the family Dematiaceae. In other words, even where Saccardo's classification is accepted, there are instances in which the nature of the pigmentation has been overlooked in favor of more fundamental taxonomic criteria. There are also numerous instances where such is not the case. Dematiaceous members of *Scopulariopsis* (Fig. 182) were included by Ôta (1928) in the genus *Phaeoscopulariopsis* and into *Masoniella* by Smith (1952). This disposition, however, was criticized by Hughes (1953), who was supported by Barron *et al.* (1961) and Morton and Smith (1963); most workers now regard this species group as *Scopulariopsis* irrespective of whether the conidia are hyaline or pigmented.

Srinivasan (1958) erected a new genus *Hyalostachybotrys* to include two *Stachybotrys*-like fungi, recovered from the rhizosphere of sugar cane, which had hyaline conidia and conidiophores. Barron (1964) disagreed with this disposition and recommended that they be considered under *Stachybotrys*. In this he was supported by Rifai (1964), who described *Stachybotrys bambusicola* found on the fallen culm sheaths of bamboo. In this species the spores are pink in the mass but hyaline to subhyaline when viewed individually.

*Acrostaphylus*, erected by Arnaud (1953), was validated by Subramanian (1956a) to include the dematiaceous *Nodulisporium*-like fungi. Here again, in the complex of species involved, we have all gradations between hyaline and pigmented to the point where assigning a given species to one or another genus becomes arbitrary and we wonder if the distinction is necessary.

As pointed out by Hughes (1953), Goos (1956) and others, most workers on Hyphomycetes are losing confidence in the major divisions based on pigmentation of the conidia and conidiophores. Hughes also cautions us, however, with regard to color: ". . . it is worthwhile remembering that large numbers of Hyphomycetes have brightly colored fructifications and mycelium, characters which we have learned to associate with the Hypocreales and some Basidiomycetes; examples are *Apiocrea chrysosperma* (Tul.) Syd. (=*Sepedonium chrysospermum* Fr.) and *Nectria inventa* Pethybr. (=*Acrostalagmus cinnabarinus* Corda). With further knowledge of the perfect states of Hyphomycetes, color may prove to be a useful character."

**Synnematous and Mononematous Forms**

The taxonomic significance of synnemata will continue to be a difficult problem in the classification of the Hyphomycetes until such time as the factors which control their development are better understood. Factors influencing the development of synnemata (coremia) have received scant attention in the past, but recent contributions by Taber and Vining (1959) on *Isaria cretacea*, Loughheed (1961) on *Hirsutella gigantea* and Carlile *et al.* (1961) on *Penicillium claviforme* are helping us to understand not only the structure and development of synnemata but also the nutritional and environmental factors which influence their formation. Loughheed (1963) gives a comparative discussion of the results obtained in the various organisms studied.

In some fungi the formation of synnemata is apparently mandatory (Loughheed, 1961) and the fungus will not sporulate without first forming synnemata. On the other hand, it is generally appreciated that, in many stilbaceous genera, formation of synnemata is, to say the least, sporadic and inconsistent. Looseness of terminology has added to the confusion in this area. The terms coremial, synnematal and funiculose have been used more or less interchangeably. The hyphal ropes (funicles) present in many species of *Cephalosporium* have frequently been referred to as synnemata and have been the basis of a number of vague and uncertain genera.

In *Penicillium claviforme* the synnemata have a sterile stalk and bear a sporulating head of aggregated penicilli. *P. claviforme* and other stilbaceous *Penicillium* species have been classified by some under *Coremium*. Raper and Thom (1949), however, properly classified these forms with the mononematous species of the genus. Similarly, the synnematal forms of *Paecilomyces* have frequently been classified under *Isaria*. *P. farinosus* may produce beautiful and elegant synnemata in culture, as may *P. fumosoroseus*. Different strains of these species show different capabilities in this regard, and in some strains formation of synnemata is suppressed. However, whether synnemata are formed or not in *Paecilomyces*, abundant mononematous conidiophores are produced from the "ground hyphae."

As pointed out by Hughes (1953), *Scopulariopsis* and *Stysanus* (= *Doratomyces*) show close morphological similarities. *Doratomyces* is essentially a coremial form of *Scopulariopsis* and all isolates of *Doratomyces* produce a *Scopulariopsis* state from the "ground hyphae." Some isolates of *Doratomyces* recovered in our laboratory eventually lost the synnematal state and the cultures appear as *Scopulariopsis*.

*Graphium* is a heterogeneous assemblage of forms in which the sporogenous cells may be annellophores, sympodulae or phialides. These forms have their mononematous counterparts in the genera *Phialocephala* (phialides), *Leptographium* (annellophores) and *Verticladiella* (sympodulae). We could, therefore, applying accepted criteria, have six genera involved in the above complex. It would seem more reasonable to neglect the synnematous or mononematous nature of the conidiophores in this case and contain these in three genera only. If the sporogenous cells are phialides, then *Phialocephala* would suffice irrespective of whether they are borne in mononematous or synnematous fructifications.

**Septation of Spores**

Considerable stress is placed in the Saccardo system on spore septation. As was pointed out by Hughes, the absence of septation or variation in septation has been given too much emphasis.

*Hormodendrum* is separated from *Cladosporium* on the basis of spore septation. But, in the development of their conidiophores, in their pigmentation and dendritic habit, in their production of branching chains of blastospores, these two genera are essentially the same. In *Hormodendrum* the conidia are considered nonseptate and in *Cladosporium* one-septate. In actual fact, no such distinction can be made. Septation is variable; in some species the conidia are mostly continuous, with occasional septate spores. In other species the spores may be mostly two-celled with occasional nonseptate spores. Between these two extremes all intermediate conditions may occur. Moreover septation is strongly influenced by age and substrate. While *Hormodendrum* is still recognized in some recent treatments on Hyphomycetes, it would be best to consider it a synonym of *Cladosporium*.

In *Cladobotryum variospermum* (Fig. 54*A*), two-celled phialospores are produced in fragile chains from verticillate, branched conidiophores. In the *Cladobotryum* state of *Hypomyces roseus* (Fig. 54*B*), the conidia are phragmospores, but this is no reason to separate this conidial state into a different form-genus.

In his treatment of the genus *Pithomyces* Ellis (1960) gives a good example of the modern approach to this problem. He includes amerospores, didymospores, phragmospores and dictyospores within this genus, giving septation of the spores only specific significance in his delimitation of taxa.

**Sporodochium and Acervulus**

The presence of a sporodochium is the hallmark of the family Tuberculariaceae of the Hyphomycetes. Ainsworth (1963) defines a sporodochium as "a mass of conidiophores tightly placed together on a stroma or mass of hyphae." The presence of an acervulus is regarded as the key diagnostic feature of the Melanconiales. Acervulus is defined in the dictionary of Ainsworth and Bisby as "an erumpent, cushion-like mass of hyphae having conidiophores and conidia, and sometimes setae" (Ainsworth, 1963). Alexopoulos (1962), on the other hand, defines sporodochium as "a cushion-shaped stroma covered with conidiophores," a definition which resembles closely that of Ainsworth and Bisby for an acervulus. Moreover, many fungi recognized as being sporodochial produce a nice fringe of setae round the margin of the fructification, as, for example, *Volutella ciliata* and *Myrothecium indicum*. The presence or absence of setae is, therefore, of little help. The fact is that, for beginner or expert, the distinction between a sporodochium and an acervulus is frequently very difficult to make. Generally, if the fructification is flat or saucer-shaped, it is referred to as an acervulus; if cushion-shaped, it is a sporodochium.

Unfortunately, many fungi which produce acervuli on their natural substrate do not do so in culture. Certain *Colletotrichum* species may produce very nice acervuli on the host plant but large, cushion-shaped, sporodochial-like masses in culture. *Metarrhizium anisopliae* and *Myrothecium verrucaria* are very closely related. They are generally considered to be sporodochial. I have seen many isolates of these species in culture, and if they had been described as acervular, I would have found it difficult to disagree. Many fungi in culture show a different growth habit than on their natural substrate, and fungi which normally produce acervuli and sporodochia may do so sporadically or not at all in culture. *Fusarium* is a good example of this. The majority of species do not produce sporodochia in culture, yet many keys to the Hyphomycetes consider *Fusarium* only under the family Tuberculariaceae.

It would not be a great inconvenience if the limits of the Hyphomycetes were stretched to include the Melanconiales.

### Wet Spores and Dry Spores

Mason (1937) emphasized the distinction between dry and slimy spores and introduced the concept of the biological spore type. This approach was accepted by Wakefield and Bisby (1941) when they listed the British Hyphomycetes under Xerosporae (dry spores) and Gloiosporae (slimy spores). The use of this "biological spore type" as a major criterion for separation of the Hyphomycetes into two major groups has not been generally accepted because it is too difficult to apply. It cuts across many of the natural associations, and species of a genus or even strains of a species may show both dry- and wet-spored forms. For example, *Fusarium moniliforme* has a microconidial state in which the spores are borne in short chains. In variety *subglutinans*, however, the spores slime down to form gloeoid balls. In *Gliomastix murorum* the globose or subglobose spores are borne in short, tortuous chains, whereas in *G. murorum* var *felina* the spores slime down to form dark, glistening balls. *Memnoniella echinata* is essentially a *Stachybotrys*, except that the spores persist in chains rather than slime down. In his studies of *Stachybotrys*, Zuch (1946) found forms intermediate between *Memnoniella* and *Stachybotrys*. Smith (1962) transferred *M. echinata* to *Stachybotrys*, and this seems a logical disposition. In *Gliocladium* the spores generally slime down to form gloeoid spore masses; in some members of the *G. roseum* series, however, and in other *Gliocladium* species, the spores form slimy columns. In *Clonostachys* the columns are rather better developed, and more persistent, but nevertheless, in view of the intermediate forms, there seems no good reason for separating this genus from *Gliocladium*. *Paecilomyces elegans* normally produces chains of conidia, but, unlike most species of this genus, the spores are borne with their long axis at a slight angle to the main axis of the chain and the conidia slip slightly downwards. In some species of *Paecilomyces* the spores may slime down completely to form a spore ball reminiscent of *Gliomastix*.

I do not necessarily advocate the policy of uniting similar dry- and wet-spored species into one genus. Sometimes there are other mitigating circumstances. *Gliocephalis hyalina* is basically similar to *Aspergillus* except that the spores slime down. These two genera, however, appear different in other respects. In *G. hyalina*, the stout conidiophore and large vesicle, the shape of the phialides and spores and the fact that it will not grow on normal laboratory media and is either a parasite, predator, or highly specialized saprophyte, all indicate differences rather than similarities. It is only in general morphology that the two genera appear similar.

If we do not regard color as a primary criterion of distinction, nor septation, nor whether the fungus is mononematous or synnematous, nor whether the spores are wet or dry, it becomes clear that we can no longer accept in principle the system of Saccardo. All or many of the criteria used by Saccardo seem to cut across the natural relationships between and within what we would like to regard as genera. It is then essentially an artificial classification.

This is not to suggest that the Saccardo system lacks merit. It has real advantages in its comprehensive nature. The criteria used for identification are relatively easy to evaluate, and this is especially important for beginning students. The system has had many critics but its simplicity will perhaps be better appreciated when we attempt to put the alternative systems into practice. The fact that there is no recent text, either generalized or specialized, which attempts to use the modern system of taxonomic evaluation in Hyphomycetes is a reflection, not only on the value of Saccardo's system, but also on the difficulties of applying the principles of the modern system to the practical problems of identification.

The first important contributions to a new approach to the classification of the *Fungi Imperfecti* were made by Vuillemin (1910, 1911), who pointed out the limitations of the term "conidium" and proposed a classification based on different spore forms. He divided the fungi into two main groups, *viz.* those which produce thallospores and those which produce *conidia vera.* In the *conidia vera*, spores are cut off from the hypha which bear them as soon as they are formed. Thallospores, on the other hand, are an integral part of the hypha which bear them. In his thallospore group, Vuillemin recognized blastospores, arthrospores and chlamydospores. He also introduced the term "aleuriospore" to describe thallospores which resemble conidia in their position, color, form, structure and dimensions, but differ in that they are not cut off from the parent mycelium by a natural means of dehiscence. Vuillemin divided his *conidia vera* into three groups according to the nature of the conidium-bearing hyphae. If the conidia are borne on undifferentiated hyphae, he classified them in the Sporotrichées, if on well developed conidiophores, then in the Sporophorées, and if at the apex of special flask-shaped structures (phialides), in the Phialidées. Vuillemin's recognition of the phialide as a specific type of sporogenous cell is one of his outstanding contributions. His concepts were discussed critically by Mason (1933), who drew attention to the heterogeneous nature of the Phialidées as conceived by Vuillemin and defined the term "phialide" more precisely.

Vuillemin's system was the basis of the classification of the *Fungi Imperfecti* presented by Langeron and Vanbreuseghem (1952), which was summarized by Goos (1956).

While the observations of Vuillemin and Mason laid the foundation for a new approach to the classification of the Hyphomycetes, their suggestions were not immediately adopted; principally because their observations were not comprehensive enough to be of use at the practical level.

The ideas proposed, however, were the basis of an extensive reappraisal of the Hyphomycetes by Hughes (1953).

### CLASSIFICATION OF HUGHES

In an outstanding contribution to the mycological literature, Hughes (1953) presented an experimental classification of the Hyphomycetes based on the different types of conidium development. The mononematous, synnematous or tuberculariaceous nature of the conidiophores, the form of the mature conidia, their dematiaceous or mucedinaceous nature, their septation and the presence or absence of slime around them were treated as subsidiary characters.

Hughes divided the Hyphomycetes into eight sections and two subsections as follows.

#### Section IA

Mycelium generally narrow. Conidia usually developing in acropetal succession as blown-out ends at the apex of simple or branched conidiophores which do not then increase in length. The basal conidia of chains, aggregated around the apical region of a conidiophore, may be morphologically different from the others, and in one instance are modified into permanent metulae which bear a terminal conidium and a number of subterminal conidia.

Into this section Hughes classified *Bispora*, *Cladosporium*, *Septonema* and similar fungi. Hughes referred to the conidia in this section as "blastospores" (arising as globular buds or blown-out ends) with their development in acropetal chains inferred by this term.

*Section IA is considered later under the Blastosporae.*

#### Section IB

Mycelium generally wide. Conidia developing in acropetal succession as blown out ends on simple or branched conidiophores; sometimes the lateral branches are modified entirely into a number of conidia or into solitary conidia, and in these instances the conidia are borne on conspicuous denticles. The solitary conidia, or short, simple or branched chains of conidia, may be aggregated on well differentiated swollen cells and arise simultaneously on them. In examples with intercalary or lateral swollen and fertile cells bearing simultaneously produced conidia, the main stalk may proliferate to develop further intercalary fertile cells or bear further lateral fertile branches.

In this section Hughes included the conidial state of *Pellicularia pruinata*, *Nematogonium*, *Oedocephalum*, *Botryosporium*, *Botrytis* and similar fungi. Where the conidia are solitary, he referred to them as "solitary blastospores"; where they are in clusters, he referred to them as "botryose solitary blastospores" and where in chains, "botryose blastospores," as in *Nematogonium*. Certain members of this group are basically similar to Section IA. As noted by Hughes in *Pellicularia pruinata*, the sparingly branched conidiophores are really of the *Cladosporium* type. This species, however, is connected through *Oidium aureum* and *Oidium conspersum* to the structures formed in *Oedocephalum* and *Nematogonium*.

Because of the similarity between forms such as the *Oidium* (= *Acladium*) fructifications of *Pellicularia pruinata* and the fructifications of *Cladosporium*, Hughes at first considered combining these two subsections. He found that more uniformity was obtained in both Sections by separating into IA those forms with predominantly narrow hyphae and into IB those forms with predominantly wide hyphae.

*Section IB includes the group considered later under Botryoblastosporae.* Intermediate forms such as the conidial state of *Pellicularia pruinata* have not been recorded from soil, and the two groups Blastosporae and Botryoblastosporae thus appear more distinctive than they actually are.

**Section II**

Conidia arising as blown-out ends of the apex of simple or branched conidiophores and the ends of successively produced new growing points developing to one side of the previous conidium. The conidiophore therefore either increases in length or becomes swollen as a result of conidium production. Acropetal chains of conidia may develop on the primary conidia.

In this section Hughes included *Dactylosporium*, *Diplorhinotrichum*, *Verticicladiella*, *Tritirachium*, *Costantinella* and similar fungi. Hughes referred to the conidia here as "terminus spores." He noted that, in many cases, as in *Ramularia*, the terminus spores bud at the apex and produce blastospores, but (in the chains produced) the basal or primary conidium is always a true terminus spore. In *Arthrobotrys*, where the conidia may develop in clusters, he referred to them as "botryose terminus spores."

*Section II is discussed later under Sympodulosporae.*

**Section III**

Conidia usually thick-walled, arising solitarily as blown-out ends of the apex of simple or branched conidiophores; a plurality of conidia may be produced, each new conidium developing as a blown out end of successive proliferations through the scars of previous conidia, so that the conidiophores in such cases become annellate.

In this section Hughes included *Trichocladium* and *Bactridium* as examples of the solitary forms and *Scopulariopsis*, *Annellophora* and *Lepto-*

*graphium* as examples of the annellated conidiophore in which a plurality of conidia develop.

Hughes referred to the conidia here as "chlamydospores" which are usually produced at the apex of a hypha. He pointed out that they were usually solitary, large, thick-walled and dark-colored, but might be small, hyaline and not particularly thick-walled, especially when developing in slimy heads. When conidia are terminal, annellations may be produced on the conidiophore and successive conidia may remain loosely attached in chains.

Hughes pointed out that the disadvantage of the character implicit in the term annellophore was the great difficulty that might be encountered in observing the actual annellations, more especially in those species whose annellophores are narrow and hyaline.

In this Section, Hughes considered two distinct spore forms. *I have considered those in which the spores are solitary under Aleuriosporae and those in which a succession of spores is formed from an annellophore under Annellosporae.*

**Section IV**

Conidia (phialospores) developing in rapidly maturing basipetal series from the apex of a conidiophore (phialide) which may or may not possess an evident collarette.

In this section were included the well known and readily recognized phialosporous genera such as *Penicillium*, *Aspergillus*, *Chalara*, *Phialophora*, *Menispora*, *Fusariella* and similar fungi.

Hughes referred to the conidia here as "phialospores." The term "polyphialide" was introduced by Hughes to include phialides which produced a series of open ends from each of which a basipetal succession of phialospores is produced.

*Section IV is considered later under Phialosporae.*

**Section V**

Conidia developing in gradually maturing basipetal series and originating by the meristematic growth of the apical region of the conidiophore in such a way that the chain of conidia merges imperceptibly with the conidiophore that gives rise to the chain.

Hughes referred to the conidia here as "meristem arthrospores," and they occur in true chains; their basipetal development, due to the meristematic growth of the conidiophore, seemed to him to necessitate a special term for them, although they may be related to arthrospores.

Included here are the conidial states of *Hysterium insidens*, *Sirodesmium diversum*, *Erysiphe polygoni*, *Trimmatostroma* and similar fungi.

*Section V is considered later under Meristem Arthrosporae.*

**Section VI**

Conidia usually thick-walled, developing from pores on conidiophores of determinate or indeterminate length; they are solitary or in whorls and may occur in acropetal chains. The conidiophore may proliferate through the terminal pore to produce a further terminal conidium or the conidiophore may develop a succession of terminal conidia on successive proliferations developing just below the previous conidium.

Hughes included here *Helminthosporium*, *Diplococcium*, *Stemphylium*, *Torula*, *Alternaria* and similar fungi. He referred to the conidia as "porospores." He noted that, in *Helminthosporium velutinum*, solitary apical and lateral porospores in verticils are produced, while in *H. teres* successive terminus porospores develop. Chains of porospores occur in some species such as *Diplococcium spicatum*. Porospores are dry.

*Section VI is considered later under Porosporae.*

**Section VII**

Conidia developing by the basipetal fragmentation of conidiophores, of determinate length, which do not possess a meristematic zone.

In this section Hughes included *Geotrichum*, *Coremiella*, *Oidiodendron* and similar fungi. He referred to the conidia as "arthrospores" and these may be slimy or dry.

*Section VII is considered later under Arthrosporae.*

**Section VIII**

Conidia borne singly at apex, or singly at the apex and laterally, often in regular whorls on conidiophores showing basal elongation. Conidia often with a longitudinal slit in the wall, but this character by no means restricted to this section.

Hughes included here *Papularia*, *Arthrinium*, *Camptoum* and similar genera. The conidia here are dry and produced as blown-out ends and are borne either singly at the apex or singly at the apex and laterally, generally in basipetal whorls. Hughes considered the basal elongation of the conidiophores, which may be called basauxic conidiophores, as the hallmark of this section. The conidia are dry.

*Section VIII is discussed later under Meristem Blastosporae.*

## ALTERNATIVE CLASSIFICATIONS

The classification of Hughes was accepted and extended by Tubaki (1958), who added a ninth Section to include *Trichothecium roseum* and divided Sections III, IV and VII into subsections. Later, Tubaki (1963) presented a more generalized scheme in which he divided the Hyphomycetes into six major groups named after the type of spore produced as follows.

Blastosporae: Hyphomycetes producing blastospores
Type genus: *Cladosporium* Link.
Radulasporae: Hyphomycetes producing radulaspores
Type genus: *Beauveria* Vuillemin.
Aleuriosporae: Hyphomycetes producing aleuriospores
Type genus: *Scopulariopsis* Bainier.
Phialosporae: Hyphomycetes producing phialospores
Type genus: *Catenularia* Grove.
Porosporae: Hyphomycetes producing porospores
Type genus: *Helminthosporium* Link.
Arthrosporae: Hyphomycetes producing arthrospores
Type genus: *Geotrichum* Link.

Tubaki classified 130 selected genera of Hyphomycetes with his revised system. He pointed out the difficulties in treating bibliographic genera and the need for many more developmental studies and noted that he omitted well known genera because of uncertainties regarding their conidial structures. Naming the sections on spore type has obvious advantages over the numerical system adopted by Hughes and Tubaki in their earlier experimental classifications.

The genus *Papularia* (= *Arthrinium*) is considered by Hughes in Section VIII. Tubaki placed this genus in his series Radulasporae. The relationship between Tubaki's revised classification and that of Hughes is indicated in Table I.

An alternative system of classification of the Hyphomycetes which merits attention was given by Subramanian (1962), who recognized six

TABLE I

*A comparison of the scheme proposed for soil Hyphomycetes with the classifications presented by Hughes (1953), Subramanian (1962), and Tubaki (1963).*

| Subramanian | Hughes | Barron | Tubaki |
|---|---|---|---|
| Torulaceae | Section IA | Blastosporae | Blastosporae |
| | Section IB | Botryoblastosporae | |
| | Section II | Sympodulosporae | Radulasporae |
| Bactridiaceae | Section III | Aleuriosporae | Aleuriosporae |
| | | Annellosporae | |
| Tuberculariaceae | Section IV | Phialosporae | Phialosporae |
| Coniosporiaceae | Section V | Meristem Arthrosporae | |
| Helminthosporiaceae | Section VI | Porosporae | Porosporae |
| Geotrichaceae | Section VII | Arthrosporae | Arthrosporae |
| | Section VIII | Meristem Blastosporae | |

morphological types of spores and, based on these, he divided the group into six families as follows.

Torulaceae: Hyphomycetes producing blastospores
Type genus: *Torula* Pers. ex Fr.
Bactridiaceae: Hyphomycetes producing gangliospores
Type genus: *Bactridium* Kunze ex Fr.
Tuberculariaceae: Hyphomycetes producing phialospores
Type genus: *Tubercularia* Tode ex Fr.
Helminthosporiaceae: Hyphomycetes producing porospores
Type genus: *Helminthosporium* Link ex Fr.
Geotrichaceae: Hyphomycetes producing arthrospores
Type genus: *Geotrichum* Link ex Sacc.
Coniosporiaceae: Hyphomycetes producing meristem arthrospores
Type genus: *Coniosporium* Link ex Fr.

Subramanian recognized a seventh spore type "the spiculospore," to include the spore produced at the tip of a pointed structure, as in *Hirsutella* or *Akanthomyces*. He suggested, however, that further work was necessary before this spore-type category could be accepted for any formal taxonomic grouping. Subramanian also considered an eighth spore type, "the chlamydospore," but regarded this as having no taxonomic significance in the delimitation of genera. He introduced the term "gangliospore" to replace the much abused "aleuriospore." The gangliospore is not precisely equivalent to the aleuriospore as used by Tubaki as it does not include conidia produced on annellophores, which Subramanian considered under "blastospores" in his family Torulaceae (Section VIII).

Within each of the spore types recognized by Subramanian, on the basis of initiation and development he noted several well-defined differences, which he summarized as follows.

1. Production of a single solitary spore, terminating the growth of the conidiophore.

2. Production of successive solitary spores by (*a*) meristematic activity of a sporogenous cell; (*b*) proliferation of the conidiophore through the scar of a fallen spore; (*c*) sympodial growth of the conidiophore and the formation of spores repeatedly from new growing points.

3. Production of successive spores in chains by (*a*) acropetal budding; (*b*) meristematic activity of the conidiophore tip or sporogenous cell such as a phialide (basipetal chains); (*c*) sympodial growth of the conidiophore and formation of spores repeatedly from new growing points; basipetal chains different in ontogeny from those of 3*b*.

On the basis of these secondary criteria, Subramanian divided the Hyphomycetes into 24 sections dispersed among his six families. Thirteen of these sections were in the Torulaceae, four in the Bactridiaceae, two in the Tuberculariaceae, two in the Helminthosporiaceae, two in the Geotrichaceae and one in the Coniosporiaceae.

It is unfortunate that Subramanian selected *Torula* as the type genus for those forms producing blastospores. Hughes included *Torula* in his Section VI (= Porosporae). The genus is somewhat anomalous. Each spore is produced through a minute pore in the darkened terminal cell of a hypha or previously formed conidium; the cells of any spore, however, develop blastogenously in acropetal succession. Hughes (1953) pointed out the similarity between the darkened terminal cells in *Torula* and the darkened, swollen apex of the sporogenous cell of *Stemphylium*. The disposition of *Torula* must be regarded as controversial; I prefer to place this genus in the Porosporae. Although the cells are produced in acropetal succession blastogenously, the entire spore originates through a pore. Subramanian's Torulaceae includes Sections IA, IB, II, III (in part) and VIII of Hughes. The relationships of Subramanian's classification and that of Tubaki and Hughes is given in Table I.

Of the classifications available, I have elected to follow that of Hughes, which was found the most suitable for the spectrum of fungi recovered from soil. In naming the sections, I have preferred to follow Tubaki and name each section according to the type of spore produced.

I have elevated Hughes' subsections IA and IB to section rank, as Blastosporae and Botryoblastosporae, respectively. As represented in the soil fungi, these two series seem to be fairly natural groupings. This arrangement alleviates the problem for beginning students of relating the acropetal blastospores of *Cladosporium* (IA) and the blastospores developing synchronously on denticles on a swollen ampulla, as in *Botrytis* (IB).

In Hughes' Section III, I have separated the forms which produce solitary or botryose aleuriospores (Aleuriosporae) from those which produce the conidia in basipetal chains from an annellated sporogenous cell (Annellosporae).

In *Echinobotryum*, *Wardomyces* and similar genera, the first spore is apical and terminal. A succession of spores is produced below and to one side of the apex. Hughes considered *Echinobotryum* in his Section II (Sympodulosporae). I have preferred to consider them as botryose forms of the Aleuriosporae. The spores have a fairly broad attachment to the sporogenous cell and do not secede readily. They do so by rupture of the sporogenous cell, part of which is left as a fringe at the base of the spore, a feature typical of the Aleuriosporae. *Echinobotryum* and similar genera are related to the solitary aleuriospore forms such as *Humicola* through transitional species such as *Mammaria echinobotryoides*.

In the classification of the Hyphomycetes proposed, I have therefore recognized ten sections in all, excluding the *Mycelia Sterilia*. The classification is essentially that of Hughes, with minor modifications as noted above. The relationship of this classification to that of Hughes is indicated in Table I. The various sections recognized are Aleuriosporae, Annellosporae, Arthrosporae, Blastosporae, Botryoblastosporae, Meristem Arthrosporae, Meristem Blastosporae, Phialosporae, Porosporae and Sympodulosporae. These sections, arranged in alphabetical order, are discussed in more detail in the following chapter.

CHAPTER

# II

# The Hyphomycetes

## SERIES ALEURIOSPORAE

In this group the conidia (aleuriospores) are produced singly (rarely in short chains) and develop terminally as blown-out ends of the sporogenous cells. A plurality of conidia is formed by the development of successive spores laterally and below the terminal spore to give an apical cluster of conidia (botryoaleuriospores).

Aleuriospores are usually thick-walled and frequently lightly or strongly pigmented, but may be thin-walled and hyaline. They may be amerospores, didymospores, phragmospores, dictyospores or helicospores, and either smooth or finely or coarsely roughened. Sometimes they have vesicular outgrowths from the spore wall. Solitary aleuriospores frequently secede with difficulty from the parental hyphae and do so by rupture of the sporogenous cell or one of the supporting cells. In such cases a trace of the parental hypha can frequently be found as a frill around the truncate attachment point of the spore. Botryoaleuriospores usually secede more readily than solitary aleuriospores.

Well developed conidiophores are not the rule in this group and most members produce the conidia either on short, undistinguished hyphal branches (pedicels), which function as the sporogenous cells, or more or less sessile on the vegetative hyphae. In a few genera the conidiophores are well differentiated.

Hughes suggested the term "chlamydospore" to describe the conidia in this group. In current mycological terminology they are frequently referred to as aleuriospores. Subramanian (1962) suggested the term "gangliospore." They are included in the Aleuriosporae of Tubaki (1963) and the Torulaceae of Subramanian (1962). The terms "chlamydospore" and "aleuriospore" have been used more or less interchangeably and somewhat loosely in the literature. In its popular concept, the term "chlamydospore" is frequently applied to the thick-walled, resistant spore formed in an intercalary position from a modified cell of a vegetative hyphae. The term has, unfortunately, been used to describe the teliospore of the Ustilaginales, which is also produced in an intercalary position. There are hundreds of fungi which produce intercalary or terminal thick-walled resting spores which are little modified from the hyphae which gave them origin. These are more or less nondescript states which have, for the most part, little taxonomic value. It would be a pity to confound these with the more or less well defined states which are being considered as genera here. I have, therefore, preferred to use the term "aleuriospore" in a more restricted sense to describe the situation in which the tip of the sporogenous cell or hyphal branch swells up to form a vesicular apex which is predestined to be a spore. Subramanian (1962) also recognized the chlamydospore as distinct from his "gangliospore" and considered it as a thick-walled thallospore formed from pre-existing elements of the vegetative hyphae (and sometimes of spores) and biologically serving for perennation and not dispersal. While the aleuriospore must be recognized as a well defined spore type, between it and the chlamydospore there are all gradations to the point where it is difficult, if possible, to make a distinction.

In many of the aleuriosporous genera, well differentiated conidiophores are lacking and the sporogenous cell subtending the spore arises more or less directly from a vegetative hypha. This is the case in *Humicola* (Fig. 1*B*). A short, cylindrical or slightly inflated, sporogenous branch arises more or less at right angles to a supporting hypha. It swells apically to form a vesicle-like cell. This enlarges, is cut off from the sporogenous cell and becomes thick-walled and dark. The spore secedes by rupture of the subtending pedicel. Some *Humicola* species recovered from soil have one or several thin spots, possibly germ pores, over the surface of the spore. Germ pores or germ slits are not uncommon in aleuriosporous fungi.

In *Trichocladium asperum* (Fig. 1*F*) the blown-out end of the sporogenous cell becomes one-septate and the resultant two-celled aleuriospore becomes dark and roughened at maturity with warty outgrowths. In *T. opacum* the aleuriospores are smooth and several-celled.

The sporogenous cell may be reduced or lacking in some members of this group. In *Mammaria echinobotryoides* (Fig. 2*B*) the mature spores are more or less sessile on the hyphae. The spores in this fungus are large

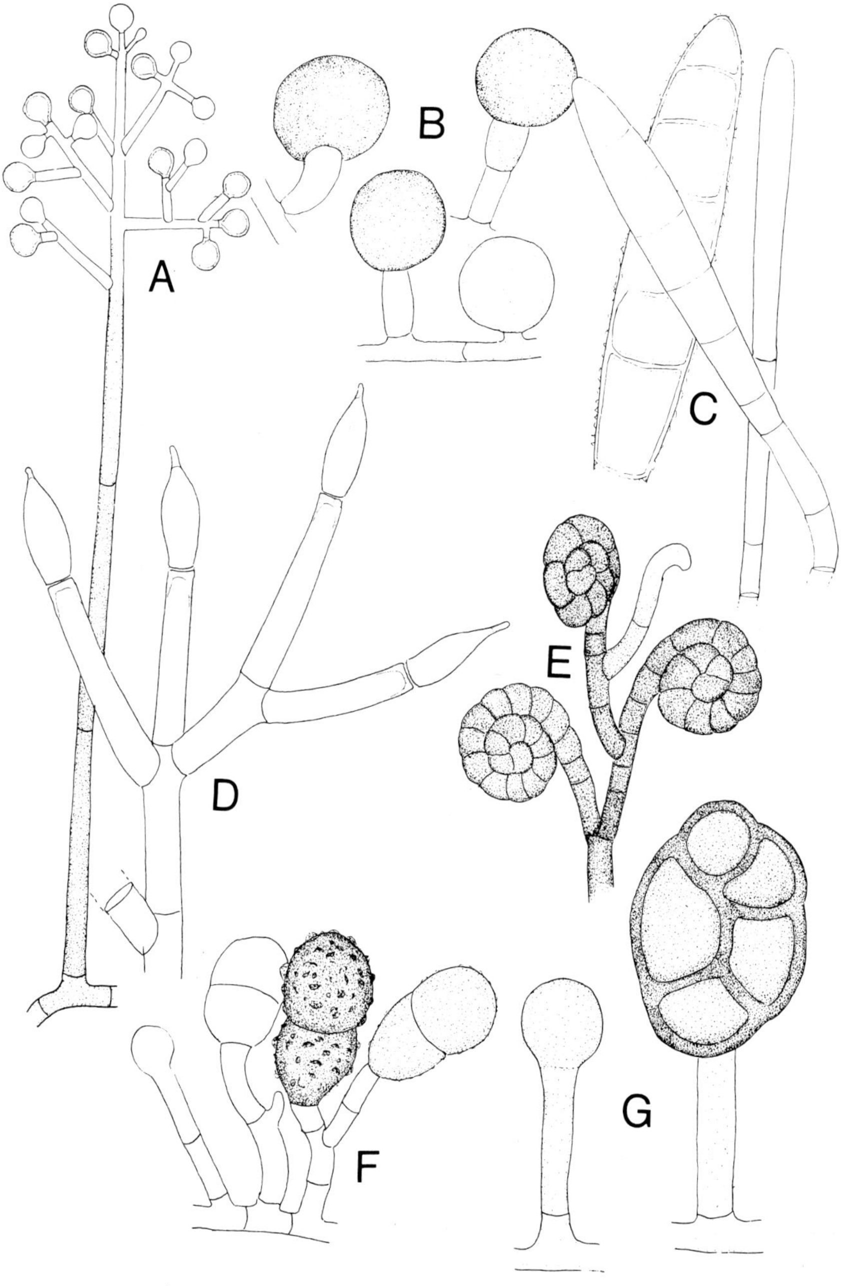


FIG. 1. Aleuriosporae. *A*, *Staphylotrichum coccosporum*. *B*, *Humicola grisea*. *C*, *Microsporum fulvum*. *D*, accessory spore state of *Sepedonium ampullosporum*. *E*, *Helicosporina veronae*. *F*, *Trichocladium asperum*. *G*, *Monodictys* sp.

and papillate or flame-shaped. Usually borne singly, they may occur in *Echinobotryum*-like clusters, hence the specific epithet. They have a well marked, longitudinal germ slit. In *Microsporum gypseum* (Fig. 1*C*) the spores are large and the conidium originates as a clavate terminal portion of a sporogenous branch. This becomes multiseptate, thick-walled and lightly pigmented. The mature, more or less fusiform spore secedes by rupture of a small cell subtending the spore.

Some aleuriospores are septate both horizontally and longitudinally. In *Monodictys* (Fig. 1*E*) such dictyospores become thick-walled and dark. In *Diheterospora* (Fig. 76) they remain hyaline to lightly pigmented and a thick, hemicellulose-like material is deposited on the inner wall, which results in a very small central lumen.

Well differentiated conidiophores are the exception in the Aleuriosporae. One of the most striking examples of conidiophore development is found in *Staphylotrichum coccosporum* (Fig. 1*A*) described by Nicot and Meyer (1956), and probably very common in high-organic soils. In this species, stout, erect, brown-pigmented conidiophores are produced. These branch irregularly at the apex to form tree-like heads. Branching is characteristically more or less at right angles to the main conidiophore axis. The ultimate branchlets bear terminal, globose or subglobose, brown, thick-walled aleuriospores. These are produced quite profusely under some conditions and may arise from the vegetative hyphae in botryoid clusters reminiscent of *Botryotrichum piluliferum*. In the latter, the aleuriospores are similar in size and pigmentation, but a well differentiated conidiophore is lacking. *Botryotrichum piluliferum* produces sterile setae, singly or in clusters, interspersed with the sporogenous hyphae.

Aleuriospores are usually solitary and terminal. In *Echinobotryum atrum* (Fig. 2*C*), a second conidium is produced by a blowing out of the sporogenous cell just below the apex. A succession of conidia produced below the apex and laterally gives rise to an apical cluster.

In *Wardomyces humicola* (Fig. 2*E*) the conidiophores are small and slightly swollen, they are simple or more often branched to form secondary and tertiary series with the ultimate branchlets functioning as the sporogenous cells. A single conidium is produced at the apex of each sporogenous cell. A succession of conidia develops below the apex and laterally to form an apical cluster of spores. A similar type of development is also found in *Gilmaniella humicola* (Fig. 2*A*).

It is very common for genera which produce aleuriospores to have an accessory spore state which belongs to one or other of the designated groups. These are frequently phialospores, but not always so. *Diheterospora* has associated *Verticillium* and *Paecilomyces* states. In *Mycogone* and *Sepedonium* the accessory states are *Verticillium* or *Verticillium*-like. In *Humicola* and *Botryotrichum* short chains of phialospores are

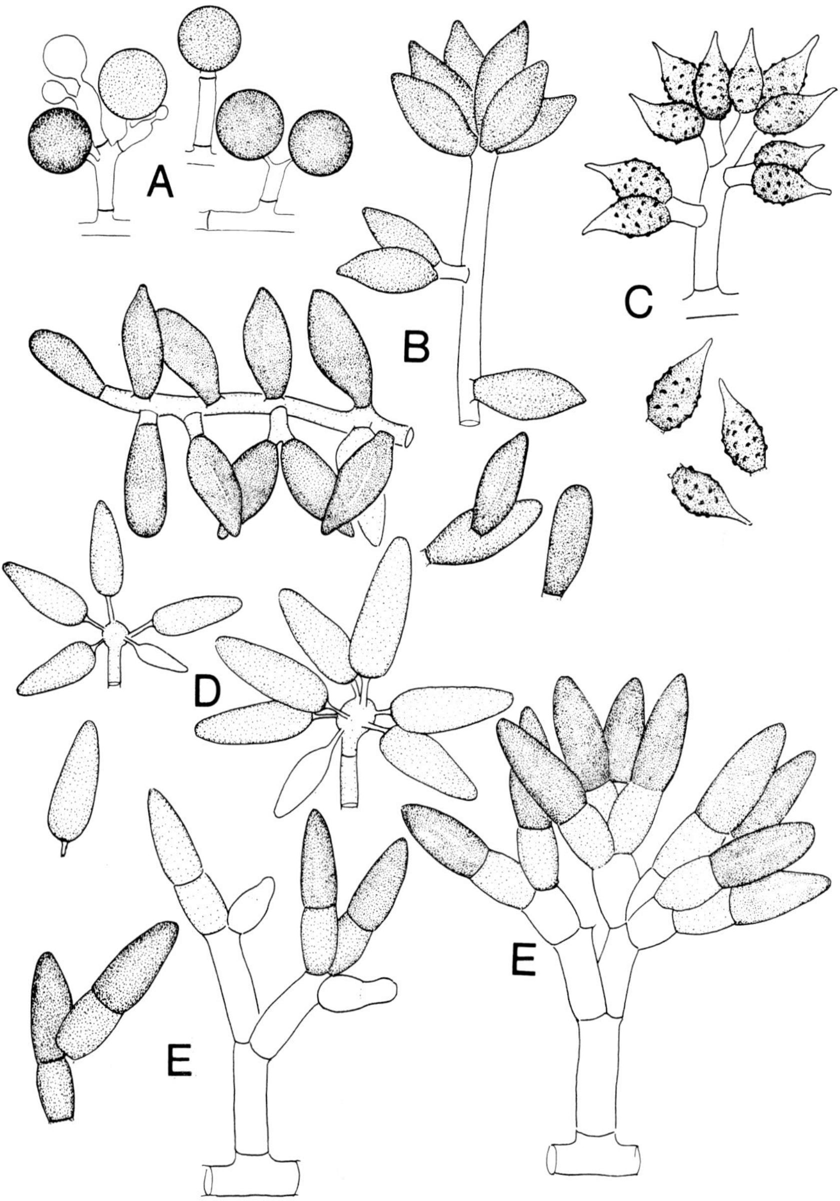


FIG. 2. Aleuriosporae. *A*, *Gilmaniella humicola*. *B*, *Mammaria echinobotryoides*. *C*, *Echinobotryum atrum*. *D*, *Asteromyces cruciatus*. *E*, *Wardomyces humicola*.

produced from solitary phialides. *Echinobotryum atrum* is the accessory state of *Doratomyces stemonitis.*

### SERIES ANNELLOSPORAE

In this series the first-formed spore is produced terminally on the sporogenous cell. Each new conidium develops as a blown-out end of a proliferation through the scar left by the previous conidium. A succession of proliferations is accompanied by an increase in length of the sporogenous cell. The scars of successive conidia give the sporogenous cell, in the spore-bearing region, an annellated appearance.

Spores in this group may be amerospores, didymospores or phragmospores. They are hyaline or pigmented and may be smooth or conspicuously roughened. They may be produced in short or long chains, or may slime down to form gloeoid balls or heads. Characteristically, the spores have a well marked, truncate attachment point corresponding in width to the conidiophore which gave them origin. In some members of the group the parental hypha appears as a delicate fringe around the base of the spore, but this is not always apparent. Annellations are not always distinct on the annellophore and the nature of the sporogenous cell is sometimes difficult to ascertain. This is especially true when the sporogenous cells are narrow and hyaline or when the spores are borne in mucus. Frequently slight irregularities in the outline of the sporogenous cells are indicative of annellophores. Shoemaker (1964) has suggested staining in a 1 per cent aqueous solution of Azur A to reveal the annellations. This technique is helpful in some cases but not in others. The fact that the young annellophores are relatively short and the older annellophores have a long, cylindrical apex is one of the most revealing diagnostic characters in this group.

The nature of the sporogenous cell in this series was explained lucidly by Hughes (1953) for *Annellophora africana.* In this species the first conidium is terminal, and, after seceding, the conidiophore proliferates through the terminal scar and the apex of the proliferation blows out into another conidium which is cut off from the proliferated conidiophore by a septum; this septum appears at a variable distance above the scar of the first conidium. A series of such proliferations, with a single conidium produced at each new level, results in a closely or distinctly annellated appearance of the conidiophore. Hughes referred to such a conidiophore as an annellophore and pointed out that this type of spore formation was far more general in the Hyphomycetes than had previously been supposed and that little attention had been given to annellations in published descriptions and illustrations.

*Scopulariopsis* is one of the most prevalent members of this series recovered from soil. Here the young annellophores are either cylindrical or flask-shaped with a short cylindrical neck. In *Scopulariopsis brumptii* (Fig. 182) the sporogenous cells are borne singly and directly from the vegetative hyphae. The annellophores are bulbous below with a narrow, cylindrical neck and give rise to short chains of dark, roughened spores.

In *Scopulariopsis brevicaulis* (Fig. 3*C*) the young annellophores are flask-shaped and are borne singly or more often in groups or in penicillate arrangements. The sporogenous cells are relatively large in this species and it is not difficult to observe either the annellations or irregularities in outline caused by successive proliferations of the sporogenous cells. The spores are large, subglobose, frequently coarsely roughened at maturity and have a broad, truncate base which indicates the attachment point to the sporogenous cell. The basal frill of parental hyphae is seen rather well in this species.

*Doratomyces* (=*Stysanus*) is to all intents and purposes a coremial form of *Scopulariopsis*. In *D. stemonitis* (Fig. 81) the stout, dark synnemata are more or less erect and branch at the top to produce an elongate, feathery, sporiferous head. The sporogenous cells are relatively small with a bulbous base and short neck. The successive annellations are very close together and give a dark, banded appearance to the upper portion of the sporogenous cell. *Doratomyces* species are usually very dark and colonies appear dark-brown to black. *Trichurus* is exactly like *Doratomyces* except that some of the terminal branches in the spore head grow out as stout, elongate appendages (Fig. 3*A*), giving the fructification a striking appearance.

In *Leptographium lundbergii* (Fig. 3*B*) conidiophores are mononematous and have a more or less unbranched main conidiophore axis bearing an apical penicillus composed of several series of metulae with the ultimate series bearing groups of slender annellophores. The spores slime down in gloeoid masses around the sporogenous cells and the apex of the conidiophores. The *Graphium* state of *Petriella setifera* produces the conidia on annellophores in a similar manner to *Leptographium lundbergii*, with the spores gathering in mucoid balls at the apex of the synnemata.

## SERIES ARTHROSPORAE

In this series are classified those Hyphomycetes in which the conidia (arthrospores) are produced after septation and breaking up of simple or branched sporogenous hyphae. The sporogenous hyphae may be morphologically and functionally identical with the vegetative hyphae or may be produced on more or less well differentiated conidiophores.

Arthrospores are, in their simplest form, similar in shape and size to the cells which gave them origin and are often relatively thin-walled and cylindric in shape, with the end wall truncate or slightly convex. Fre-

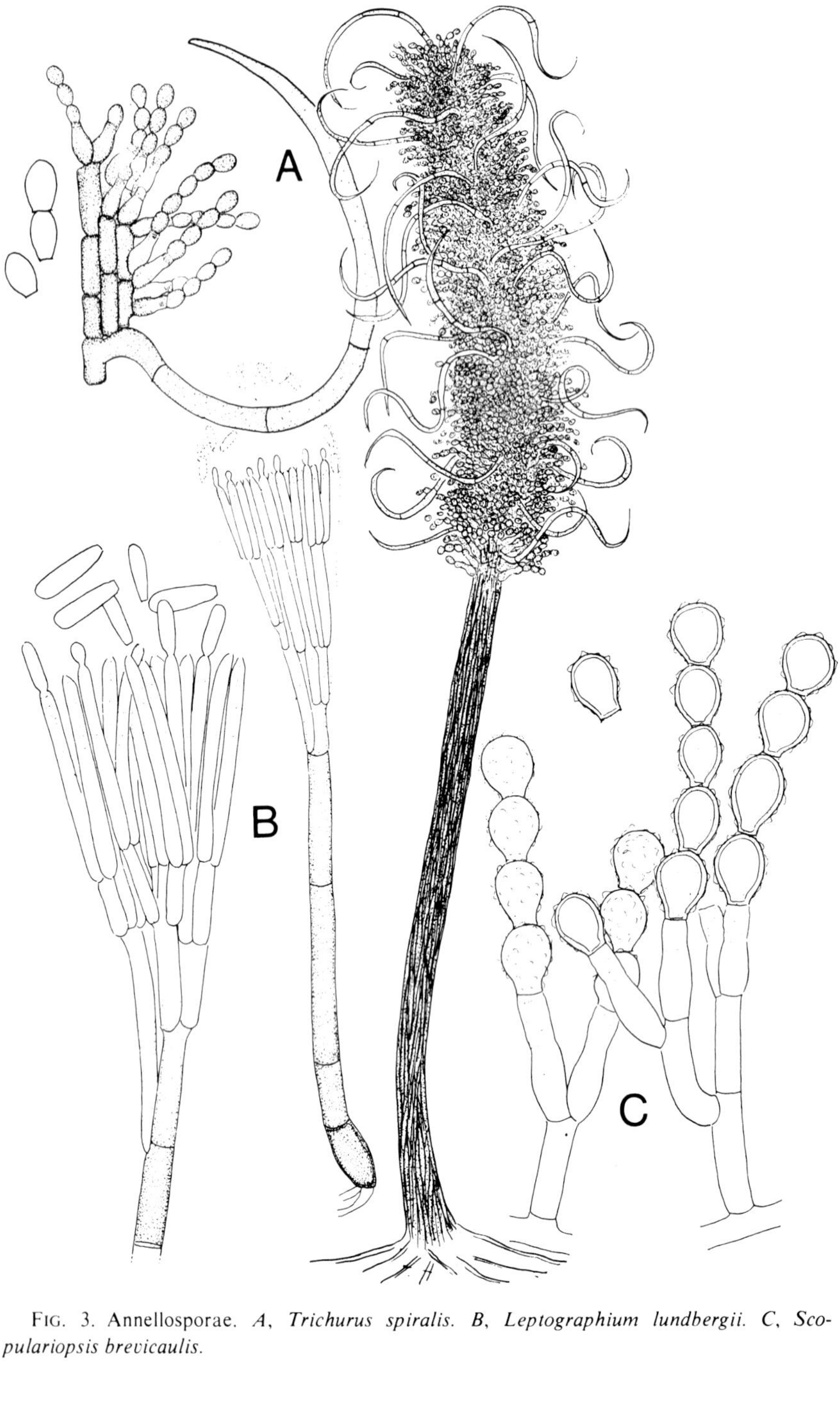


FIG. 3. Annellosporae. *A*, *Trichurus spiralis*. *B*, *Leptographium lundbergii*. *C*, *Scopulariopsis brevicaulis*.

quently, the cells may become considerably differentiated with thick walls and are globose, subglobose, ellipsoid or even helmet-shaped. They may be smooth or roughened, pale or dark, and in the mature state their "arthrospore" origin is not always obvious.

In *Geotrichum candidum* (Fig. 4*C*) more or less undifferentiated simple or branched sporogenous hyphae develop septa in basipetal succession. Each cell becomes thick-walled and secedes by the breaking down of the outer wall and a separation of the common transverse wall. It seems that an inner wall is laid down on each side of the transverse wall which either ruptures along its length or gelatinizes. The spores are short-cylindric and truncate, or more often slightly convex, at the end walls.

I have isolated a *Geotrichum*-like fungus (Fig. 4*A*) from soil on two occasions in which the sporogenous hyphae are very delicate, unbranched and grow suberect from the substratum. These aerial sporogenous hyphae may be produced in a dense stand and become septate and mature basipetally in typical arthrospore fashion. At maturity a simple chain of ellipsoid spores is produced.

The conidial state of *Monascus purpureus* (Figs. 4*E* and 117*E*) shows an interesting type of arthrospore development. The apical cell is cut off from a more or less erect vegetative branch. This cell swells up and becomes helmet-shaped. Cells are cut off in basipetal succession and mature to form short chains of spores. The final chains show a remarkable similarity to those of *Scopulariopsis*. In some cases the spores mature from both ends of the hyphae at the same time and their arthrosporic nature is then self evident.

In *Coremiella ulmariae* (Fig. 5), as described by Hughes (1953), the conidiophores are in tufts and each grows apically and branches irregularly or somewhat dichotomously to become arborescent towards the apex. Transverse septa are laid down in basipetal succession in the upper branches and neighboring or alternate cells develop thicker walls, a process which also takes place basipetally. The intermediate cells lose their contents entirely and the lateral walls remain thin, collapse inwards and finally break readily to form the oblong or cuboid conidia. The conidium has developed within the outer wall of the original hypha and the characteristic papillae and minute pore of the end walls are clearly visible. Only the terminal branches of the conidiophores become modified into conidia.

In *Oidiodendron maius* (Fig. 149) the main conidiophore axis is more or less erect and branches irregularly at the apex. The primary, pigmented branches give rise to delicate, interlacing, sporogenous hyphae. The sporogenous hyphae are hyaline, filamentous, narrow and frequently sinuous. These become septate basipetally from their apex back to the main conidiophore axis or one of the primary, pigmented branches. After sep-

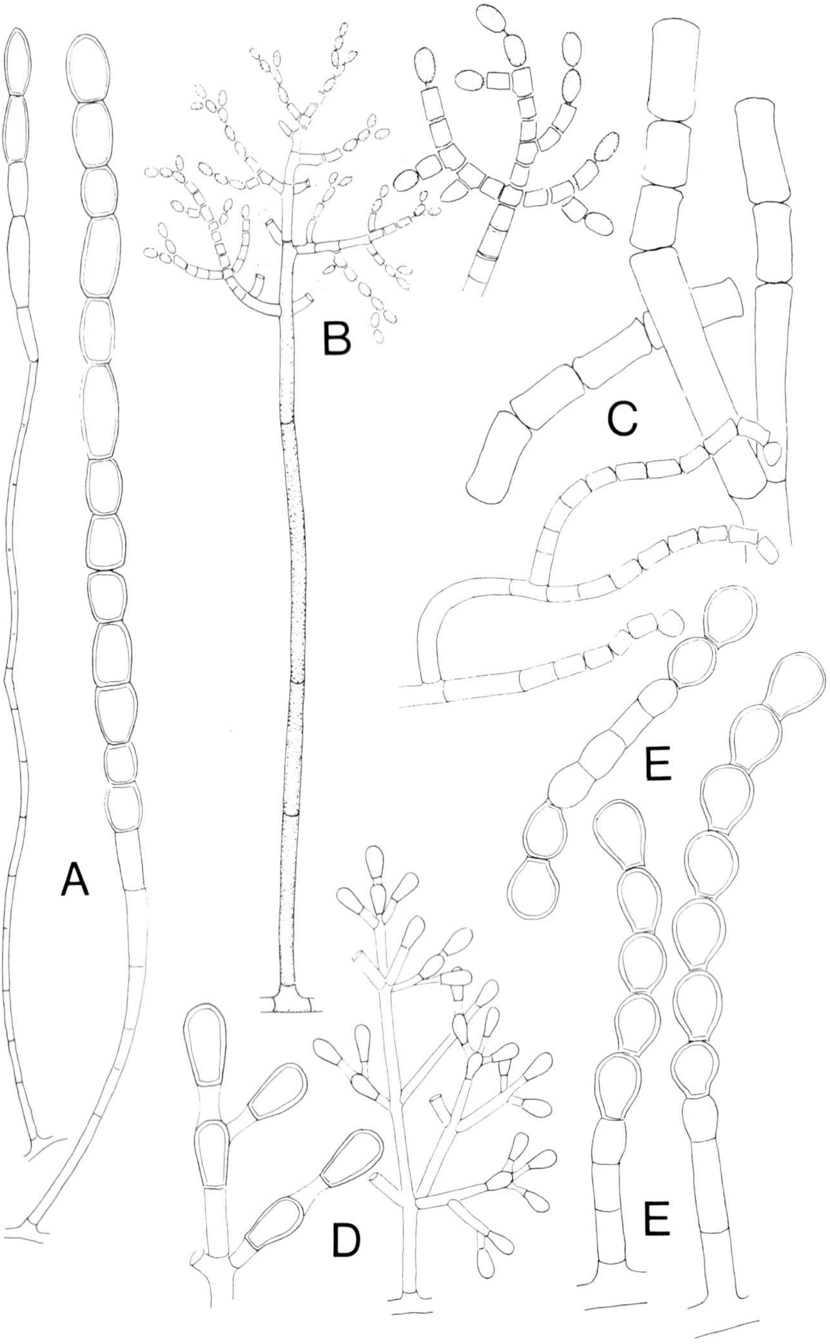


FIG. 4. Arthrosporae. *A*, *Geotrichum* sp. *B*, *Oidiodendron griseum*. *C*, *Geotrichum candidum*. *D*, *Chrysosporium pannorum*. *E*, conidial state of *Monascus purpureus*.

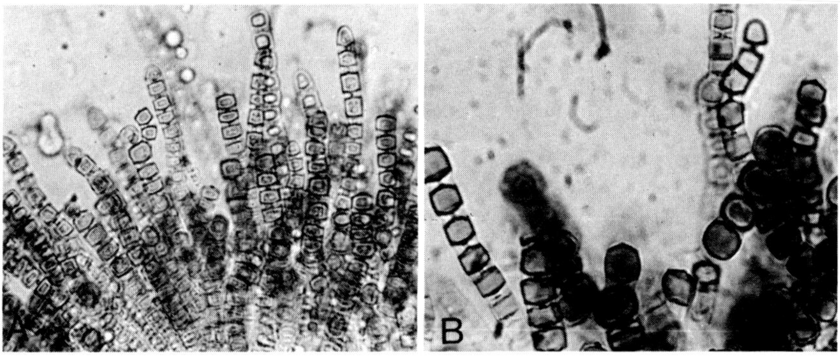


FIG. 5. *Coremiella ulmariae*. *A*, thick-walled arthrospores developing in chains. *B*, the parental hyphae can still be seen enveloping the arthrospores. The isthmus-like connectives between spores are characteristic of this species. (From type material, IMI 83012).

tation, an area around each septum becomes clear. As the spores mature, the clear area takes the form of a gelatinous plug between developing spores. The original septum can sometimes be seen traversing the plug. As the spores mature and round off, the plug diminishes, and finally all that remains is a narrow connective between adjacent spores; the general appearance is that of beads on a string.

*Chrysosporium* is generally regarded as an aleuriospore-forming genus (Carmichael, 1962) and is placed in Section III of Hughes' system. In *C. pannorum* (Fig. 4*D*) sporogenous hyphae with verticillate branching are produced in clumps from the vegetative hyphae. The terminal cell differentiates to form a spore, and subsequently spores may be produced in basipetal succession to form short, branched chains. In *Chrysosporium* sp. (Fig. 59*C*) the sporogenous hyphae are branched and sinuous and very similar to those found in *Oidiodendron*. The spores again are cut off and mature in basipetal succession to form branching chains of spores. The method of spore formation in these two species of *Chrysosporium* is essentially the same as other members of this series, and I include them in the arthrosporic forms rather than the aleuriosporic forms. In old cultures of *Chrysosporium pannorum* the sporogenous hyphae may be produced on more or less erect conidiophores (Fig. 52), as in *Oidiodendron*.

In *Chrysosporium aureum* the conidia are produced singly and terminally on the vegetative hyphae or branches and would be classified as aleuriospores.

## SERIES BLASTOSPORAE

The conidia in this series are solitary or in chains. Conidia frequently develop in acropetal succession as blown-out ends of simple or branching conidiophores. The apical branches of the conidiophore may frequently be delimited and secede to function as conidia (ramoconidia). Sometimes conidiophores are lacking and solitary conidia develop blastogenously from the hyphal cells. In forms lacking hyphae, blastospores are produced by budding from pre-existing conidia. In some forms the blastospores bud out successively at the apex of a more or less short sporogenous cell to form an apical cluster of blastospores.

Structurally, the simplest members of this group are the so-called asporogenous yeasts. These imperfect states of the Hemiascomycetidae may lack mycelium completely, as in *Cryptococcus*, in which each spore arises from a pre-existing spore by budding. A succession of blastospores may, under certain conditions, hold together to form a weak hyphal development referred to as pseudomycelium. More often, the attachment between the parent and daughter cells is a tenuous one and the spores come apart readily on mounting. In *Candida* (Fig. 6*C*) the mycelium is consistently present and the conidia are produced either by direct budding to produce solitary spores which arise at right angles to the vegetative hyphae or in short fragile chains.

In *Monilia cinerea* var. *americana* (=conidial *Monilinia fructicola*) (Fig. 6*F*), the causal agent of brown rot of stone fruits, the conidiophores are little distinguished from the vegetative hyphae. Branching chains of blastospores are produced and these arise in complex tufts from the substratum. The mature chains of conidia have a distinctly "beaded" appearance and fragment readily into nonseptate, hyaline conidia.

In *Cladosporium cladosporioides* (Fig. 6*A*) a dense stand of more or less erect brown conidiophores is produced. These branch irregularly at the apex and each branchlet gives rise to a branching chain of blastospores, to give the mature fructification an arborescent appearance. Many members of this genus have very dark connectives between adjacent spores, which gives them a distinctive appearance. In most species of the genus the spore chains are so fragile that it is very difficult to make a mount in which the conidia remain together in chains. Their catenulate nature, however, is evident from the attachment points at both ends of the spores (Fig. 55*A*).

*Pycnostysanus resinae* (Fig. 170) is a coremial form of this series. The conidiophores aggregate to form dark-brown synnemata which produce branching chains of blastospores at the apex. The spores are similar to those found in *Cladosporium*.

In *Hyalodendron* (Fig. 122 and 123) the conidia and chains are similar to *Cladosporium* save that they are hyaline. In the *Hyalodendron* species I have recovered from soil, the main conidiophore axis is usually poorly developed, with the branching chains of conidia arising either directly from the vegetative hyphae or from very short conidiophores. As in

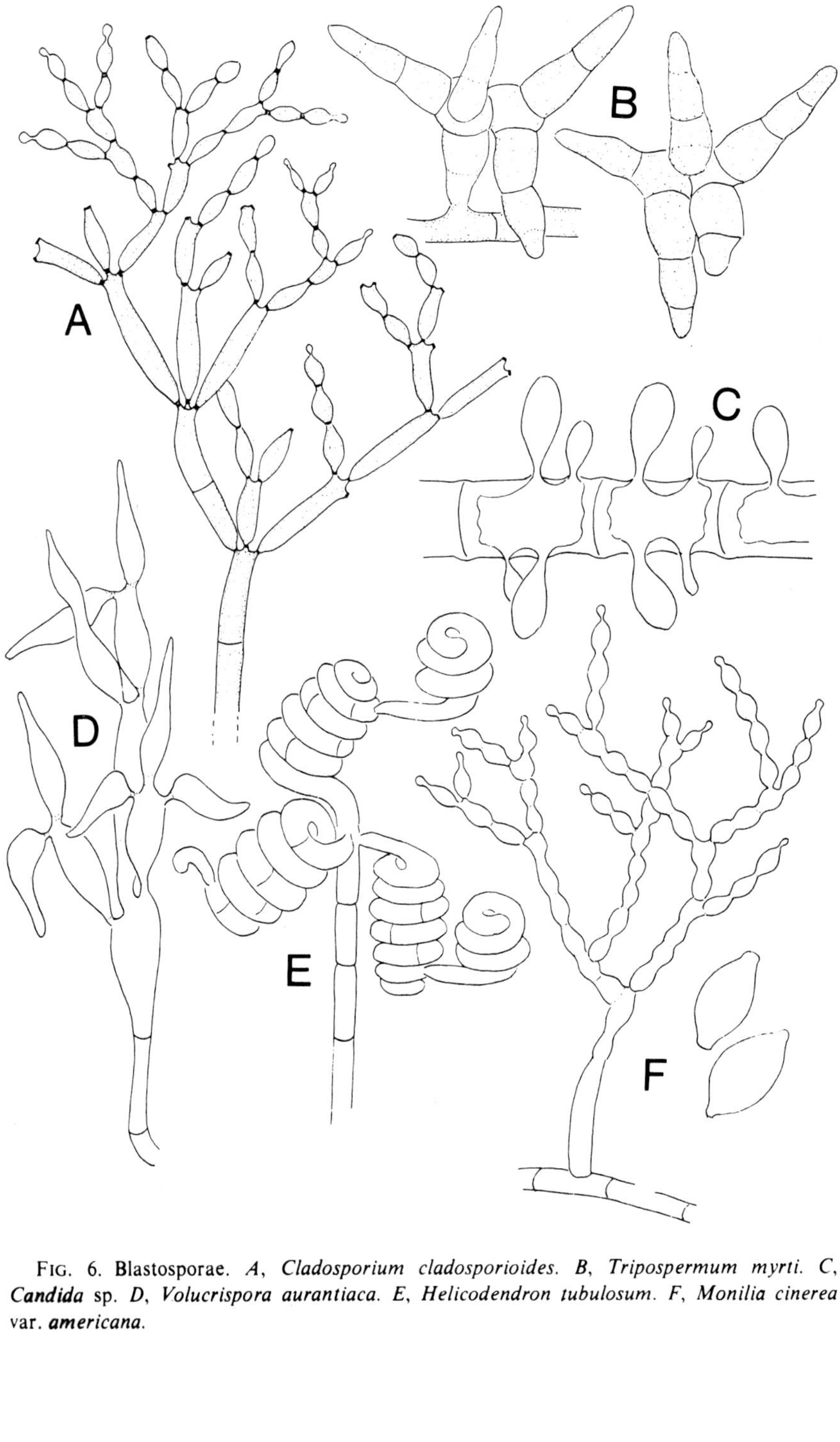


FIG. 6. Blastosporae. *A*, *Cladosporium cladosporioides*. *B*, *Tripospermum myrti*. *C*, ***Candida*** sp. *D*, *Volucrispora aurantiaca*. *E*, *Helicodendron tubulosum*. *F*, *Monilia cinerea* var. ***americana***.

*Cladosporium*, the conidial chains of *Hyalodendron* are very fragile and are best studied in slide culture. The conidia have dark connectives at the attachment points.

In *Bispora* (Fig. 34), unlike most of the other members considered, the acropetal series of spores forms simple chains. The spores here are dark, two-celled and have a conspicuous central septum.

In *Sphaeridium candidum* (Fig. 186), another coremial member of the Blastosporae, the ultimate branches produce long, branching chains of hyaline, nonseptate conidia which appear like little tufts of cotton on the dark synnemata, giving the colonies a striking appearance.

*Periconia macrospinosa* is typical of the genus *Periconia*, and the spores develop in acropetal succession but mature in basipetal succession. The blastogenous nature of the conidia is therefore not as obvious, since the oldest spores appear to be terminal. The conidia may be produced on a stout, dark, unbranched, erect conidiophore, but frequently in culture the conidiophores are suppressed and the branching chains of conidia arise more or less directly from the vegetative hyphae. In *P. macrospinosa* the ornamentation of the conidia is unusual, with large laminated or fibrillose outgrowths, which give the spores a shaggy rather than spiny appearance (Fig. 157*B*).

## SERIES BOTRYOBLASTOSPORAE

In this series the conidia are produced on well differentiated, swollen, sporogenous cells. The swollen apex of the sporogenous cell bears the conidia and is referred to as an ampulla. Each ampulla produces a large number of conidia more or less simultaneously over the surface. Primary conidia may remain solitary or give rise to simple or branching acropetal chains of blastospores. The conidiophores may be determinate or indeterminate. In determinate conidiophores, growth of the conidiophore or branch ceases with the production of the ampulla. In indeterminate conidiophores, the main axis may continue growth repeatedly from the ampulla, to give a noded appearance to the conidiophores, with the conidia borne at the "nodes." Indeterminate conidiophores may, on the other hand, produce an acropetal succession of lateral ampullae or fertile branches which bear solitary or clustered ampullae at the ends.

The term "ampulla" was adopted by Hughes (1953) for the conidium-bearing swellings in this group, following its introduction by Klebahn (1930) and subsequent use by Whetzel and Drayton (1932) for the conidium-bearing swellings in *Botrytis*. It is important to emphasize here that the ampullae of the Botryoblastosporae are distinct in kind from similar swellings found in certain of the Sympodulosporae such as *Arthrobotrys*. The ampulla of the Botryoblastosporae is formed prior to spore production. The vesicle in the Sympodulosporae is produced, in most cases, as a result of successive spore formation.

In *Botrytis cinerea* (Fig. 8*C*) the conidiophores are tall, stout, and dark-brown, with the main axis dichotomously or subdichotomously branched. Near the apex of each branch a number of short sporogenous branches are produced, each terminating in a solitary ampulla on which the conidia develop simultaneously. The spores become lightly pigmented and appear grey to the eye, hence giving the common name of "grey mold" to this and other species of the genus. In *Botrytis*, growth of the conidiophore ceases with the production of the terminal ampullae, and this is therefore an example of a determinate type of conidiophore.

In *Oedocephalum* sp. (Fig. 8*B*) the conidiophores are simple, hyaline and more or less erect. Each conidiophore terminates in a solitary ampulla in this species, with the spores being blown out simultaneously over the surface. Mature spores secede readily and the ampullae appear minutely echinulate, a feature characteristic of the Botryoblastosporae but by no means restricted to this series.

In *Cephaliophora tropica* (Fig. 7*C*) conidiophores are not well developed and the subglobose to clavate sporogenous ampullae arise more or less sessile and at right angles to the vegetative hyphae. The conidia are confined to the upper two-thirds of each ampulla and at maturity are multiseptate, thick-walled and lightly pigmented. An optical section through the mature fructification can sometimes reveal a striking symmetry (Fig. 42*B*).

In *Gonatobotrys simplex* (Fig. 7*B*) the conidiophores produce a solitary, terminal ampulla as in *Oedocephalum*, but, subsequent to spore production, the main conidiophore axis continues growth from the apex of the ampulla and, after a period of sterile growth, produces a second terminal ampulla at a higher level. This process is repeated and results in a succession of ampullae, each bearing a cluster of spores, which gives the mature fructification a noded appearance.

In *Gonatobotryum apiculatum* (Fig. 7*A*) the main conidiophore axis is simple, with each ampulla borne singly and terminally. Ampullae are only slightly broader than the width of the conidiophore. A primary series of conidia is blown out from the ampulla. Conidia of the primary series produce simple or branching chains of blastospores. The conidiophore then continues growth and produces a second ampulla at a higher level. This process is repeated several times, to give an elongate, simple conidiophore with radiating chains of blastospores at each "node."

An interesting member of this series was described by Rai and Tewari (1963) as *Acremoniella serpentina*. The main conidiophore axis is indeterminate, more or less erect and sinuous. Below the septa, lateral branches may develop which are similar in appearance to the main axis, and these may also give rise to secondary and tertiary branching systems. The mature fructification consists of complex tufts of intermingled conidiophores. Below certain septa, lateral ampullae are blown out. These

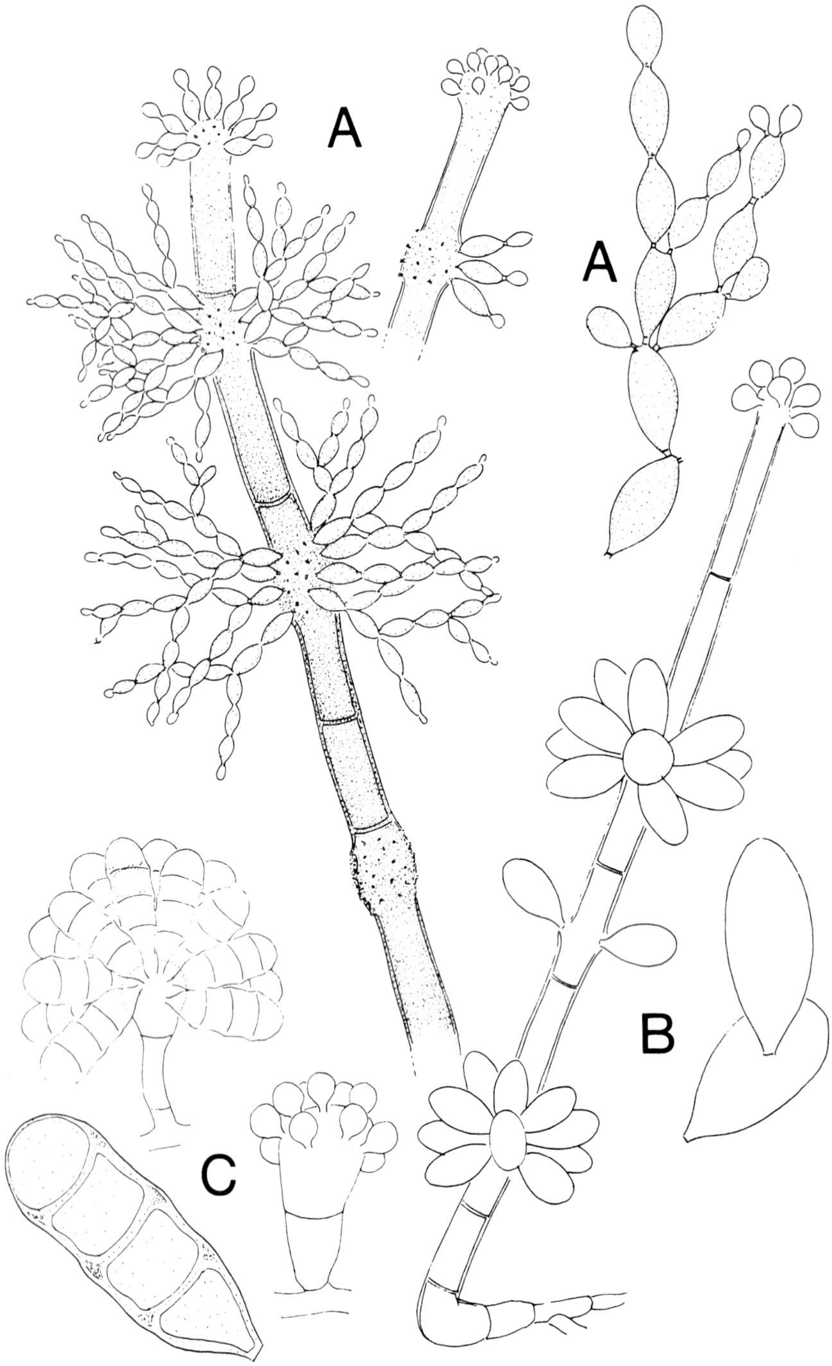


FIG. 7. Botryoblastosporae. *A*, *Gonatobotryum apiculatum*. *B*, *Gonatobotrys simplex*. *C*, *Cephaliophora tropica*.

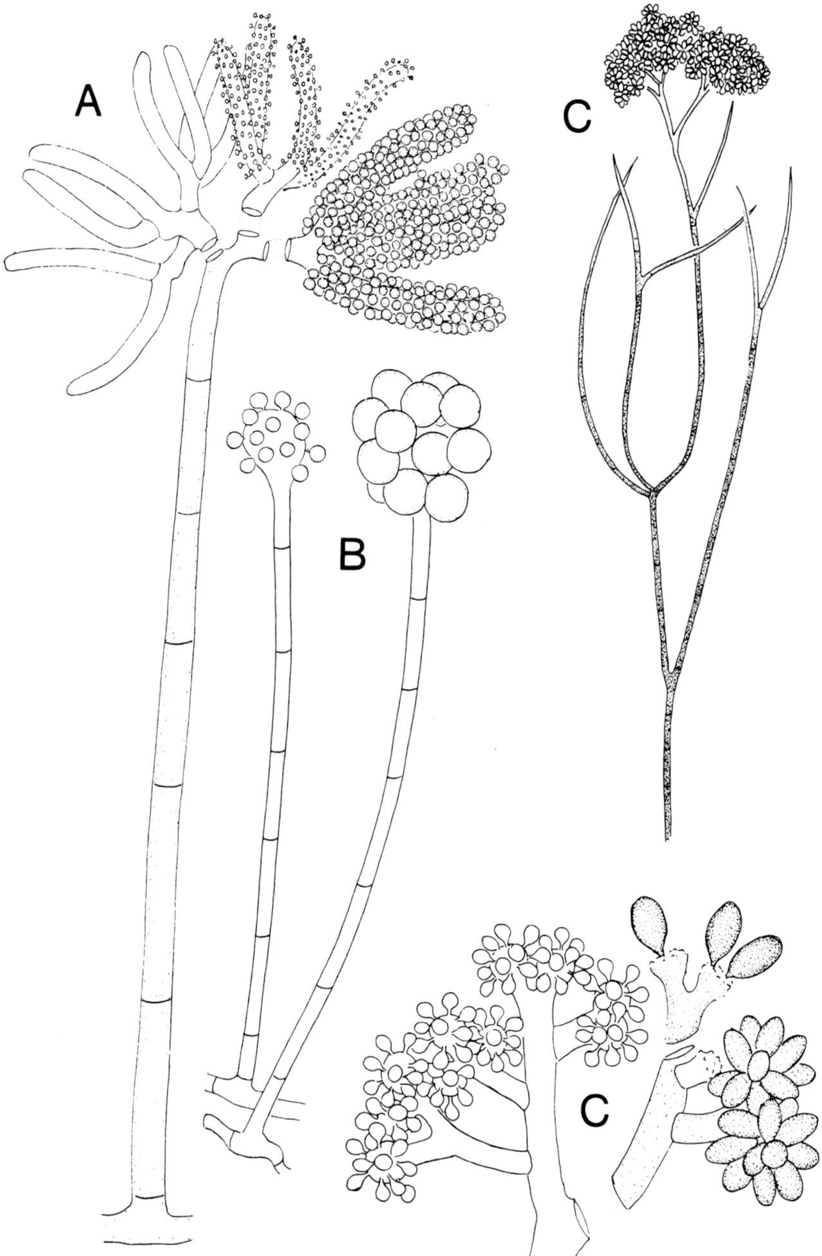


FIG. 8. Botryoblastosporae. *A*, *Ostracoderma* state of *Peziza ostracoderma*. *B*, *Oedocephalum* sp. *C*, *Botrytis cinerea*.

develop in more or less acropetal succession on the conidiophore, to give a series of ampullae along the length of the conidiophore at different stages of development. Ampullae may be globose, subglobose or somewhat elongate, and form conidia simultaneously in the manner characteristic of the group. From the above description it is clear that *A. serpentina* is not congeneric with *A. atra*, the type species of *Acremoniella*, and must eventually be redisposed.

In *Botryosporium longibrachiatum* (Fig. 35) the conidiophores are more or less erect, dichotomously or subdichotomously branched and indeterminate in growth. The conidiophores bear lateral, fertile branches in acropetal succession. Each fertile branch is somewhat clavate, being narrower at the attachment end and broader near the apex, where it bears a cluster of up to six ampullae. The small, ovoid conidia develop synchronously on minute denticles all over the surface of the vesicle. The mature fructification is pure white and quite striking in appearance.

## SERIES MERISTEM ARTHROSPORAE

The conidia develop in a gradually maturing, basipetal series and originate by meristematic growth of the apical region of the conidiophore in such a way that the chain of conidia merges imperceptibly with the conidiophore that gives rise to the chain.

According to Hughes (1953) the fungi classified in this series develop basipetal chains from poorly developed conidiophores which possess a generative or meristematic region towards the apex, with the result that the conidiophore merges imperceptibly with the chain of conidium initials which exhibit a gradual maturation towards the distal end of the chain.

Hughes included here the conidial states of *Hysterium insidens* and *Sirodesmium diversum* and also the *Acrosporium* (*Oidium*) state of *Erysiphe polygoni*.

Members of this series are uncommon in soil. The only species isolated from soil is *Coniothecium betulinum* (see *Trimmatostroma*). In *Coniothecium betulinum* the fructification is pustular. A section through a pustule shows the conidia to be dictyospores, developing in basipetal chains from more or less undifferentiated hyphae of the stroma (Fig. 9). Hughes pointed out that branched chains may occur, but these are not formed acropetally. The origin of a branch is at a triradiate conidium or at a conidium bearing a short lateral outgrowth at right angles, but at no regular position with regard to the linear part of that particular conidium from which it originates. Hughes believed that lateral branches have arisen by a lateral (as well as longitudinal) swelling of a conidium initial; septation is followed by the longitudinal expansion of one cell, and the lateral expansion of the other to produce a basipetal chain similar to the main chain on which it is borne.

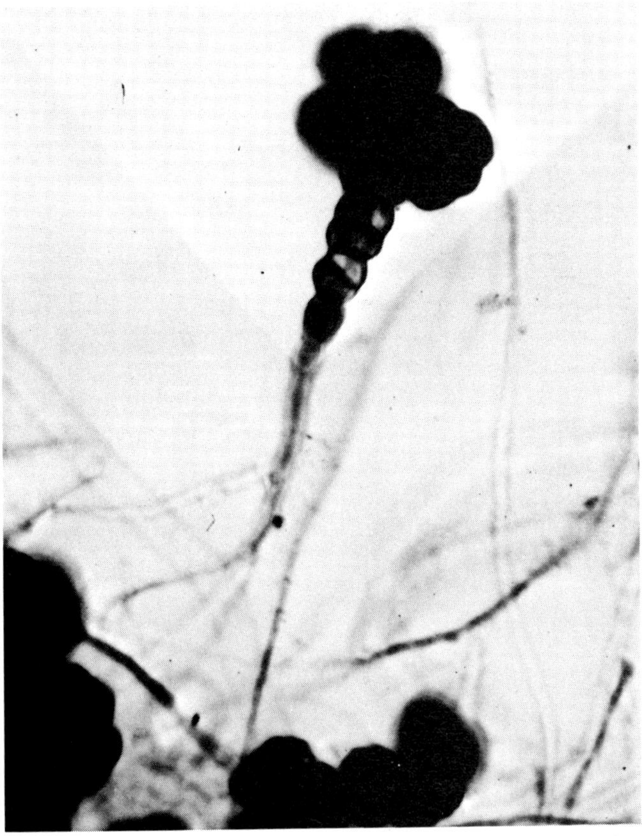

FIG. 9. Meristem Arthosporae. *A*, *Trimmatostroma betulinum* (Courtesy C. Dickinson).

## SERIES MERISTEM BLASTOSPORAE

Conidia are borne singly at the apex, or singly at the apex and laterally, and are often in regular whorls on conidiophores which elongate from the base. The conidia frequently have a longitudinal slit in the wall, but this character is not restricted to this series.

In this section Hughes included those species in which the conidiophores elongate from a basal meristem. The conidiophore itself may arise from a barrel-shaped or flask-shaped conidiophore mother cell which usually bears a terminal (often atypical) conidium before development of the conidiophore. The conidia of mature conidiophores are either solitary at the apex or borne apically and irregularly on short stalks along the increasing length of the conidiophore. They may also be produced apically and in whorls between thickened septa of the conidiophore. All the conidia arise as blown-out ends, and the oldest conidia are usually towards the apex and the youngest towards the elongating base of the conidiophore.

The various species produce fructifications which are usually pustular, superficial and with dry, powdery conidia. The conidia are mostly colored and may be amerospores or septate in such a way that they are composed of a flattened or irregular plate of 4, 8, or 16 cells. The amerospores are usually flattened, almost lenticular and bivalvate with a hyaline rim or germ slit.

In *Arthrinium phaeospermum* (= *Papularia sphaerosperma*) (Fig. 10) the fructifications may be superficial or immersed then erumpent. In culture the mycelium produces a white or dirty-white, cottony growth with funiculose strands of mycelium. Conidiophores are borne in clusters on the aerial hyphae. The conidiophore arises directly from the vegetative hypha as a swollen outgrowth, with a slender tubular tip which blows out terminally to form the first conidium. Subsequent conidia are borne laterally on an elongating tubular extension of the sporogenous cell. The youngest conidia are nearest the base. Conidia are lenticular, with a conspicuous, hyaline rim.

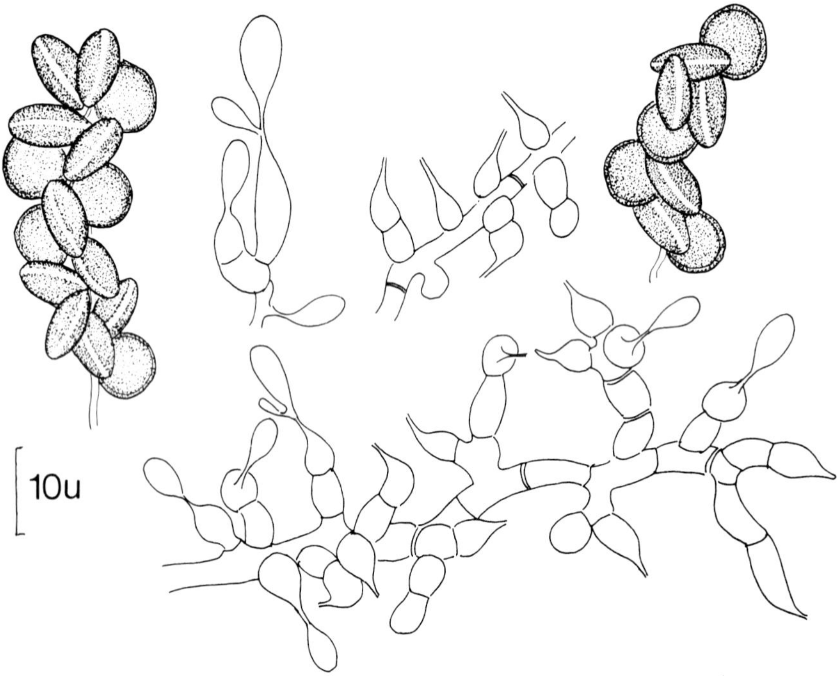


FIG. 10. Meristem Blastosporae. *Arthrinium phaeospermum* (= *Papularia sphaerosperma*).

In *Spegazzinia tessarthra* (Fig. 185) the conidiophores arise from conidiophore mother cells and bear the single, terminal conidium prior to elongation. At maturity the conidiophores are of two kinds, micro- and macroconidiophores, each bearing different kinds of conidia, a single one at the apex of each. The macroconidiophores are dark-brown and do not possess wide, black, transverse septa nor a germ slit.

### SERIES PHIALOSPORAE

In the Phialosporae the sporogenous cells remain more or less constant in length and the conidia (phialospores) are abstricted successively in basipetal succession from an open growing point at the apex of the sporogenous cell.

The term "phialide" is restricted by Hughes to those unicellular structures which are usually terminal on simple or branched conidiophores. They are oval to subcylindrical to flask-shaped or subulate, often with a well differentiated basal swelling and a narrower distal neck, with or without a terminal collarette. From the apex, each phialide develops a basipetal succession of phialospores without an increase in the length of the phialide itself. If the phialide does proliferate by extension of the sporogenous cell through the mouth of the phialide, then a plurality of conidia develops at each level (with rare exceptions). Sometimes a phialide will possess two or three functional collarettes, in which case the term "polyphialide" is applied to it.

In some genera the sporogenous cells arise more or less directly from the vegetative hyphae or from hyphal ropes, as in *Cephalosporium*, *Torulomyces*, *Gliomastix* and certain *Paecilomyces* species. In *Torulomyces lagena* (Fig. 11*C*) a solitary, swollen phialide is borne on a short stalk and gives rise to a long chain of spherical, slightly roughened conidia. Collarettes are not found in *Torulomyces* or *Paecilomyces* but are evident in some *Gliomastix* species such as *G. guttuliformis* (Fig. 98).

Collarettes are particularly conspicuous in *Catenularia*, *Chloridium* and most species of *Phialophora*. In *P. verrucosa* (Fig. 12*B*) the phialides are short and swollen, solitary or in clusters and have a large darkly pigmented collarette. In *Phialophora* the spores differentiate singly within the cup and slime down at the mouth to form a ball. In *Catenularia* the spores also differentiate singly, but successive spores here persist to form short chains as in *C. cuneiformis* (Fig. 41). In *Catenularia* the sporogenous cells are solitary and borne on short, septate, pigmented conidiophores. In *Chloridium chlamydosporis* (Fig. 11*F*) the sporogenous surface of the phialide is close to the mouth of the collarette, and in this species several spores of different ages may develop concurrently, to form either a spore-ball or short column. In both *Chloridium* and *Catenularia*

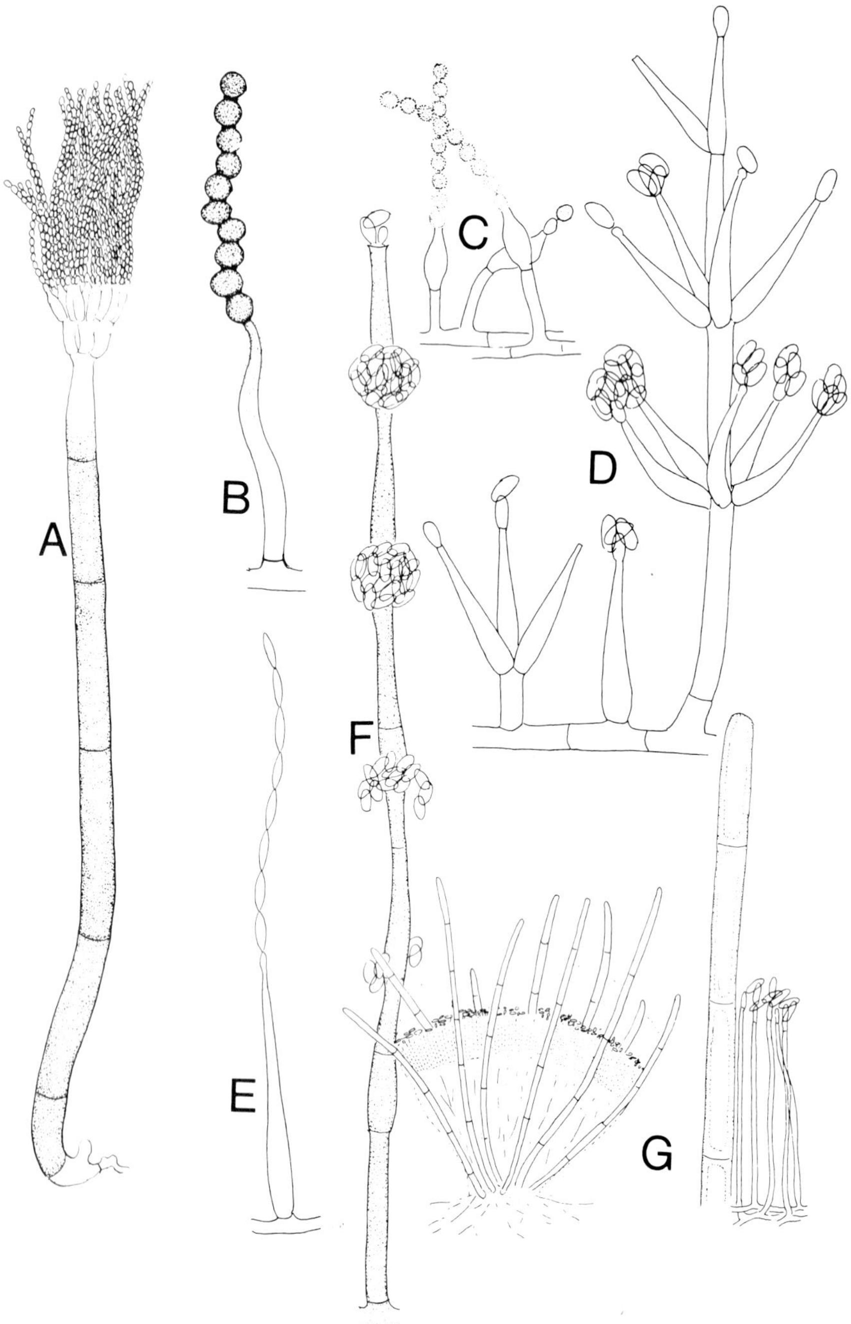


FIG. 11. Phialosporae. *A*, *Thysanophora penicillioides*. *B*, *Gliomastix murorum*. *C*, *Torulomyces lagena*. *D*, *Verticillium lateritium*. *E*, *Paecilomyces terricola*. *F*, *Chloridium chlamydosporis*. *G*, *Volutella ciliata*.

the phialides may proliferate repeatedly through the cup to form secondary, tertiary or higher order sporogenous cells. Proliferation of the phialide through the collarette is not common in *Phialophora*.

In *Thielaviopsis basicola* (Fig. 12*H*) the spores are differentiated within an elongate, tubular sporogenous cell. Such spores are frequently referred to as "endospores," but this term is not restricted to such genera and is frequently applied to any phialospore produced within a collarette. In *Chalara* (Fig. 49*B*) and similar genera the spores are short-cylindric and truncate at both ends; they are pushed out through the mouth of the phialide to form long, fragile chains. The sporogenous cell frequently has a slightly swollen base and a long, tubular barrel and is referred to as a "rifle cell" in some of the literature.

In the above mentioned genera, the phialides arise more or less directly from the vegetative hyphae or hyphal ropes; less often, a few phialides may arise in a group from a short conidiophore, particularly in *Phialophora*. In a large number of the common phialosporous Hyphomycetes, however, the phialides arise from a well defined specialized main axis, the phialophore. The phialides may arise directly from the main axis or indirectly on branches or prophialides. In *Verticillium lateritium* (Fig. 11*D*) and some species of *Paecilomyces*, the phialides are borne in whorls of three or more. Such whorls are referred to as verticils and the arrangement is said to be verticillate. In *Gonytrichum macrocladum* (Fig. 12*E*) the sporogenous cells do not arise directly from the main axis but from a specialized "collar hypha" which grows out from below a septum on the main axis. Hughes (1951c) referred to these whorls of phialides as "false verticils."

In *Penicillium*, *Gliocladium*, *Phialocephala*, *Thysanophora* and some species of *Paecilomyces* the conidiophore branches apically to form a *penicillus*. The main axis of the conidiophore is unbranched; near the apex primary, secondary and tertiary branching occurs to form a tight, brush-like head (Figs. 11*A* and 12*G*). The ultimate branchlets, which bear the phialides in groups, are referred to as prophialides or metulae.

In *Aspergillus* (Fig. 26) and *Gliocephalis* (Fig. 94) the apex of the main conidiophore axis swells up to form a vesicle on which the phialides are borne directly, as in *Aspergillus fumigatus*, or a series of prophialides is formed from which phialides are borne in clusters, as in *A. niger* and *Gliocephalis hyalina*. In *Aspergillus* the phialospores persist in long chains, while in *Gliocephalis* and *Goidanichiella* (Fig. 12*C*) they slime down to form gloeoid heads.

Most phialospores are nonseptate, but in a number of genera they are one- or several-septate. In *Cladobotryum variospermum* the spores are one-septate and persist in fragile chains (Fig. 54). In *Fusariella* (Fig. 86) the conidia are phragmospores and are attached "tip to tail" in short chains. In the majority of genera which produce phragmospores from

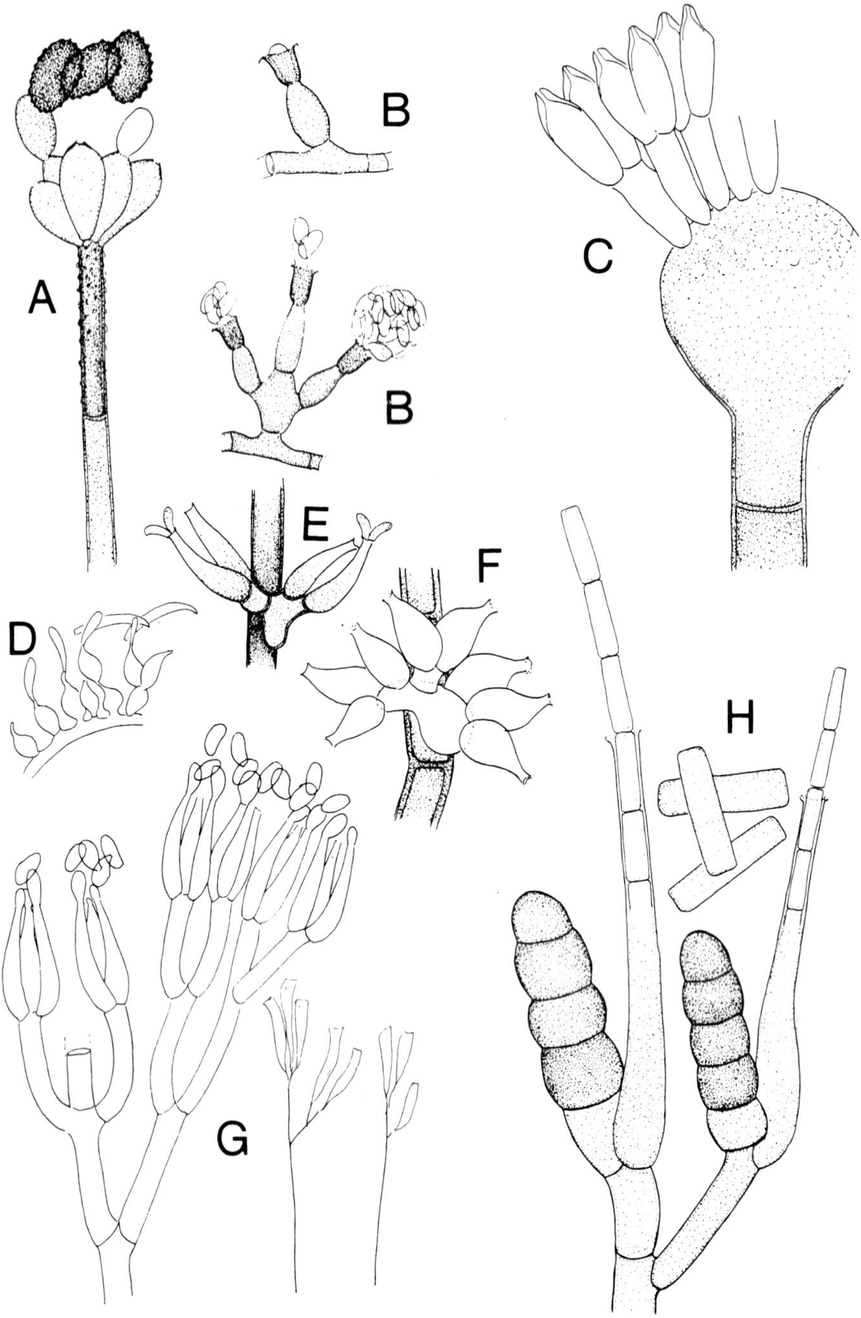


FIG. 12. Phialosporae. *A*, *Stachybotrys atra*. *B*, *Phialophora verrucosa*. *C*, *Goidanichiella* sp. *D*, *Harposporium lilliputianum*. *E*, *Gonytrichum chlamydosporium*. *F*, *Gonytrichum* sp. *G*, *Gliocladium roseum*. *H*, *Thielaviopsis basicola*.

phialides, the spores slime down to form gloeoid balls or spore masses as in *Fusarium*, *Cylindrocarpon* and *Cylindrocladium*.

In *Menispora ciliata* (Fig. 132) the phialospores have a single setula at each end. Not all species of this genus, however, have setulate phialospores. In *Leptodiscus* (Fig. 128) the two-celled, setulate phialospores are borne in sporodochium-like fructifications. It was demonstrated by McVey and Gerdemann (1960) that the setulae have a function in spore dispersal. In the mucous mass the setulae are folded back along the spore, and when the mucus is diluted the setulae spring violently outwards. This same phenomenon was also reported by Hughes and Kendrick (1963) for *Menispora*.

Phialospores may be divided into "wet-spored" and "dry-spored." These biological spore types do not have great taxonomic significance, but they are convenient in some instances for distinguishing between genera. In some genera the spores of certain species may slime down, whereas those of others persist in dry chains. In a few cases, both wet and dry spores are found within a single species. *Gliomastix murorum* (Fig. 11*B*) has globose to subglobose darkly pigmented spores borne in short, tortuous chains. In *G. murorum* var. *felina* the spores slime down to form dark, gloeoid balls on the simple sporogenous cells. Similarly, in *F. moniliforme* the microspores are produced in chains, whereas in *F. moniliforme* var. *subglutinans* the spores slime down to give the appearance of a *Cephalosporium*. In *Chloridium chlamydosporis* (Fig. 11*F*) the spores slime down at the apex of the phialide to form balls, but in *C. caudiger* the spores adhere to form short columns.

In most genera the phialospores arise singly and successively from the mouth of the phialide. In some species of *Chloridium* several spores (at different stages of development) may develop concurrently at the sporogenous apex.

In most cases the phialide terminates in a single mouth through which a succession of spores is produced. In some genera, however, a succession of mouths is produced at the apex of a single sporogenous cell with one or a succession of phialospores being produced from each mouth. Such a structure has been referred to as a polyphialide as described for *Chaetopsis wauchii* (Fig. 48) by Hughes (1951c). Tubaki (1963) has described *Chloridium*, *Catenularia* and several other genera as having polyphialides. Here, however, after a number of spores have been produced, the growing point proliferates *through* the phialide and a second sporogenous cell is produced at a higher level. The first collarette is immediately nonfunctional. I do not regard these as polyphialides in the sense described by Hughes.

In *Fusarium chlamydosporum* (Fig. 16*C*) the microspores are produced on a very distinctive type of polyphialide. In some strains a succession of spores is produced from each mouth, but in others, so far as I can determine, only a single conidium is produced. If the macrospores are lacking or sporadic, this species is very difficult to identify as a *Fusarium*. It has been described as *Dactylium fusarioides* and *Nodulisporium didymosporum*.

### SERIES POROSPORAE

The conidia are usually thick-walled and develop through pores in the wall of the conidiophores. The conidiophores are of determinate or indeterminate length. If the latter, then increase in length is either by proliferation of the conidiophore through the apical pore or by sympodial extension of the conidiophore by renewal of growth from a point lateral to and just below the apex.

There is a wide variation in spore septation in this group and conidia may be amerospores, didymospores, phragmospores or dictyospores. They may be solitary or in simple or branched acropetal chains or sometimes capitate. Pores may be single and apical or scattered randomly over the conidiophore or located in groups just below the septa, to give the conidia a whorled arrangement.

As indicated by Hughes, the main character of this group is the development of the conidia through minute pores in the wall of the conidiophore. Hughes noted that the apex of the conidiophore is nearly always rounded, and the outer and inner walls come to an end abruptly, thus delimiting a more or less cylindrical pore. The base of the conidium is also usually rounded, and a morphological scar such as a break or tear in the wall is absent. The base of the conidium may be truncate; nevertheless, the wall of the conidium is continuous round the base except for the basal pore.

In *Diplococcium spicatum* (Fig. 78) the conidiophores are straight and more or less erect. They bear a number of primary and sometimes secondary lateral branches. Long chains of two-celled porospores are produced at random over the branches and the upper part of the main conidiophore axis. The chains originate through small pores in the conidiophore wall which become apparent when the fragile chains of spore are dislodged.

In *Stemphylium botryosum* (conidial *Pleospora herbarum*) (Fig. 194) the solitary conidia are produced apically at the ends of the conidiophore or its branches. The apex of the conidiophore is distinctly swollen and more darkly pigmented than the remainder of the conidiophore. The conidiophore proliferates through the pore and a second pore is produced at the apex of the proliferation. A series of proliferations give a noded appearance to the conidiophore. The spores here are sarciniform (box-shaped) and, as in many of the Porosporae, readily dislodged.

In *Ulocladium* (Fig. 216) the spores are very similar to *Stemphylium* in appearance. Here, however, the conidiophore increases in length by sympodial growth from a point lateral to and just below the first formed spore, which may be pushed over to take up a lateral position. A succession of spores is produced acropetally and the spores remain attached, apparently lateral, on a more or less geniculate conidiophore.

In *Alternaria* (Fig. 13*C*) the first formed spore is terminal. A succession of spores is formed in acropetal succession through the apex of the first formed spore. A pore may arise laterally near the apex of a conidium and originate branching chains of porospores. In *Alternaria* the conidiophore may also be indeterminate and increase by sympodial extension in the same way as *Ulocladium*.

In *Dichotomophthora portulacae* the conidiophore branches repeatedly in a dichotomous fashion with the ultimate branchlets bearing solitary porospores. In *Dichotomophthora indica* (Figs. 13*A* and Fig. 74), isolated from soil and *Portulaca oleracea* in Canada, the main conidiophore axis is unbranched and more or less erect. The ultimate dichotomies are borne in a tight series on a slightly swollen apex of the main axis, giving a head of sporogenous cells with the porospores radiating out.

In *Torula herbarum* (Fig. 13*B*) the conidiophore is poorly developed and appears as a slightly swollen tip of a vegetative hypha. The tip is slightly darkened and gives rise to branching chains of porospores produced in acropetal series. The apical cell of each conidium is darker than the others and with slightly thicker walls, reminiscent of the apex of the conidiophore in *Stemphylium*.

### SERIES SYMPODULOSPORAE

The conidia arise as blown out ends at the apex of simple or branched conidiophores and the ends of successively new growing points which develop to one side and below the previous conidium. The conidiophore, therefore, either increases in length or becomes swollen as the result of conidium production. Acropetal chains of conidia may develop on the primary conidia.

As pointed out by Hughes, the conidia are produced as blown-out ends of successive new growing points which develop just to one side of the previous terminal conidium. At maturity, therefore, a conidiophore or sporogenous cell which produces conidia in this way possesses a number of scars; each one of these was in turn terminal before being pushed aside by the development of the new growing point. Hughes noted that such conidiophores usually show a perceptible increase in length with the development of each conidium, but that in some cases a succession of growing points results, not in a longitudinal extension of the conidiophore or sporogenous cell, but in a swelling, as in *Arthrobotrys*. The distance between successive scars is often fairly constant, but in some species it

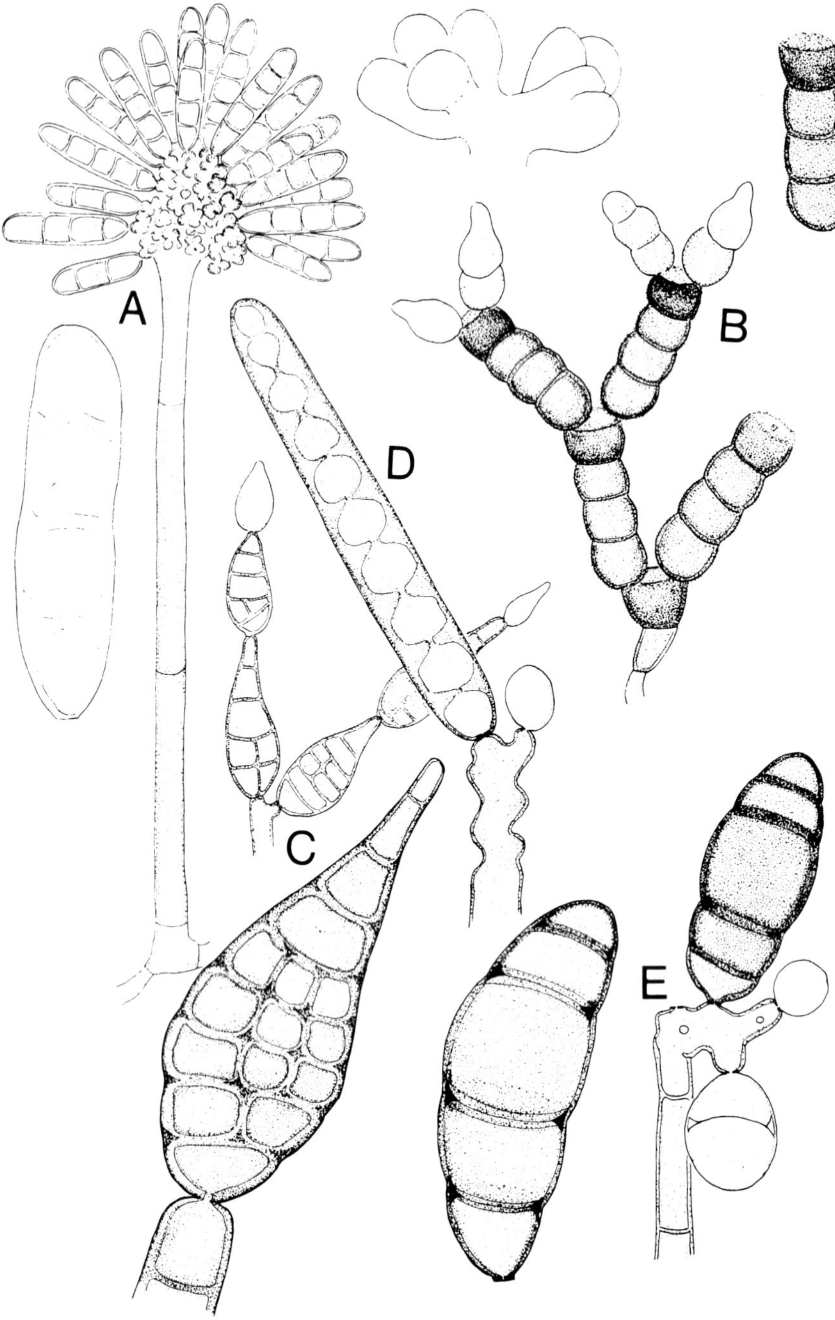


FIG. 13. Porosporae. *A*, *Dichotomophthora* sp. *B*, *Torula herbarum*. *C*, *Alternaria tenuis*, *D*, *Drechslera* sp. *E*, *Curvularia geniculata*.

varies considerably, depending on the activity of the new growing point before it blows out into a new spore. In some species of *Rhinocladiella* groups of conidial scars along the length of the conidiophore alternate with smooth areas.

In *Beauveria bassiana* (Fig. 14*B*) the sporogenous cells arise directly from the vegetative hyphae, singly, or more often in dense, radiating clusters. The sporogenous cell is inflated below and tapers to a slender apex which produces a single terminal spore. The sympodial growth of the sporogenous cell begins a little way back from the apex, and, after a succession of conidia are produced, each conidium appears to be borne on a short, slender peg.

*Tritirachium* is very close morphologically to *Beauveria*. Here, however, the sporogenous cells are less markedly inflated at the base and are borne in primary or secondary verticils from a more or less erect main conidiophore axis. In *Tritirachium roseum* (Fig. 214) the sympodial extension of the sporogenous cell originates very near the previous spore. The final sporogenous cell has therefore a markedly rachis-like appearance.

In *Verticicladiella abietina* (Fig. 219) the mononematous conidiophores are stout, tall and unbranched along most of their length. At the apex they bear a verticillate arrangement of up to four series of metulae with the final series bearing groups of sporogenous cells to form a compact sporogenous layer. In his treatment of *Verticicladiella*, Kendrick (1962) first introduced the term "sympodula" to describe the type of sporogenous cell discussed in this series. The conidia accumulate in slime to form cream-colored heads which may darken to yellow-brown in age.

In *Graphium ulmi* the method of conidium formation is exactly as in *Verticicladiella*. Here, however, mononematous conidiophores are lacking and the conidiophores are aggregated into dark, stout synnemata. In *Verticicladiella* and *Graphium ulmi* the sporogenous cells are slender, hyaline and somewhat tapered towards the distal end. Projections or scars left by the seceding conidia are not prominent.

In *Rhinocladiella atrovirens* and other *Rhinocladiella* sp. (Fig. 173) the scars are very pronounced, as they are in *Nodulisporium hinnuleum* (Fig. 14*C*) and, indeed, in most of the genera in which the spores are produced in dry heads.

In *Scolecobasidium constrictum* (Fig. 179) the sporogenous cell may be short and swollen or longer and cylindrical. In either case the spore arises from a delicate, tubular prolongation of the apex of the sporogenous cell. When the spore secedes, it leaves this prolongation as a delicate, tubular appendage on the sporogenous cell.

In *Arthrobotrys oligospora* (Fig. 15*B*) the main axis and branches of the conidiophores are stout and hyaline. Spores are produced in acropetal

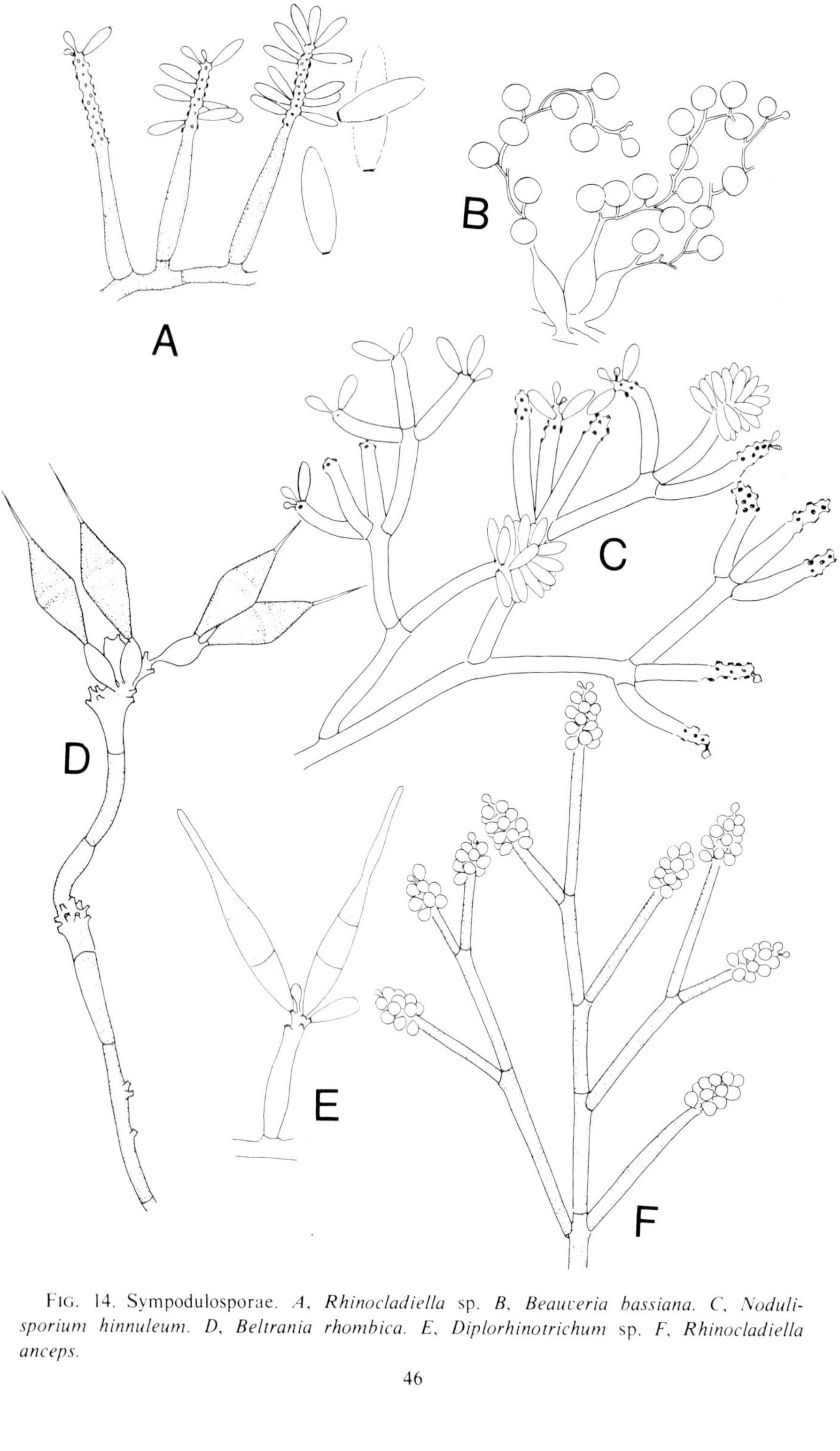


FIG. 14. Sympodulosporae. *A*, *Rhinocladiella* sp. *B*, *Beauveria bassiana*. *C*, *Nodulisporium hinnuleum*. *D*, *Beltrania rhombica*. *E*, *Diplorhinotrichum* sp. *F*, *Rhinocladiella anceps*.

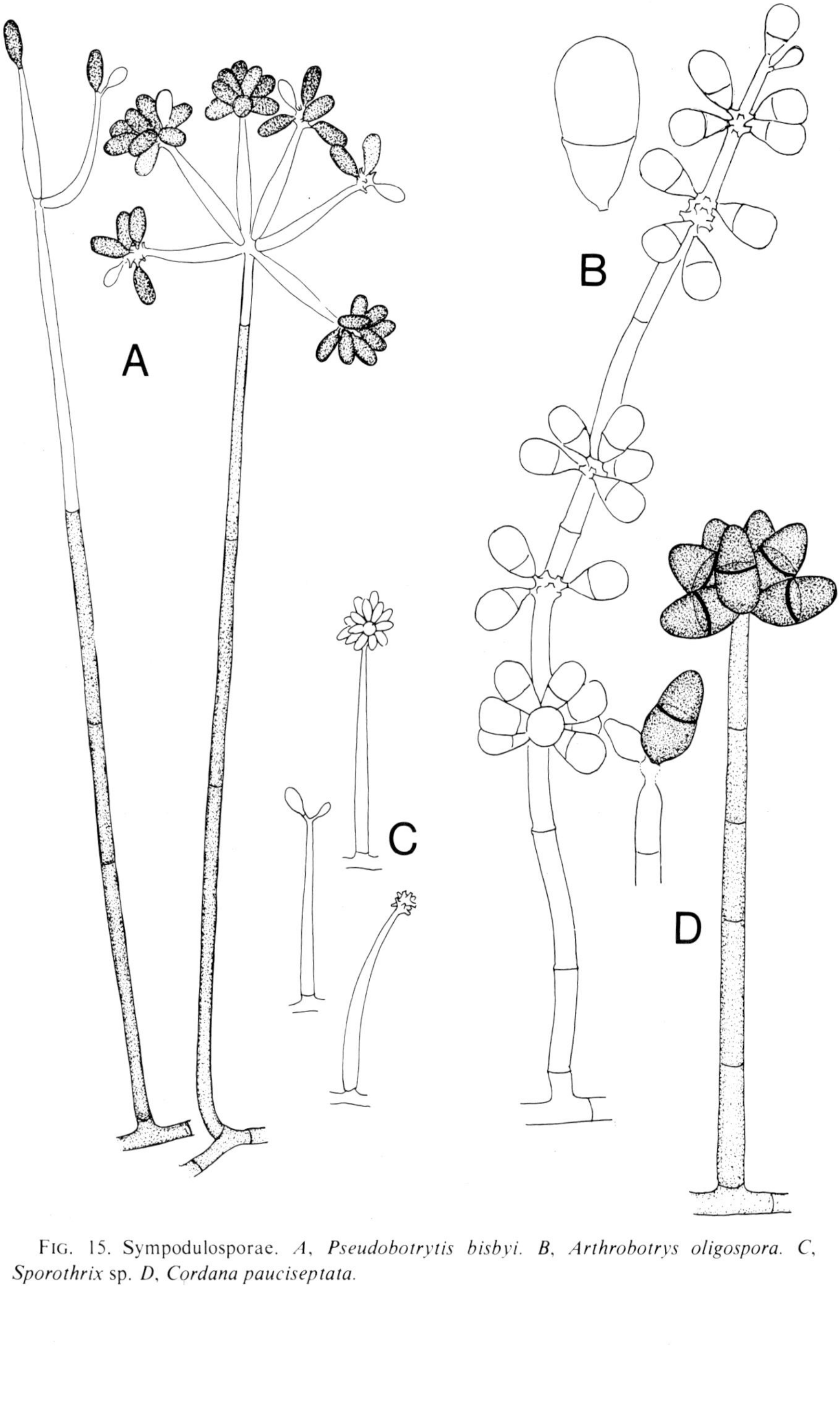


FIG. 15. Sympodulosporae. *A*, *Pseudobotrytis bisbyi*. *B*, *Arthrobotrys oligospora*. *C*, *Sporothrix* sp. *D*, *Cordana pauciseptata*.

succession at the apex, but there is no marked elongation of the sporogenous cell; rather, it swells up slightly to form an apical cluster of spores on a slightly swollen apex. The main conidiophore axis may then proliferate through the apex and, after a period of more or less sterile growth, produce a second acropetal cluster. Successive proliferations give a noded appearance to the final conidiophore, with each node bearing a cluster of spores, all apparently of the same age except for the terminal sporogenous apex on which spores are still developing. On casual inspection, the mature state of *Arthrobotrys* seems very similar in appearance to *Gonatobotrys* of the Botryoblastosporae, but in the latter the spores on the vesicle develop synchronously rather than in acropetal succession.

In *Pseudobotrytis bisbyi* (Fig. 15*A*) the sporogenous cells are borne in groups of up to eight, radiating out from the apex of an unbranched, pigmented conidiophore. The spores are one-celled, darkly pigmented, and borne acropetally at the apex of the sporogenous cell which swells up, as in *Arthrobotrys*, so that in the mature state a cluster of spores is borne on a bulbous, terminal swelling.

### ANOMALOUS SPORE FORMATION

A number of genera do not fit readily into the classification proposed above. The method of spore production may be controversial, with different interpretations being put forward by the various workers concerned. In some cases a superficial examination will suggest one type of spore origin, while a more detailed study reveals another. In a few cases, the method of spore formation seems to be a hybrid between two recognized types. This is not surprising, and one might expect to find not only anomalous methods of spore formation which do not agree with any of the suggested categories, but also a certain amount of overlap between categories.

In *Wallemia ichthyophaga* (Fig. 16*B*) the conidiophores are produced in a dense, more or less erect stand from the substratum. Each conidiophore terminates in a solitary, apical, phialide-like structure which is slightly constricted near the apex and bears what appears as a pigmented collarette. From the mouth of the collarette a filamentous sporogenous hypha grows out. This is nonseptate at first, but septates and matures basipetally into unicellular, arthrospore-like units which become spherical, roughened and lightly pigmented. A casual inspection indicates that this genus belongs in the Arthrosporae, but the phialide-like sporogenous cell indicates affinities to the Phialosporae and a possible origin in this group. It seems likely that the "phialide" acts as a meristem from which the sporogenous hypha continues to grow while the arthrospores develop above.

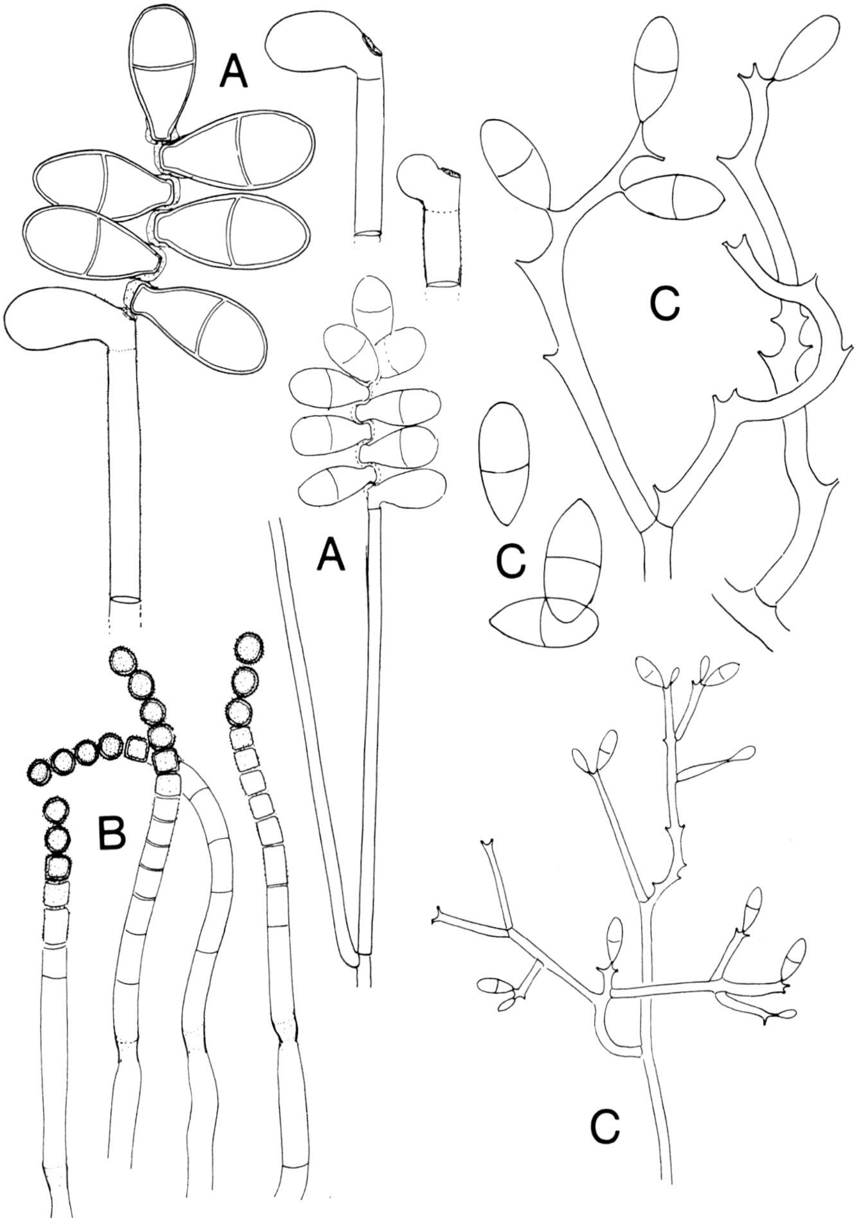


FIG. 16. Miscellaneous. *A*, *Trichothecium roseum*. *B*. *Wallemia ichthyophaga*. *C*, *Fusarium chlamydosporum*.

The method of conidium development in *Trichothecium roseum* was described in detail by Ingold (1956). The conidiophore is usually erect, 100 to 250 $\mu$ long and sparingly septate in the lower half. At maturity it is more or less cylindrical and slightly roughened with minute granules. The first spore is cut off terminally, with its long axis in line with that of the conidiophore. After the first spore is formed, the apex of the conidiophore immediately below blows out laterally to form a second spore which is cut off by a cross wall when it is close to maturity. A third spore arises below the second in a similar manner but on the other side of the conidiophore. A two-ranked chain of conidia is thus produced, with all conidia being horizontal except the first (Fig. 16*A*). Ingold noted that, in producing basipetal chains of spores without any change in length, the conidiophore resembles a phialide, but lacks a distal swelling and apical neck. He suggested that the conidiophore of *T. roseum* represents a modification of the annellophore type found in *Scopulariopsis*, but it differs in that the level at which each spore is produced is the same and therefore no annular scars are produced. He considered the conidia as terminal thallospores.

My own observations on *T. roseum* indicate that a part of the conidiophore is incorporated into the base of each spore as it is blown out sideways. The conidiophore therefore must either *decrease* in length as a succession of spores is formed or *increase* in length by meristematic growth from below. The conidia in *T. roseum* might be modified arthrospores.

In *Fusarium chlamydosporum* the microspore state has an unusual range of development. The sporogenous cell produces a terminal mouth on the phialide from which a ball of one-septate phialospores is produced. The sporogenous cell then proliferates sympodially and produces a second mouth at a higher level. Repeated proliferations give a polyphialide, but of rather an unusual shape. In some strains only one spore is produced from each mouth and the sporogenous cell appears denticulate. The macrospore state may be rare or lacking and such a strain of *F. chlamydosporum*, which produces solitary conidia on "denticles" from a sympodially extending sporogenous cell, appears as a member of the Sympodulosporae (Fig. 16*C*). If one was not familiar with the range of strains found in this species, then an isolate could easily be misidentified. Nicot (1956) described *F. chlamydosporum* as *Nodulisporium didymosporum*. This species is interesting in indicating possible relationships between the Phialosporae and Sympodulosporae.

CHAPTER

# III

# Mycelia Sterilia

Many fungi do not produce any recognizable sexual or asexual conidial state in culture. Such forms are frequently classified for convenience in the *Mycelia Sterilia*. This group is a catchall which may include a few well defined and easily recognizable genera, but more often is a repository for a large number of nondescript mycelial isolates.

Some genera like *Papulaspora* (Fig. 155) produce irregular clusters of cells called "bulbils" which in this genus are frequently orange to red-brown in color. Other genera such as *Sclerotium* and *Rhizoctonia* produce sclerotia. These may be hard and black and more or less regular in size and shape, as in *Sclerotium rolfsii* (Fig. 178). In *Rhizoctonia solani* (Fig. 175) the sclerotia are less well-defined, irregular in size and shape, brown to dark-brown in color and tend to be spongy in texture.

A large number of soil isolates produce neither spores, bulbils, nor sclerotia in culture. It cannot be hastily assumed, however, that these species belong properly in the *Mycelia Sterilia* in the strict sense. Many fungi fail to sporulate under the conditions imposed upon them in the laboratory. A number of the saprophytic and parasitic species of the Porosporae fruit poorly or not at all on normal laboratory media. Certain *Botrytis* species may develop only the sclerotial phase in culture with the conidial state being suppressed. A large number of Ascomycetes and Basidiomycetes which lack an asexual state will not produce their sporocarps with the use of normal laboratory techniques and will therefore be classified as *Mycelia Sterilia*. Warcup and Talbot (1962) recovered 31 mycelial cultures of Basidiomycetes which formed clamp connections and a further 11 isolates not forming clamps which later proved to be Basidiomycetes. The work of Warcup and Talbot (1962, 1963) encompassing both Basidiomycetes and Ascomycetes points out the need for more sophisticated and elaborate techniques to establish the identity of *Mycelia Sterilia* from soil. All too often mycelial fungi which may have great significance in the soil processes are dismissed lightly for no other reason than that they cannot be readily identified. The solution to this problem presents an interesting challenge to the soil ecologists.

CHAPTER

# IV

# Key to Series

| | |
|---|---|
| **1.** Conidia lacking, with units of reproduction or perennation consisting of irregular groups of cells (bulbils) or sclerotia . . | **Mycelia Sterilia** (see *Papulaspora, Rhizoctonia* and *Sclerotium* |
| **1.** Conidia present . . . . . . . . . . . . . . . . . . . . | **2** |
| **2.** Conidia phialospores, produced basipetally in chains or balls from a phialide which does not increase in length with successive spore production; a succession of mouths may be produced by sympodial proliferation to form a polyphialide . . . . . . . . . . . . . . . . . . . . . . . . . . . . . . . . | **Phialosporae** *p.* 68 |
| **2.** Conidia not as above . . . . . . . . . . . . . . . . | **3** |
| **3.** Conidia aleuriospores, solitary or in botryose clusters, produced by a blowing out of the terminal portion of a hypha or sporogenous cell or conidiophore, seceding with difficulty and then by rupture of the parental hypha which remains as a fringe round the base of the spore . . . | **Aleuriosporae** *p.* 56 |
| **3.** Conidia not as above . . . . . . . . . . . . . . . . | **4** |

| | |
|---|---|
| **4.** Conidia sympodulospores, never in chains, produced on denticles from a sympodially extending sporogenous cell which elongates to form a rachis or enlarges to form a vesicle with successive spore production . . . . . . . . . . . . . . . . . . . | **Sympodulosporae** *p.* 77 |
| **4.** Conidia not as above . . . . . . . . . . . . . . . . | **5** |
| **5.** Conidia blastospores, produced apically or laterally by budding from a hypha, sporogenous cell, conidiophore or previously formed conidium; conidia solitary or in acropetal chains, seceding readily . . . . . . . . . . . . . . . . . . . . . . . . . . . . | **Blastosporae** *p.* 62 |
| **5.** Conidia not as above . . . . . . . . . . . . . . . . | **6** |
| **6.** Conidia botryoblastospores, developing simultaneously from the swollen apex (ampulla) of a sporogenous cell, on short denticles, borne singly or in acropetal chains . . . . . . . . . . . . . . . . . . . . . . . . . . . . | **Botryoblastosporae** *p.* 66 |
| **6.** Conidia not as above . . . . . . . . . . . . . . . . | **7** |
| **7.** Conidia porospores, produced through minute pores in the wall of the conidiophore or sporogenous cell or previously formed conidium, septate, pigmented, solitary or in acropetal chains . . . . . . . . | **Porosporae** *p.* 75 |
| **7.** Conidia not as above . . . . . . . . . . . . . . . . | **8** |
| **8.** Conidia annellospores, produced in basipetal succession in balls or chains from the apex of a flask-shaped or cylindrical sporogenous cell which increases in length with successive spores to form an elongate, annellated apex; conidia frequently truncate at the base . . . . . . . . . . | **Annellosporae** *p.* 60 |
| **8.** Conidia not as above . . . . . . . . . . . . . . . . | **9** |
| **9.** Conidia arthrospores, produced by basipetal septation and fragmentation of vegetative or sporogenous hyphae, nonseptate, in simple or branching chains . . | **Arthrosporae** *p.* 61 |
| **9.** Conidia not as above . . . . . . . . . . . . . . . . | **10** |

**10.** Conidia meristem arthrospores, produced in basipetal chains from a conidiophore which increases in length by meristematic growth from the base . . . **Meristem Arthrosporae** (see *Trimmatostroma* and *Spegazzinia*, only genera in this series from soil)

**10.** Conidia meristem blastospores, lens-shaped with conspicuous hyaline rim, darkly pigmented . . . . . . . . . . . . . . . . . . . **Meristem Blastosporae** (see *Arthrinium*, only genus of this series from soil)

CHAPTER

# V

# Keys to Genera

## KEY TO ALEUREOSPORAE

In this group the conidia (aleuriospores) are produced singly (rarely in short chains) and develop terminally as blown out ends of the sporogenous cells. A plurality of conidia is formed by the development of successive spores laterally and below the terminal spore to give an apical cluster of conidia (botryoaleuriospores).

| | |
|---|---|
| **1.** Conidia coiled, palmate or stellate .... | **2** |
| **1.** Conidia not as above; globose, subglobose, ovoid or cylindric etc. ........... | **4** |
| **2.** Conidia coiled ........................ | **Helicosporina** |
| **2.** Conidia not coiled .................... | **3** |
| **3.** Conidia like a hand with finger-like extensions ............................ | **Dictyosporium** |
| **3.** Conidia stellate, composed of four equal cells with spiny outgrowths ........... | *see* **Spegazzinia** |
| **4.** Conidia septate ..................... | **5** |
| **4.** Conidia non-septate ................... | **17** |

| | | |
|---|---|---|
| **5.** | Conidia muriform (with longitudinal and transverse septa) . . . . . . . . . . . . . . . . . . . | **6** |
| **5.** | Conidia with transverse septa only . . . . | **9** |
| **6.** | Conidia hyaline or light-colored . . . . . . . | **Diheterospora** |
| **6.** | Conidia darkly pigmented . . . . . . . . . . . | **7** |
| **7.** | Conidia spherical, rough, clustered in sporodochium-like masses . . . . . . . . . . . | **Epicoccum** |
| **7.** | Conidia not in sporodochia . . . . . . . . . . . | **8** |
| **8.** | Conidia with prominent apical cell . . . . | **Acrospeira** |
| **8.** | Apical cell not conspicuously enlarged . . | **Pithomyces** or **Monodictys** |
| **9.** | Conidia one-septate . . . . . . . . . . . . . . . . . | **10** |
| **9.** | Conidia several-septate . . . . . . . . . . . . . . | **13** |
| **10.** | Conidia consisting of a very large, rough, pigmented, apical cell and a small, hyaline, smooth, basal cell . . . . . . . . . . . . . | **11** |
| **10.** | Both cells of conidium pigmented . . . . . | **12** |
| **11.** | Accessory spore state *Verticillium*-like . . . . . . . . . . . . . . . . . . . . . . . . . . . . . . | **Mycogone** |
| **11.** | Accessory spore state aspergilliform . . . | **Chlamydomyces** |
| **12.** | Conidia solitary, one per sporogenous cell . . . . . . . . . . . . . . . . . . . . . . . . . . . . . . | **Trichocladium** |
| **12.** | Conidia borne in clusters on hyaline branched conidiophores . . . . . . . . . . . . . | **Wardomyces** |
| **13.** | Conidia hyaline or lightly pigmented . . | **14** |
| **13.** | Conidia brown to dark-brown or olivaceous . . . . . . . . . . . . . . . . . . . . . . . . . . . | **15** |
| **14.** | Conidia fusiform, thick-walled, lightly pigmented . . . . . . . . . . . . . . . . . . . . . . . . | **Microsporum** |
| **14.** | Conidia cylindric to fusiform, thin-walled, hyaline . . . . . . . . . . . . . . . . . . . . . | **Trichophyton** |
| **15.** | Conidia very large 50 to 150 $\mu$ long . . . . | **Clasterosporium** |
| **15.** | Conidia usually less than 50 $\mu$ long . . . . | **16** |

| | |
|---|---|
| **16.** Conidia broadly fusiform with small central lumen, very thick-walled, 3- to 7-septate, *Bipolaris*-like in appearance . . | **Murogenella** |
| **16.** Conidia not as above, central lumen much wider than the walls are thick . . . | **Trichocladium** or **Pithomyces** |
| **17.** Conidia very large (15 to 30 by 14 to 22 $\mu$), sporogenous cells sharply tapered at the apex . . . . . . . . . . . . . . . . . . . . . . . . . . | *see* **Acremoniella** |
| **17.** Conidia smaller, usually less than 20 $\mu$ in diameter . . . . . . . . . . . . . . . . . . . . . . . . | **18** |
| **18.** Conidia hyaline to lightly pigmented . . | **19** |
| **18.** Conidia brown to dark-brown . . . . . . . . . | **26** |
| **19.** Conidia spherical or almost so . . . . . . . . | **20** |
| **19.** Conidia distinctly longer than broad . . . | **25** |
| **20.** Conidia with conspicuous protuberances arising from the wall . . . . . . . . . . . . . . . . | **21** |
| **20.** Conidia lacking prominent, hyaline protuberances . . . . . . . . . . . . . . . . . . . . . . . . . . | **22** |
| **21.** Protuberances finger-like . . . . . . . . . . . . | **Histoplasma** |
| **21.** Protuberances large, hemispherical vesicles . . . . . . . . . . . . . . . . . . . . . . . . . . . . . . | **Stephanoma** |
| **22.** Conidia distinctly warty . . . . . . . . . . . . . | **Sepedonium** |
| **22.** Conidia smooth . . . . . . . . . . . . . . . . . . . . | **23** |
| **23.** Conidia thick-walled, golden-brown . . . | **Botryotrichum** |
| **23.** Conidia thin-walled . . . . . . . . . . . . . . . . . | **24** |
| **24.** Conidia borne on short, tapering pedicels . . . . . . . . . . . . . . . . . . . . . . . . . . . . . . | *see* **Umbelopsis** |
| **24.** Conidia more or less sessile, fructification a sporodochium . . . . . . . . . . . . . . . . | *see* **Beniowskia** |
| **25.** Conidia solitary, borne laterally or terminally on hyphae little distinguished from the vegetative hyphae . . . . . . . . . . | **Sporotrichum** |
| **25.** Conidia borne in short chains . . . . . . . . . | **Chrysosporium** |

**26.** Conidia produced mostly singly at the apex of the sporogenous cell .......... **27**
**26.** Conidia produced in clusters at the apex of each sporogenous cell ............. **34**

**27.** Conidia globose to subglobose or triangulate ............................... **28**
**27.** Conidia distinctly longer than broad ... **33**

**28.** Conidia triangulate or broader than long .............................. **29**
**28.** Conidia globose to subglobose ........ **30**

**29.** Conidia black, slightly broader than long .............................. *see* **Nigrospora**
**29.** Conidia dark-brown, thick-walled, triangulate, with prominent germ pores.. **Rhinocladium**

**30.** Conidia rough ....................... **Thermomyces**
**30.** Conidia smooth .................... **31**

**31.** Conidia golden-brown to brown ....... **32**
**31.** Conidia dark-brown .................. **Humicola**

**32.** Conidiophores tall, stout, brown, branching irregularly at the top, aleuriospores borne singly at the ends of the ultimate branchlets ........................ **Staphylotrichum**
**32.** Conidiophores inconspicuous, aleuriospores borne on short sporogenous branches arising from the vegetative hyphae ............................. **Botryotrichum**

**33.** Conidia dark-brown, smooth, ovoid to flame-shaped, with a longitudinal germ slit ............................... **Mammaria**
**33.** Conidia pale-brown or subhyaline, formed laterally and apically on more or less undifferentiated hyphae .......... **Sporotrichum**

**34.** Conidia bivalvate (lens-shaped) with hyaline rim ......................... **35**
**34.** Conidia not as above, globose, subglobose, flame-shaped, etc. ............. **36**

**35.** Sterile setae present ................. *see* **Cordella**
**35.** Sterile setae absent .................. *see* **Arthrinium**

**36.** Conidiophores consisting of several series of inflated hyaline cells .............. **37**
**36.** Conidiophores not conspicuous, with cylindrical sporogenous cells arising more or less directly from the vegetative hyphae ............................ **38**

**37.** Conidia spherical with apical germ pore ................................ **Gilmaniella**
**37.** Conidia ovoid to short-cylindric, sometimes papillate ...................... **Wardomyces**

**38.** Conidia with broad attachment ....... **Echinobotryum**
**38.** Conidia borne on short, slender stalks from the sporogenous cells ........... **Asteromyces**

KEY TO ANNELLOSPORAE

In this series the first formed spore is produced terminally on the sporogenous cell. Each new conidium develops as a blown out end of successive proliferations through the scar left by the previous conidium. A succession of proliferations is accompanied by an increase in length of the sporogenous cell. The scars of successive conidia give the sporogenous cell, in the spore-bearing region, an annellated appearance.

All the Annellosporae so far recovered from soil have nonseptate spores except those producing sporodochial fructifications, in which case they may be multiseptate, dark-brown and setulate.

**1.** Conidiophores aggregated into synnemata or borne on sporodochia or acervuli ..... **2**
**1.** Conidiophores not as above, solitary or inconspicuous ....................... **7**

**2.** Conidiophores aggregated into synnemata; conidia nonseptate .............. **3**
**2.** Conidiophores on sporodochia or acervuli; conidia dark, septate, setulate ..... **6**

| | |
|---|---|
| **3.** Spores sliming down to form gloeoid heads . . . . . . . . . . | **4** |
| **3.** Spores in dry heads . . . . . . . . . . | **5** |
| **4.** Synnemata darkly pigmented . . . . . . . . . . | *see* **Graphium** |
| **4.** Synnemata lightly pigmented . . . . . . . . . . | *see* **Stilbum, Heterocephalum** |
| **5.** Prominent curved or straight appendages arising from the spore-bearing region of the synnemata . . . . . . . . . . | **Trichurus** |
| **5.** Appendages lacking . . . . . . . . . . | **Doratomyces** |
| **6.** Conidia with a single apical setula . . . . . | **Monochaetia** |
| **6.** Conidia with several or branched setulae . . . . . . . . . . | **Pestalotia** |
| **7.** Sporogenous cells in an apical penicillus on stout, brown conidiophores . . . . . . . . . | **Leptographium** |
| **7.** Well developed conidiophores with apical penicillus lacking . . . . . . . . . . | **8** |
| **8.** Sporogenous cells with cylindrical stalk and vesicle in the upper part, arising singly from the hyphae; conidia in short chains, hyaline . . . . . . . . . . | *see* **Monocillium** |
| **8.** Sporogenous cells not markedly inflated in the upper part, borne singly or in groups or in penicillate arrangements . . . | **Scopulariopsis** |

## KEY TO ARTHROSPORAE

In this series the conidia are produced after septation and breaking up or rounding off of simple or branched sporogenous hyphae. The sporogenous hyphae may be morphologically and functionally identical with the vegetative hyphae or may be produced on more or less well developed conidiophores.

Arthrospores are, in their simplest form, short-cylindric with truncate ends, but may be considerably differentiated with thick walls and may be globose, subglobose, ellipsoid or even helmet-shaped. They may be smooth or roughened, pale or darkly pigmented and in their mature state their arthrospore origin is not always obvious.

Arthrospores are always in chains and, with the exception of *Polyscytalum*, nonseptate.

| | | |
|---|---|---|
| **1.** | Conidia arthrospores and blastospores . . | **Trichosporon** |
| **1.** | Conidia arthrospores only . . . . . . . . . . . . . | **2** |
| **2.** | Conidiophores well developed, more or less erect and usually slightly or markedly pigmented . . . . . . . . . . . . . . . . . . . . . . . . . | **3** |
| **2.** | Conidiophores lacking or not obvious . . . | **6** |
| **3.** | Conidia in short, simple chains, arising from simple, brown, sympodially extending conidiophores . . . . . . . . . . . . . . . . . . . . | *see* **Sympodiella** |
| **3.** | Conidia in branching chains . . . . . . . . . . . | **4** |
| **4.** | Conidiophores stout, more than 5 $\mu$ wide and hyaline or lightly pigmented . . . . . . . | **Amblyosporium** |
| **4.** | Conidiophores narrow, less than 5 $\mu$ wide, and brown . . . . . . . . . . . . . . . . . . . . . . . . | **5** |
| **5.** | Conidia spherical to ovoid or short-cylindric . . . . . . . . . . . . . . . . . . . . . . . . . . . . . . . . | **Oidiodendron** |
| **5.** | Conidia several times as long as broad (long-cylindric) . . . . . . . . . . . . . . . . . . . . . . | **Polyscytalum** |
| **6.** | Conidia never spherical, usually truncate at one or both ends . . . . . . . . . . . . . . . . . . . . | **7** |
| **6.** | Conidia spherical or helmet-shaped . . . . | **8** |
| **7.** | Conidia truncate at both ends, in long chains . . . . . . . . . . . . . . . . . . . . . . . . . . . . . . | **Geotrichum** |
| **7.** | Conidia truncate at one end, in short chains . . . . . . . . . . . . . . . . . . . . . . . . . . . . . . | **Chrysosporium** |
| **8.** | Conidia spherical, lightly pigmented, tan-brown in mass . . . . . . . . . . . . . . . . . . . . . . . | **Wallemia** |
| **8.** | Conidia hyaline, helmet-shaped to ovoid . . . . . . . . . . . . . . . . . . . . . . . . . . . . . . | conidial **Monascus** |

## KEY TO BLASTOSPORAE

Conidia develop in acropetal succession as blown out ends of simple or branching conidiophores which do not then increase in length. The apical branches of the conidiophore frequently secede and may then function as conidia (ramoconidia). Sometimes conidiophores are lacking and solitary conidia or chains of conidia may originate more or less directly from the vegetative hyphae. In a few genera, hyphae are lacking and the colonies are made up of masses of blastospores.

1. Conidia regularly or irregularly coiled . . . **2**
1. Conidia not coiled . . . **3**

2. Conidia solitary . . . **Helicoon**
2. Conidia in chains . . . **Helicodendron**

3. Conidia triradiate, tetraradiate, stellate, or irregularly branched . . . **4**
3. Conidia not as above, globose to ovoid or cylindric . . . **7**

4. Conidia brown, stellate, with five septate arms . . . **Tripospermum**
4. Conidia hyaline . . . **5**

5. Conidiophores lacking, conidia with elongate main axis and primary and secondary branches . . . **Varicosporium**
5. Conidiophores present, bearing apical cluster or triradiate or tetraradiate conidia . . . **6**

6. Fructification a synnema, conidia symmetrical, with four, petal-like arms . . . **Riessia**
6. Conidia bird-like, with main axis and one to three wing-like arms . . . **Volucrispora**

7. Colonies yeast-like, consisting largely of masses of globose to ellipsoid conidia, hyphae lacking or restricted . . . **8**
7. Colonies not yeast-like . . . **14**

8. Hyphae present . . . **9**
8. Hyphae lacking . . . **11**

9. Hyphae dark-brown, conidia borne on small denticles . . . **Aureobasidium**
9. Hyphae hyaline, conidia not borne on denticles . . . **10**

| | |
|---|---|
| **10.** Arthrospores present ................ | **Trichosporon** |
| **10.** Arthrospores absent, blastospores only .. | **Candida** |
| | |
| **11.** Colonies pink, red, or yellow ......... | **12** |
| **11.** Colonies white to cream, conidia surrounded by capsule ................. | **13** |
| | |
| **12.** Conidia borne on denticles, shot off .... | **Sporobolomyces** |
| **12.** Conidia not borne on denticles ........ | **Rhodotorula** |
| | |
| **13.** Starch produced in capsule round conidium ............................ | **Cryptococcus** |
| **13.** No starch produced in capsule ........ | **Torulopsis** |
| | |
| **14.** Conidia formed on synnemata or sporodochia ............................ | **15** |
| **14.** Synnemata and sporodochia lacking ... | **18** |
| | |
| **15.** Sporodochium produced ............. | **16** |
| **15.** Synnema produced .................. | **17** |
| | |
| **16.** Sporodochium setose, conidia multiseptate ............................ | *see* **Abgliophragma** |
| **16.** Sporodochium not setose, conidia nonseptate, produced sessile on anastomosing sporogenous hyphae ............. | **Beniowskia** |
| | |
| **17.** Synnemata simple, conidia brown, *Cladosporium*-like .................. | **Pycnostysanus** |
| **17.** Synnemata branching, conidia hyaline to lightly pigmented, in cottony tufts ..... | **Sphaeridium** |
| | |
| **18.** Conidia borne singly (rarely two) on each sporogenous cell ..... .............. | **19** |
| **18.** Conidia in chains or gloeoid masses ... | **21** |
| | |
| **19.** Sporogenous cells globose to ovoid, arising from strongly reflexed branch (falx) of the conidiophore .................. | **Zygosporium** |
| **19.** Sporogenous cells not as above ........ | **20** |

**20.** Conidia black, slightly longer in horizontal axis, arising from very short, tapering sporogenous cells .................... *see* **Nigrospora**
**20.** Conidia brown, ovoid, borne on long, sharply tapered sporogenous cells ..... *see* **Acremoniella**

**21.** Conidia septate .................... **22**
**21.** Conidia nonseptate .................. **24**

**22.** Conidia regularly several-septate ...... **Septonema**
**22.** Conidia with one or two septa ......... **23**

**23.** Conidia dark brown with very dark central septum, in simple chains .......... **Bispora**
**23.** Conidia continuous or once-septate, pale-brown with very dark connectives, in branching chains ................. **Cladosporium**

**24.** Conidia hyaline to lightly pigmented .. **25**
**24.** Conidia brown to dark-brown ........ **29**

**25.** Conidiophores brown, conidia short-cylindric to cylindric ................ **Polyscytalum**
**25.** Conidiophores hyaline or inconspicuous .............................. **26**

**26.** Conidia with pronounced black truncate scars .............................. **Hyalodendron**
**26.** Conidia lacking dark scars ........... **27**

**27.** Conidia globose to subglobose or ovoid.. **Monilia**
**27.** Conidia short-cylindric to cylindric .... **28**

**28.** Conidiophores present, conidia with isthmus-like connections between spores ............................ **Hormiactis**
**28.** Conidiophores inconspicuous .......... **Fusidium**

**29.** Conidia sliming down, conidiophores stout, erect, brown .................. **Haplographium**
**29.** Conidia in dry chains ................ **30**

| | |
|---|---|
| **30.** Conidia ovoid to long-ovoid .......... | **31** |
| **30.** Conidia spherical .................... | **32** |
| **31.** Conidia with prominent scars at the attachment points, conidiophores irregularly branched at the apex ........... | **Cladosporium** |
| **31.** Conidia lacking attachment scars, long ovoid, rough; conidiophores, where present, forming apical penicillus ..... | *see* **Periconiella** |
| **32.** Setae present ...................... | **33** |
| **32.** Setae absent ....................... | **34** |
| **33.** Conidiophores with pronounced, dark, apical swelling ..................... | **Lacellina** |
| **33.** Conidiophores lacking apical swelling.. | **Lacellinopsis** |
| **34.** Conidiophores with pronounced dark apical swelling ..................... | **Haplobasidion** |
| **34.** Conidiophores lacking apical swelling.. | **35** |
| **35.** Conidia produced acropetally but maturing basipetally ................... | **Periconia** |
| **35.** Conidia developing and maturing acropetally ........................... | **Cladosporium** |

## KEY TO BOTRYOBLASTOSPORAE

In this series the conidia are produced on well differentiated, swollen sporogenous cells called ampullae. The conidia are solitary or in simple or branched chains. From each ampulla the conidia arise more or less simultaneously, all initiating and developing synchronously. In the case of chains of conidia, the first series from the ampullae are also synchronous. The conidiophores may be determinate with the conidiophores or branches terminating in ampullae. If indeterminate, the main axis may proliferate repeatedly through the apex to form sterile and fertile areas, giving a noded appearance to the mature conidiophore, or the conidiophore gives rise to an acropetal succession of lateral fertile branches which bear solitary or clustered ampullae.

In the mature state, *Arthrobotrys* and certain of the other genera of the sympodulosporae may be confused with the Botryoblastosporae. In the sympodulosporae, however, the conidia are not produced simultaneously on a vesicle; rather they form successively, and continued conidium production forms the vesicle.

1. Conidia produced on a sporodochium or synnema .......... **Abgliophragma**
1. Sporodochium lacking .......... **2**

2. Conidia in chains, conidiophores indeterminate with spore-bearing "nodes" .......... **3**
2. Conidia not in chains .......... **4**

3. Conidia and conidiophores dark-brown .......... **Gonatobotryum**
3. Conidia and conidiophores hyaline .......... **Nematogonium**

4. Conidia septate .......... **5**
4. Conidia nonseptate .......... **6**

5. Conidia several-septate with solitary, sessile ampullae .......... **Cephaliophora**
5. Conidia one-septate, conidiophores with spore-bearing nodes .......... *see* **Arthrobotrys**

6. Conidiophores determinate, growth ceasing with production of terminal ampullae .......... **7**
6. Conidiophores indeterminate, continuing growth to give successive new levels of conidium production .......... **10**

7. Terminal vesicles (ampullae) more or less spherical .......... **8**
7. Terminal vesicles (ampullae) elongate .......... **9**

8. Conidiophores hyaline, ampullae large .......... **Oedocephalum**
8. Conidiophores pigmented, branching, ampullae small .......... **Botrytis**

9. Conidiophores terminating in a radiating cluster of elongate sporogenous cells .......... *see* **Ostracoderma**
9. Conidiophores terminating in a solitary or linear series of sporogenous cells .......... **Olpitrichum**

10. Conidiophores simple, sporogenous cells terminal and eventually intercalary .......... **Gonatobotrys**
10. Conidiophores branching, with sporogenous cells borne on lateral fertile branches .......... **Botryosporium**

## KEY TO PHIALOSPORAE

In this section the conidia are phialospores and are produced in rapidly maturing, basipetal succession from the apex of the sporogenous cell (phialide) which may or may not possess a collarette.

| | |
|---|---|
| **1.** Conidiophores aggregated into synnemata or borne on sporodochia or acervuli . . . . . . . . . . | **2** |
| **1.** Conidiophores not as above, solitary . . . | **27** |
| **2.** Conidiophores aggregated into synnemata . . . . . . . . . . | **3** |
| **2.** Conidiophores on acervuli or sporodochia . . . . . . . . . . | **12** |
| **3.** Synnemata determinate with sterile stalk and sporiferous head . . . . . . . . . . | **4** |
| **3.** Synnemata indeterminate, continuing growth from the apex . . . . . . . . . . | **11** |
| **4.** Spores sliming down to form gloeoid heads . . . . . . . . . . | **5** |
| **4.** Spores persisting in chains or dry masses . . . . . . . . . . | **9** |
| **5.** Synnemata darkly pigmented . . . . . . . . . | **6** |
| **5.** Synnemata lightly pigmented . . . . . . . . . | **8** |
| **6.** Spores globose to ovoid, nonsetulate . . . | *see* **Graphium** |
| **6.** Spores falcate to crescent-shaped, with a setula at each end . . . . . . . . . . | **7** |
| **7.** Synnema consisting of central sterile seta with cortex of fused phialides at the base . . . . . . . . . . | **Menisporopsis** |
| **7.** Synnema dark-brown, stout, flaring at the apex and bearing slimy masses of spores in heads . . . . . . . . . . | **Thozetellopsis** |
| **8.** Phialides long and slender, borne singly at the apex of the conidiophore . . . . . . . | **Stilbum** |
| **8.** Phialides aggregated into groups or verticils . . . . . . . . . . | **Dendrostilbella** |

| | |
|---|---|
| **9.** Spores persisting in chains . . . . . . . . . . . | *see* **Doratomyces** |
| **9.** Spores produced in dry masses . . . . . . . . | **10** |
| **10.** Sporogenous head yellow-orange, bearing interwoven sterile branches to form a basket-like container with radiating sterile setae . . . . . . . . . . . . . . . . . . . . . . . . | **Heterocephalum** |
| **10.** Sporogenous head consisting of a dry spore mass on top of a sterile, somewhat clavate synnema . . . . . . . . . . . . . . . . . . . . | **Ciliciopodium** |
| **11.** Spores borne in chains . . . . . . . . . . . . . . . | **Paecilomyces** |
| **11.** Spores borne singly or in small balls, individual spores enclosed in a mucous sheath . . . . . . . . . . . . . . . . . . . . . . . . . . | **Hirsutella** |
| **12.** Conidia bearing setulae or membranous appendage . . . . . . . . . . . . . . . . . . . . . . . . | **13** |
| **12.** Conidia lacking setulae or membranous appendage . . . . . . . . . . . . . . . . . . . . . . . | **16** |
| **13.** Conidia with setula at each end . . . . . . | **14** |
| **13.** Conidia with membranous appendage at each end . . . . . . . . . . . . . . . . . . . . . . . . | **Starkeyomyces** |
| **14.** Conidia nonseptate with single setula at each end . . . . . . . . . . . . . . . . . . . . . . . . . | **Thozetellopsis** |
| **14.** Conidia septate . . . . . . . . . . . . . . . . . . . . | **15** |
| **15.** Conidia once-septate, hyaline, with single setula at each end . . . . . . . . . . . . . | **Leptodiscus** |
| **15.** Conidia several septate, darkly pigmented with one or more setulae at the ends . . . . . . . . . . . . . . . . . . . . . . . . . . . . | *see series* **Annellosporae** |
| **16.** Conidia septate . . . . . . . . . . . . . . . . . . . . | **17** |
| **16.** Conidia nonseptate . . . . . . . . . . . . . . . . | **18** |
| **17.** Conidia rounded at both ends . . . . . . . . . | **Cylindrocarpon** |
| **17.** Conidia with a foot cell at the attachment end . . . . . . . . . . . . . . . . . . . . . . . . | **Fusarium** |
| **18.** Fructifications with setae . . . . . . . . . . . . | **19** |
| **18.** Fructifications lacking setae . . . . . . . . . . | **21** |

**19.** Setae stout and darkly pigmented . . . . . **Colletotrichum**
**19.** Setae hyaline or lightly pigmented . . . . **20**

**20.** Spore mass dark, olivaceous . . . . . . . . . . **Myrothecium**
**20.** Spore mass white, cream, or pink . . . . . **Volutina**

**21.** Conidia short-cylindric, adhering in tall green columns . . . . . . . . . . . . . . . . . . . . . **Metarrhizium**
**21.** Conidial sliming down to form gloeoid masses . . . . . . . . . . . . . . . . . . . . . . . . . . **22**

**22.** Conidia dark olive-green in mass, ovoid to short-cylindric . . . . . . . . . . . . . . . . . . . **Myrothecium**
**22.** Conidial masses light-colored . . . . . . . . . **23**

**23.** Phialides minute, globose to ovoid, conidia asymmetric . . . . . . . . . . . . . . . . . . . **Harposporium**
**23.** Phialides cylindrical to filiform or flask-shaped . . . . . . . . . . . . . . . . . . . . . . . . . . **24**

**24.** Phialides in whorls . . . . . . . . . . . . . . . . . . **Dendrodochium**
**24.** Phialides not in whorls . . . . . . . . . . . . . . **25**

**25.** Conidia many times longer than broad (filiform) . . . . . . . . . . . . . . . . . . . . . . . . . **Libertella**
**25.** Conidia ovoid, short-cylindric, or allantoid . . . . . . . . . . . . . . . . . . . . . . . . . . . . . **26**

**26.** Conidia short-cylindric to bacillar . . . . . **Tubercularia**
**26.** Conidia lunate or allantoid . . . . . . . . . . . **Hainesia**

**27.** Conidiophores well developed with the main axis bearing a number of phialides . . . . . . . . . . . . . . . . . . . . . . . . . . . **28**
**27.** Conidiophores with main axis bearing a single apical phialide or sometimes groups of several phialides on short sparingly branched conidiophores . . . . . . . . . **52**

**28.** Conidia persisting in chains . . . . . . . . . . . **29**
**28.** Conidia sliming down to form gloeoid balls . . . . . . . . . . . . . . . . . . . . . . . . . . . . . . **39**

| | |
|---|---|
| **29.** Conidia nonseptate . . . . . . . . . . . . . . . . . | **30** |
| **29.** Conidia septate . . . . . . . . . . . . . . . . . . . . | **38** |
| **30.** Phialides numerous, borne directly from a vesicle or produced on prophialides arising from a vesicle . . . . . . . . . . . . . . . | **Aspergillus** |
| **30.** Phialides in a penicillus or if in an apical cluster then not exceeding eight in number . . . . . . . . . . . . . . . . . . . . . . . . . | **31** |
| **31.** Phialides in apical cluster of less than eight . . . . . . . . . . . . . . . . . . . . . . . . . . . | **32** |
| **31.** Phialides in penicillus . . . . . . . . . . . . . . | **36** |
| **32.** Phialides distinctly swollen, broader near the apex, spores dark, rough, globose . . . . . . . . . . . . . . . . . . . . . . . . . . . . . . . | **Stachybotrys** |
| **32.** Phialides bottle-shaped . . . . . . . . . . . . . . | **33** |
| **33.** Spores very large (16 to 20 by 20 to 26 $\mu$) dark, lemon-shaped . . . . . . . . . . . . . . | **Phialomyces** |
| **33.** Spores small (less than 10 $\mu$ in diameter) . . . . . . . . . . . . . . . . . . . . . . . . . . . . . | **34** |
| **34.** Spore chains short (up to five spores), spores dark, roughened . . . . . . . . . . . . . | **Eladia** |
| **34.** Spore chains long . . . . . . . . . . . . . . . . . . . | **35** |
| **35.** Spores globose to ovoid . . . . . . . . . . . . . . | **Penicillium** |
| **35.** Spores fusiform to lemon-shaped . . . . . | **Paecilomyces** |
| **36.** Conidiophores dark . . . . . . . . . . . . . . . . . | **Thysanophora** |
| **36.** Conidiophores light or hyaline . . . . . . . | **37** |
| **37.** Conidia globose to ovoid . . . . . . . . . . . . . | **Penicillium** |
| **37.** Conidia lemon-shaped to fusiform . . . . | **Paecilomyces** or **Phialotubus** |
| **38.** Conidia hyaline . . . . . . . . . . . . . . . . . . . . | **Cladobotryum** |
| **38.** Conidia pigmented . . . . . . . . . . . . . . . . . . | **Fusariella** |
| **39.** Conidia septate . . . . . . . . . . . . . . . . . . . . | **40** |
| **39.** Conidia non-septate . . . . . . . . . . . . . . . . . | **43** |

| | |
|---|---|
| **40.** Conidiophores brown, sporogenous cells borne laterally on the conidiophore or its branches ........................ | **41** |
| **40.** Conidiophores hyaline, sporogenous cells aggregated into an apical penicillus on each conidiophore ............... | **42** |
| **41.** Phialides with collarette ............ | **Menispora** |
| **41.** Sporogenous cells polyphialides (many collarettes on each sporogenous cell), conidia cylindric, septate ............ | **Chaetopsis** |
| **42.** Conidia long-cylindric, penicillus with sterile appendage .................. | **Cylindrocladium** |
| **42.** Sterile appendage lacking ........... | **Cylindrocarpon** |
| **43.** Conidiophore with unbranched, pointed main axis and thick collar of slimy spores in lower half ................. | **Chaetopsina** |
| **43.** Conidiophores not as above, with phialides alternate or verticillate along the length of the conidiophores, or in an apical penicillus ...................... | **44** |
| **44.** Phialides in whorls along the length of the conidiophore or its branches (verticillate) ......................... | **45** |
| **44.** Phialides not verticillate ............. | **47** |
| **45.** Whorls of phialides arising from special collar hyphae ...................... | **Gonytrichum** |
| **45.** Phialides arising directly from the conidiophore or its branches ............ | **46** |
| **46.** Phialides swollen below, tapering towards the apex ..................... | **Verticillium** |
| **46.** Phialides cylindrical, rounded at the tip, roughened ..................... | **Stachylidium** |
| **47.** Phialides in an apical penicillus of several series of metulae ............... | **48** |
| **47.** Phialides in an apical cluster (monoverticillate) or arising directly or indirectly from the swollen apex of the conidiophore (aspergilliform) .......... | **50** |

**48.** Conidiophores brown, phialides with pronounced collarette ............... **Phialocephala**
**48.** Conidiophores hyaline or lightly pigmented ........................... **49**

**49.** Conidiophores with several stout appendages arising from just below the apical penicillus ...................... **Gliocephalotrichum**
**49.** Conidiophores lacking appendages at the apex ........................... **Gliocladium**

**50.** Phialides swollen, in apical clusters of less than 10, borne on simple conidiophores ............................. **Stachybotrys**
**50.** Phialides numerous, borne on prophialides arising from a vesicle as in *Aspergillus* ............................. **51**

**51.** Conidiophores hyaline, nonseptate .... **Gliocephalis**
**51.** Conidiophores pigmented, septate .... **Goidanichiella**

**52.** Conidia septate ..................... **53**
**52.** Conidia nonseptate ................. **58**

**53.** Conidia darkly pigmented, in chains ... **Fusariella**
**53.** Conidia hyaline or lightly pigmented .. **54**

**54.** Conidiophores brown, conidia two-celled, sterile setae present .......... **Cylindrotrichum**
**54.** Conidiophores hyaline .............. **55**

**55.** Conidia ovoid to long-ovoid, phialides arising more or less directly from the vegetative hyphae .................. **56**
**55.** Conidia cylindric or naviculate ....... **57**

**56.** Conidiophores long, cylindric; conidia two-celled, clavate, with truncate base, in short chains ..................... *see* **Trichothecium**
**56.** Conidia ellipsoid to long-ellipsoid, phialides tapering, arising directly from hyphae ........................... **Cephalosporiopsis**

**57.** Conidia with rounded ends ........... **Cylindrocarpon**
**57.** Conidia with lowest cell modified to form "foot cell" .................. **Fusarium**

**58.** Conidia endospores, produced within a cylindrical phialide ................ **59**
**58.** Conidia produced at the apex of the phialide, with or without a collarette .... **61**

**59.** Aleuriospores absent ................ **Chalara**
**59.** Aleuriospores present ............... **60**

**60.** Aleuriospores septate, solitary ........ **Thielaviopsis**
**60.** Aleuriospores nonseptate, solitary or in chains .......................... **Chalaropsis**

**61.** Conidia produced in pronounced collarette ........................... **62**
**61.** Collarette, if present, not conspicuous .. **65**

**62.** Conidia falcate or crescent-shaped .... **Menisporella**
**62.** Conidia globose to ovoid, or triangulate ............................. **63**

**63.** Phialospores more than 10 $\mu$ long, thick-walled, dark-brown, with several germ pores ............................. **Catenularia**
**63.** Phialospores less than 10 $\mu$ long, hyaline to pale-brown ..................... **64**

**64.** Phialides proliferating repeatedly and monopodially, solitary on the conidiophores.......................... **Chloridium**
**64.** Phialides not proliferating, frequently in clusters ....................... **Phialophora**

**65.** Phialides globose to ovoid ........... **66**
**65.** Phialides flask-shaped or long and tapering.......................... **67**

**66.** Phialides arising directly from the vegetative hyphae, conidia flattened or convex on one side ..................... **Harposporium**
**66.** Phialides arising singly or in groups from irregularly branched hyaline conidiophores ........................ **Trichoderma**

**67.** Conidia in chains .................... **68**
**67.** Conidia sliming down to form gloeoid balls .................... **71**

**68.** Phialides with cylindrical stalk and swollen vesicle in the upper part ...... **69**
**68.** Phialides broader near the base and tapering towards the apex ............ **70**

**69.** Conidia spherical, in long chains ...... **Torulomyces**
**69.** Conidia ovoid to pyriform, with truncate base, in short chains ............ **Monocillium**

**70.** Phialides frequently sinuous and darkened or roughened near the apex, conidia pigmented, globose to ovoid or fusiform .......................... **Gliomastix**
**70.** Phialides hyaline, smooth, broader below and tapering to a narrow apex, conidia fusiform to lemon-shaped, sometimes short-cylindric ................. **Paecilomyces**

**71.** Spores pigmented .................. **72**
**71.** Spores hyaline ..................... **73**

**72.** Conidia bright-green, conidiophores irregularly branched .................. **Trichoderma**
**72.** Conidia pale-olive to black, phialides mostly arising directly from the vegetative hyphae ...................... **Gliomastix**

**73.** Conidiophores irregularly branched, phialides short, arising singly or in groups **Trichoderma**
**73.** Phialides long and tapering, arising singly more or less directly from the vegetative hyphae .................. **Cephalosporium** (*see also microconidial* **Fusarium, Cylindrocarpon,** *and* **Verticillium**)

### KEY TO POROSPORAE

Conidia usually thick-walled, developing through minute pores in the wall of the conidiophore, seceding readily. Conidiophores of determinate or indeterminate length. If the latter, then increase in length is either by proliferation of the conidiophore through the apical pore (monopodial) or by repeated renewal of growth lateral to and just below successive apices (sympodial).

*Note:* All members of the Porosporae recorded from soil have dark-brown, septate conidia.

| | |
|---|---|
| **1.** Conidiophores lacking, with conidia developing in simple or branched chains from the apex of a more or less undifferentiated vegetative hypha .......... | **Torula** |
| **1.** Conidiophores present, although sometimes short ........................ | **2** |
| **2.** Conidia with both longitudinal and transverse septa .................... | **3** |
| **2.** Conidia with transverse septa only ... | **6** |
| **3.** Conidia in simple or branching chains ............................ | **Alternaria** |
| **3.** Conidia not in chains ................ | **4** |
| **4.** Conidia with long, apical, cellular prolongation ........................ | **Piricauda** |
| **4.** Conidia without prolongation ........ | **5** |
| **5.** Conidia solitary; conidiophore elongating monopodially by proliferation through the scar of the previous conidium ............................ | **Stemphylium** |
| **5.** Conidiophore elongating by sympodial growth from a point to one side and below the apical conidium .............. | **Ulocladium** |
| **6.** Conidia one-septate, frequently somewhat indented at the septum; conidia solitary or in simple chains ........... | **Diplococcium** |
| **6.** Conidia multiseptate ................ | **7** |
| **7.** Conidiophores determinate, ceasing growth with production of the apical conidia ............................ | **8** |
| **7.** Conidiophores indeterminate continuing growth sympodially .................. | **9** |

| | |
|---|---|
| **8.** Conidiophores long, terminating in clusters of conidia borne on dichotomously branched sporogenous cells . . . . . . . . . . . | **Dichotomophthora** |
| **8.** Conidiophores relatively short and simple or, if sparingly branched, then not dichotomous . . . . . . . . . . . . . . . . . . . . . . . | **Helminthosporium** |
| **9.** Conidia with elongate, apical, cellular prolongation . . . . . . . . . . . . . . . . . . . . . . . | **Phaeotrichonis** |
| **9.** Conidia lacking apical prolongation . . . | **10** |
| **10.** Conidia curved, with paler end cells . . . | **Curvularia** |
| **10.** Conidia cylindrical or fusiform, sometimes curved but then concolorous . . . . . | **11** |
| **11.** Conidiophores branched irregularly at the apex . . . . . . . . . . . . . . . . . . . . . . . . . . | **Dendryphion** |
| **11.** Conidiophores simple or sparingly branched . . . . . . . . . . . . . . . . . . . . . . . . . . | **12** |
| **12.** Conidia cylindric, germinating from all cells . . . . . . . . . . . . . . . . . . . . . . . . . . . . | **Drechslera** |
| **12.** Conidia fusiform, germinating from end cells only . . . . . . . . . . . . . . . . . . . . . . . . | **Bipolaris** |

## KEY TO SYMPODULOSPORAE

Conidia arising as blown out ends at the apex of simple or branched conidiophores. The conidiophore increases in length sympodially or becomes swollen after conidium production. Acropetal chains of conidia may develop on the primary conidia.

| | |
|---|---|
| **1.** Conidiophores aggregated into synnemata or on sporodochia . . . . . . . . . . . . . . | **2** |
| **1.** Conidiophores not as above (mononematous) . . . . . . . . . . . . . . . . . . . . . . . . | **6** |
| **2.** Conidiophores on sporodochia, conidia filiform, consisting of several cells connected by narrow isthmuses . . . . . . . . . . | *see* **Abgliophragma** |
| **2.** Conidiophores aggregated into synnemata . . . . . . . . . . . . . . . . . . . . . . . . . . . . . . | **3** |
| **3.** Conidia septate . . . . . . . . . . . . . . . . . . . . . | **Isariopsis** |
| **3.** Conidia nonseptate . . . . . . . . . . . . . . . . . . | **4** |

**4.** Conidia gathering in gloeoid masses .. *see* **Graphium**
**4.** Conidia in dry heads ................ **5**

**5.** Conidia falcate to crescent-shaped .... *see* **Harpographium**
**5.** Conidia globose to subglobose ......... *Isaria cretacea*

**6.** Sporogenous cells swelling at the apex with successive conidia which are in apical or intercalary clusters .......... **7**
**6.** Sporogenous cells elongating with conidium production, or if swollen then *before* conidium production .......... **15**

**7.** Conidiophores brown ................ **8**
**7.** Conidiophores hyaline or lightly pigmented .......................... **11**

**8.** Conidia biconic, setae present ........ **Beltrania**
**8.** Conidia globose, ovoid, cylindrical or Y-shaped ........................ **9**

**9.** Conidiophores short (same length as conidia), arising directly from the vegetative hyphae ...................... **Scolecobasidium**
**9.** Conidiophores stout and erect ........ **10**

**10.** Conidiophores simple, terminating in an apical cluster of dark-brown, two-celled spores .......................... **Cordana**
**10.** Conidiophores terminating in an apical cluster of radiating elongate sporogenous cells each bearing an apical cluster of conidia ........................ **Pseudobotrytis**

**11.** Conidia septate, in terminal and intercalary clusters ...................... **12**
**11.** Conidia nonseptate, in terminal clusters only .......................... **13**

**12.** Denticles on sporogenous cells short .... **Arthrobotrys**
**12.** Denticles on sporogenous cells elongated to form a candelabrum-like arrangement ........................ **Candelabrella**

**13.** Sporogenous cells arising in whorls from a main axis . . . . . . . . . . . . . . . . . . . . . . . **14**
**13.** Sporogenous cells arising directly from the vegetative hyphae . . . . . . . . . . . . . . . **Sporothrix**

**14.** Sporogenous cells long and tapering, conidia ovoid to clavate . . . . . . . . . . . . . . **Calcarisporium**
**14.** Sporogenous cells short with reflexed apex, conidia globose . . . . . . . . . . . . . . . . **Costantinella**

**15.** Conidia coiled in two dimensions, hyaline . . . . . . . . . . . . . . . . . . . . . . . . . . . . . . . **Helicosporium**
**15.** Conidia not coiled . . . . . . . . . . . . . . . . . . **16**

**16.** Conidia septate . . . . . . . . . . . . . . . . . . . . . **17**
**16.** Conidia nonseptate . . . . . . . . . . . . . . . . . . **30**

**17.** Conidia longitudinally and transversely septate . . . . . . . . . . . . . . . . . . . . . . . . . . . . **Dactylosporium**
**17.** Conidia transversely septate . . . . . . . . . . **18**

**18.** Conidia and conidiophores hyaline or light-brown . . . . . . . . . . . . . . . . . . . . . . . . . **19**
**18.** Conidiophores or conidia dark-brown . . **25**

**19.** Conidia one-septate . . . . . . . . . . . . . . . . . **20**
**19.** Conidia multiseptate . . . . . . . . . . . . . . . . **22**

**20.** Conidiophores short (at most a few times longer than the spores), conidia long-ovoid, sometimes with apical cell prolongated . . . . . . . . . . . . . . . . . . . . . . . . **22**
**20.** Conidiophores long, conidia obovate to clavate with apical cell larger than basal cell . . . . . . . . . . . . . . . . . . . . . . . . . . . . . . **21**

**21.** Conidia borne in clusters on long denticles . . . . . . . . . . . . . . . . . . . . . . . . . . . . . . . . **Candelabrella**
**21.** Conidia solitary, borne singly on a sympodially extending conidiophore . . . . . . **Genicularia**

**22.** Conidia cylindric, multiseptate, not produced on conspicuous denticles ........ **Cercosporella**
**22.** Conidia produced on conspicuous denticles ........ **23**

**23.** Conidia multiseptate ........ **24**
**23.** Conidia one-septate or if two-septate then with long, apical prolongation .... **Diplorhinotrichum**

**24.** Conidia clavate or cylindric ........ **Dactylaria**
**24.** Conidia fusiform with central cells usually much wider than end cells ........ **Dactylella**

**25.** Conidia hyaline, two-celled ........ **Diplorhinotrichum**
**25.** Conidia pigmented ........ **26**

**26.** Conidia borne on slender stalks which are left as delicate, tubular appendages on the sporogenous cells ........ **Scolecobasidium**
**26.** Conidia secede to leave a scar or short denticle ........ **27**

**27.** Conidia large, dark-brown with long, hyaline, apical, cellular appendage .... *see* **Phaeotrichonis**
**27.** Conidia lacking apical, cellular appendage ........ **28**

**28.** Conidia top-shaped (obclavate to pyriform), multiseptate ........ **Pyricularia**
**28.** Conidia ellipsoid to long-ovoid or cylindrical ........ **29**

**29.** Conidia multiseptate (rarely one-septate), sporogenous cell bearing pronounced dark scars ........ **Pleurophragmium, Cercospora**
**29.** Conidia one-septate or continuous ..... **Fusicladium**

**30.** Sporogenous cells solitary, arising directly from the vegetative hyphae or singly on top of short conidiophores ... **31**
**30.** Several or many sporogenous cells arising singly or in clusters from a well developed simple or branched conidiophore ........ **37**

| | |
|---|---|
| **31.** Conidia asymmetric, crescent-shaped to kidney-shaped . . . . . . . . . . . . . . . . . . . . . . | **32** |
| **31.** Conidia globose to ovoid or cylindric or biconic . . . . . . . . . . . . . . . . . . . . . . . . . . . | **33** |
| **32.** Conidia hyaline, crescent-shaped to falcate . . . . . . . . . . . . . . . . . . . . . . . . . . . . . | **Idriella** |
| **32.** Conidia dark-brown, kidney-shaped . . . | **Virgaria** |
| **33.** Conidia biconic with apical end spicate . . . . . . . . . . . . . . . . . . . . . . . . . . . . . . . | **Beltrania** |
| **33.** Conidia globose to ovoid or cylindric . . . | **34** |
| **34.** Sporogenous cells hyaline . . . . . . . . . . . . | **35** |
| **34.** Sporogenous cells brown . . . . . . . . . . . . . | **36** |
| **35.** Sporogenous cells arising from the hyphae in radiating clusters, inflated base tapering to a long, narrow rachis . . . . . . | **Beauveria** |
| **35.** Sporogenous cells solitary, tapering gradually to a swollen or elongate sporogenous apex . . . . . . . . . . . . . . . . . . . . . . | **Sporothrix** |
| **36.** Sporogenous cell extending sympodially, bearing short chains of arthrospores at each new level . . . . . . . . . . . . . . . . . . . . . | **Sympodiella** |
| **36.** Sporogenous cells bearing a succession of solitary conidia . . . . . . . . . . . . . . . . . . . . | **Rhinocladiella** |
| **37.** Conidia sliming down . . . . . . . . . . . . . . . | **Verticicladiella** |
| **37.** Conidia in dry heads . . . . . . . . . . . . . . . . | **38** |
| **38.** Conidiophore branching irregularly . . . . | **39** |
| **38.** Conidiophore branching in whorls or confined to apical head . . . . . . . . . . . . . . | **40** |
| **39.** Base of conidia slightly concave . . . . . . . | **Geniculisporium** |
| **39.** Base of conidia flat . . . . . . . . . . . . . . . . . . | **Rhinocladiella, Hansfordia** |
| **40.** Sporogenous cells with recurved denticulate apex . . . . . . . . . . . . . . . . . . . . . . . . . . | **Costantinella** |
| **40.** Sporogenous cells with elongate sporogenous apex . . . . . . . . . . . . . . . . . . . . . . . . . | **41** |

**41.** Verticils of sporogenous cells borne at the apex of a stout conidiophore ...... **Verticicladium**
**41.** Verticils borne along the length of the conidiophore ........................ **42**

**42.** Sporogenous cells long and slender with elongate rachis ...................... **Tritirachium**
**42.** Sporogenous cells cylindrical, elongate, but not rachis-like .................. **Nodulisporium**

CHAPTER

# VI

# Generic Descriptions

## ABGLIOPHRAGMA Roy and Gujarati

Type Species: *Abgliophragma setosum* Roy and Gujarati.

**Generic Description:** Hyphae black, tuberculate, bearing sporodochia with black, septate setae; conidiophores subhyaline, unbranched at the tips; conidia hyaline, several-celled, cells connected by short isthmi, produced singly. (Description from Roy and Gujarati, 1966).

**Diagnostic Features:** *Abgliophragma* is characterized by setose sporodochia bearing large, hyaline phragmospores with the cells separated by narrow isthmi (Fig. 17).

**Notes:** *Abgliophragma* was erected by Roy and Gujarati (1966) for *A. setosum*, a fungus recovered from the rhizosphere of *Dichanthium annulatum* in grassland. The authors did not commit themselves as to the precise method of conidium production in *Abgliophragma*, but they noted that the fungus resembled *Gliophragma setosum*, described by Subramanian and Lhoda (1964) from horse manure. In *G. setosum* the conidia are blastospores and it is presumed that this is also true for *Abgliophragma*. It is not clear whether the spores develop successively or simultaneously from the sporogenous cells in either genus. In their drawings of *Gliophragma*, Subramanian and Lhoda depicted all of the conidia at more or less the same stage of development; this might indicate that the spores develop simultaneously and that *Gliophragma* and *Abgliophragma* belong in the Botryoblastosporae.

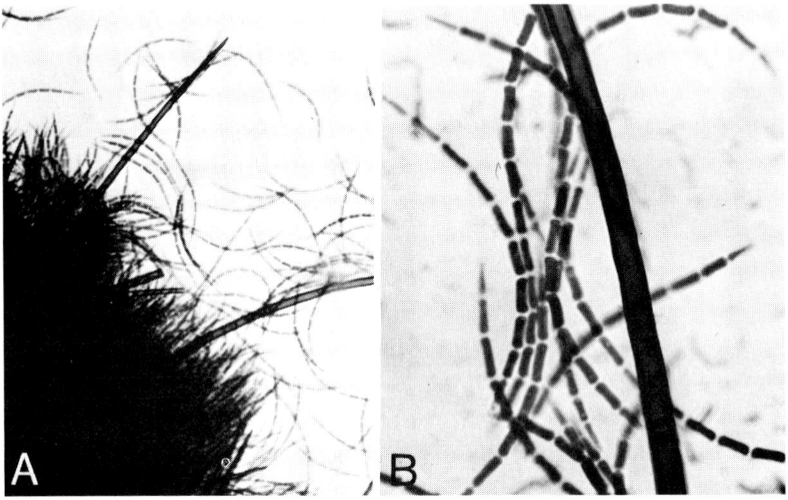

FIG. 17. *Abgliophragma* sp. *A*, the fructifications have a fringe of stout, dark setae. *B*, the conidia have narrow connectives between cells.

The fructification in *Gliophragma* is described as a synnema and in *Abgliophragma* as a sporodochium. From the figures and descriptions of *Abgliophragma* it appears that the fructification type falls into the "grey zone" and either term may be applied. In both genera the fructifications are setose, but in *Gliophragma* the entire fructification is produced in relation to a single, stout seta. In *Gliophragma* the spores are slimy; in *Abgliophragma* they are dry.

Roy and Gujarati drew attention to the similarities between *Abgliophragma* and *Schizotrichum* McAlpine but did not elaborate on the differences between these two genera.

## ACREMONIELLA Sacc.

Type Species: *Acremoniella atra* Sacc.

**Generic Description:** Conidiophores short and inconspicuous, upright or procumbent, simple or sparingly branched, hyaline, septate, with main axis and branches terminating in sharply tapered sporogenous cells; each sporogenous cell bearing a large, solitary conidium at the tip; sporogenous cells proliferate sympodially from a point behind the apex to produce one or more secondary sporogenous cells similar to the first; repeated sympodial development of the sporogenous cells gives rise to complex intertwining masses which appear as tufts; conidia large, globose to ovoid, nonseptate, pigmented, rough or smooth, thick-walled; an aspergilliform phialospore state sometimes produced. (Description based on *A. atra* and *A. verrucosa*.)

**Diagnostic Features:** The large, brown, ovoid conidia borne singly on sharply tapered sporogenous cells distinguish the common species, *A. atra* (Fig. 18*B*) and *A. verrucosa* (Fig. 18*A*).

**Notes:** An excellent account of the morphology and taxonomy of this genus was given by Mason (1933), who reported that in culture the conidiophores are not usually erect but scramble over the surface, turning up at the apex so that the spores are just free of the surface of the substratum. Isolates of *A. verrucosa* and *A. atra* from Ontario soil produced erect, complex tufts of "conidiophores" on oatmeal agar (Fig. 19). Mason noted that the conidia of *A. atra* measured 16 to 30 by 14 to 22 $\mu$ and that, in the *Aspergillus*-like phialospore state, a short conidiophore produces a globose vesicle up to 12 $\mu$ in diameter. A primary series of phialides which arise from the vesicle produces short chains of minute, spherical phialospores about 2 $\mu$ in diameter.

Tubaki (1963) suggested that the spores of *Acremoniella* are produced in succession and that the sporogenous cells are in fact annellophores. The conidia of *Acremoniella* have a tenuous relationship with the sporogenous cell and secede very readily, and may in fact be blastospores rather than aleuriospores. *Acremoniella* shows interesting similarities to *Olpitrichum* (*Rhinotrichum*) *macrosporum*. The relationships between *Acremoniella* and *Olpitrichum* are worth further study.

*Acremoniella* is infrequently reported from soil (Gilman, 1957). In our laboratory *A. atra* has been recovered only once. In Herbarium IMI there are three records of *Acremoniella* from soil. These include *A. atra*, *A. verrucosa* and an unassigned species of the genus, all from England. Miss A. Mangan (private communication) has recorded *A. atra* from rhizosphere of wheat in Ireland. *Acremoniella* is frequently associated with the seeds of higher plants (Groves and Skolko, 1946).

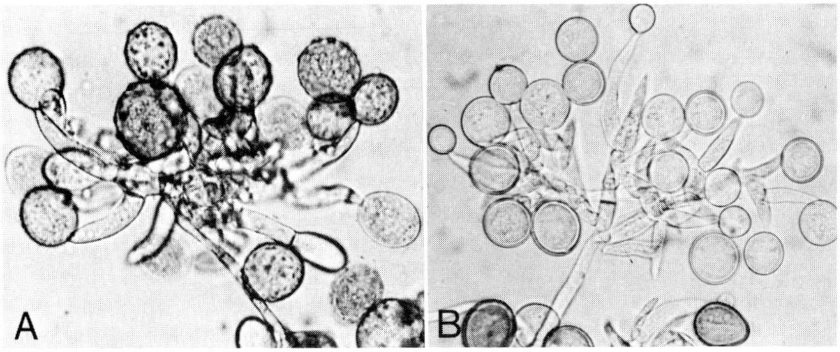


FIG. 18. *A*, *Acremoniella verrucosa*. Rough-walled conidia are borne on sharply-tapered sporogenous cells. *B*, *A. atra*. Conidia are smooth-walled in this species.

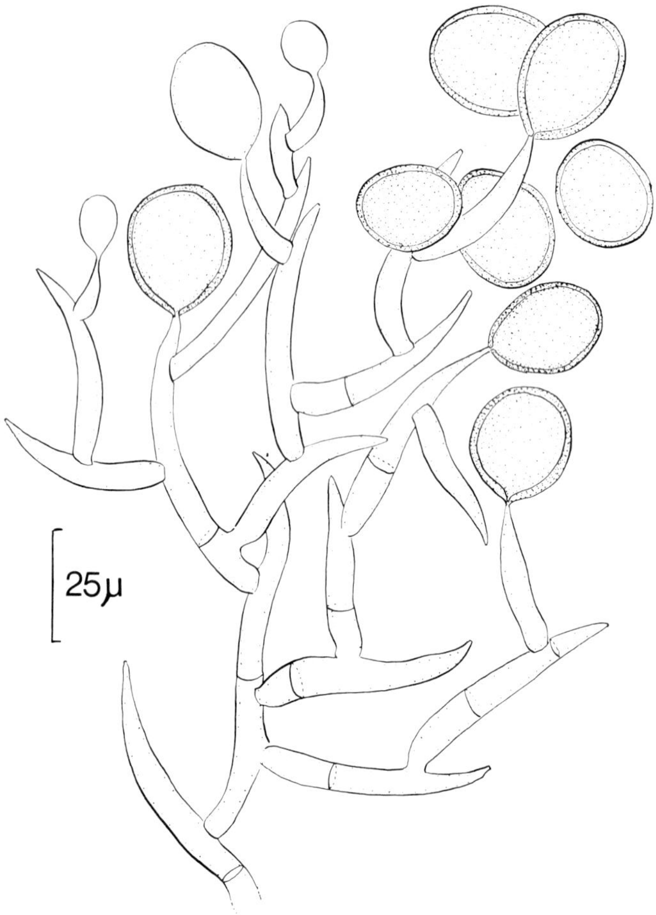


FIG. 19. *Acremoniella atra.* The sporogenous cells branch sympodially to form complex intertwining tufts (OAC 10468).

*A. serpentina* has been described from India by Rai and Tewari (1963), but this fungus is not an *Acremoniella.*

### ACROSPEIRA Berkeley and Broome

Type Species: *Acrospeira mirabilis* Berkeley and Broome.

**Generic Description:** Conidiophores inconspicuous; conidia aleuriospores, borne on short pedicels arising from the vegetative hyphae; aleuriospores several-celled; cells of irregular arrangement with apical cell larger and more darkly pigmented, roughened in some species; an *Aspergillus*-like phialospore state sometimes produced; phialospores in long chains with phialides arising more or less directly from the vegetative hyphae or from swollen ends of the hyphae or branches. (Description based on Wiltshire, 1938.)

**Diagnostic Features:** The dark, irregular aleuriospore with prominent apical cell serves to distinguish *Acrospeira* (Fig. 20).

**Notes:** The morphology of *Acrospeira* was considered in some detail by Wiltshire (1938). The genus was originally erected by Berkeley and Broome (1861), based on *A. mirabilis* found on Spanish chestnuts. According to Wiltshire, the characteristic feature of this species is the spiral coiling of the young conidium primordia. The end of a branch of the conidiophore swells slightly and becomes coiled, the coil often occurring in more than one plane. Transverse divisions develop in the coiled portion of the aleuriospore which cuts off three cells. The terminal cell increases in size and becomes globular, brown and warted. The two lower cells lie tightly adpressed to the large terminal cell. The spore is released by rupture. A small germ pore is often recognizable on the enlarged cell. Wiltshire also considered *A. levis* and *A. macrosporoidea* in his treatment, but more recently Hughes (1958) transferred these species into *Monodictys*.

*Acrospeira* is rarely reported from soil and the only record is that of

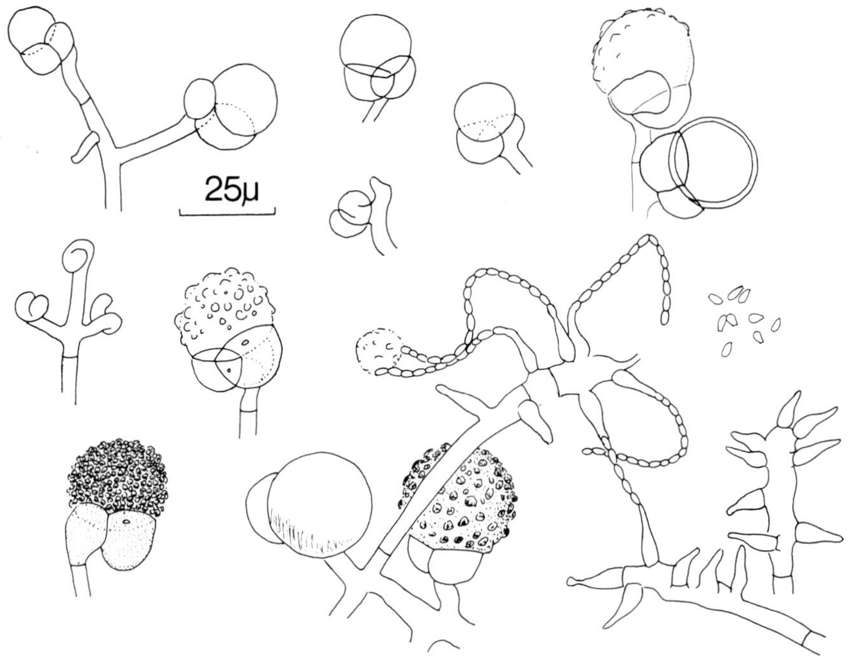


FIG. 20. *Acrospeira mirabilis*. Showing stages in the development of the aleuriospores. (Copied from Wiltshire, *Trans. Brit. Mycol. Soc.*, *21:* 211, 1938. Fig. 16)

Pugh *et al.* (1963), who recorded *A. macrosporoidea*. Following Hughes, this record would now be under *Monodictys castaneae* (Wallr.) Hughes.

## ACROTHECIUM

Ainsworth (1963) following Hughes (1958) considered *Acrothecium* as a *nomen ambiguum*.

Members of the genus *Acrothecium* are frequently reported from soil (Gilman, 1957). *Acrothecium robustum* is the most commonly recorded species and other species are reported only rarely. In its popular concept *Acrothecium* includes those fungi with dark phragmospores arising in clusters at the apex of erect, simple conidiophores. There are a number of alternative genera which would fit this somewhat generalized description. These would include *Curvularia*, *Helminthosporium*, *Drechslera*, *Bipolaris* and *Dichotomophthora*.

*A. arenarium*, described from sand dunes by Moreau and Moreau (1941), is now disposed as *Curvularia inaequalis* (Ellis, 1966).

## ALTERNARIA Nees ex Wallr.

Type Species: *Alternaria tenuis* Nees.

**Generic Description:** Conidiophores dark, septate, sometimes inconspicuous, simple or branched, bearing conidia at the apex; porospores solitary or more often produced in acropetal succession to form simple or branched chains, muriform, darkly pigmented, ovate to obclavate, tapering abruptly or gradually towards the distal end, smooth or roughened.

**Diagnostic Features:** The distinctive obclavate or beaked, muriform

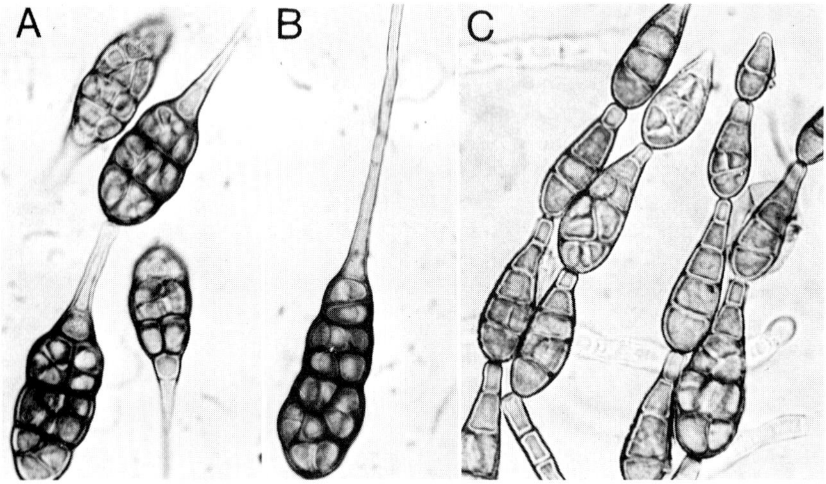


FIG. 21. *A* and *B*, *Alternaria porri*. Prolongation of the terminal cell is more pronounced in pathogenic forms. *C*, *A. tenuis*. Showing acropetal development of conidia.

porospores of *Alternaria* make it readily identifiable in most cases (Fig. 21).

**Notes:** In certain species the longitudinal septa are rare or lacking; this is particularly true of pathogenic forms. Many of the pathogenic forms fail to fruit in culture except under special conditions. For the soil microbiologist, therefore, such forms are mostly self eliminating being classified as sterile mycelia. The mycelium of some *Alternaria* species, however, is distinctive enough to encourage further experimentation to initiate sporulation and confirm their identity.

The taxonomy of *Alternaria* is difficult due to the variability in shape, size and septation of spores even within a species. The most extensive treatments of the genus are those of Neergaard (1945) and Joly (1964).

*Alternaria* species are frequently recorded from soil, organic debris and as primary and secondary invaders of higher plants.

## AMBLYOSPORIUM Fres.

Type Species: *Amblyosporium botrytis* Fres.

**Generic Description:** Conidiophores erect or suberect, stout, septate, hyaline to lightly pigmented, branching irregularly at the apex to form a radiating series of sporogenous branches; branches become septate basipetally and mature to form branching chains of arthrospores; arthrospores fall away at maturity to leave the conidiophore as a naked trunk with the concertina-like stumps of the sporogenous branches still remaining. (Description based on *A. botrytis.*)

**Diagnostic Features:** Stout conidiophores branch tree-like at the apex and finally form large arthrospores (Figs. 22 and 139*E*). The spores are frequently barrel-shaped and truncate at the ends.

**Notes:** The genus by description resembles *Oidiodendron*. It differs

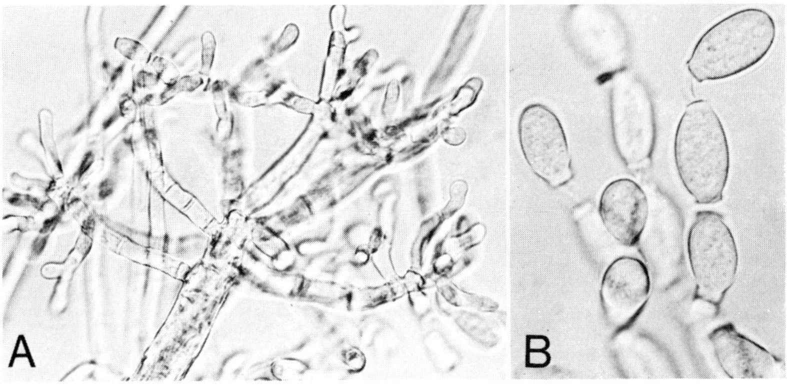


FIG. 22. *Amblyosporium botrytis*. *A*, apex of conidiophore showing primary branches with arthrospores beginning to develop from sporogenous hyphae. *B*, higher magnification with barrel-shaped arthrospores still connected by enveloping parental hyphae (IMI 1691).

from the latter in growth habit, pale coloration, large size of conidia and conidiophores and also in the precise method of spore formation. In culture, *A. botrytis*, the common species, produces a rapidly spreading, cottony, orange-colored mycelium.

Members of the genus are rarely reported from soil. Gilman (1957) reported records of *A. echinulatum* from China and Jensen (1931) recorded this genus from North American soils. The figure of *A. echinulatum* drawn by Oudemans and Koning (1902) is not *Amblyosporium* and is probably *Aspergillus*. An isolate, reputed to be this species, was considered by Raper and Fennell (1965) to be *Aspergillus ruber*.

## ARTHRINIUM Kunze ex Fr.

Type Species: *Arthrinium caricicola* Kunze ex Ficinus and Schubert.

**Generic Description:** Conidiophore mother cells subspherical, ampulliform, barrel-shaped, or broadly clavate, arising from cells of a superficial or erumpent mycelial mat; conidiophores arising singly from conidiophore mother cells, simple, often narrow, more or less cylindrical, usually colorless except for the thick transverse septa which may be highly refractive and are often brown or dark brown, growth in length takes place only at the base; conidia lateral, sometimes terminal, usually formed on very short pegs, one-celled, distinctly shaped, often flattened and with a hyaline rim or germ slit, brown or dark-brown, smooth in most species. (Description from Ellis, 1965.)

**Diagnostic Features:** *Arthrinium* is distinguished by the dark, one-celled, distinctively shaped conidia with a pronounced hyaline rim or germ slit (Figs. 10 and 23). Highly refractive septa are also characteristic of the genus.

**Notes:** An excellent and well illustrated monograph on the genus *Arthrinium* has been presented by Ellis (1965). Few members of the genus have been isolated from soil and the best known are *A. phaeospermum*, previously listed in the literature on soil fungi as *Papularia sphaerosperma*, and the *Arthrinium* state of *Apiospora montagnei*, previously found in much of the literature as *Papularia arundinis*. This latter fungus was connected to its perfect state by Hudson (1963).

## ARTHROBOTRYS Corda

Type Species: *Arthrobotrys superba* Corda.

**Generic Description:** Conidiophores more or less erect, arising from the substratum or from the aerial hyphae, hyaline, septate; first conidium apical, a succession of conidia produced in acropetal succession results in enlargement of the apex to form a vesicle on which a cluster of spores is borne; conidiophore continues growth from the apex, and after a period of sterile growth forms a second spore cluster in a similar manner to the first; process repeated to give an elongate conidiophore with apical and intercalary swellings bearing spore clusters; conidia borne on denticles over the surface of the swellings, large, hyaline, obovate, one-septate. (Description based on *A. superba*.)

**Diagnostic Features:** The large, two-celled, obovate spores are borne on node-like swellings (Fig. 24) on more or less simple conidiophores and make *Arthrobotrys* readily identifiable (Fig. 15*B*).

**Notes:** Members of this genus are commonly found as predators of nematodes and are consequently more readily recovered from soils when special techniques are used. In the writer's experience, *Arthrobotrys* species come up on dilution or soil plates from time to time but may escape the attention of the investigator. They may produce only a very thin mycelial growth, which is mostly subsurface, with the only aerial development being the sporadic formation of conidiophores. Members of this genus are usually fairly rapid growers, however, and purification is not difficult once a colony has been located.

The species most frequently isolated from soil are *A. oligospora* and *A. superba*. Records of the genus from soil include those of Guillemat and

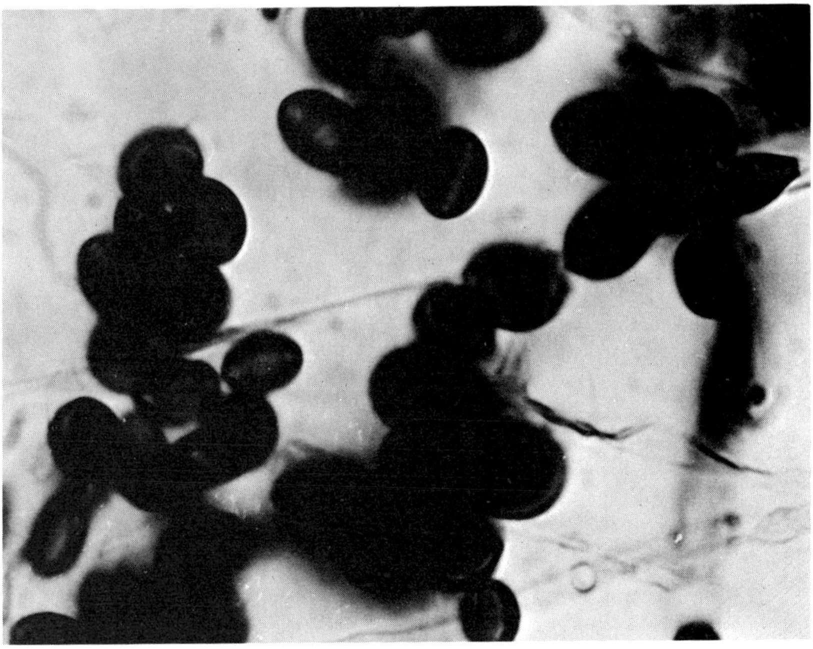

FIG. 23. *Arthrinium phaeospermum*. Conidia are lens-shaped with a hyaline rim (OAC 10155).

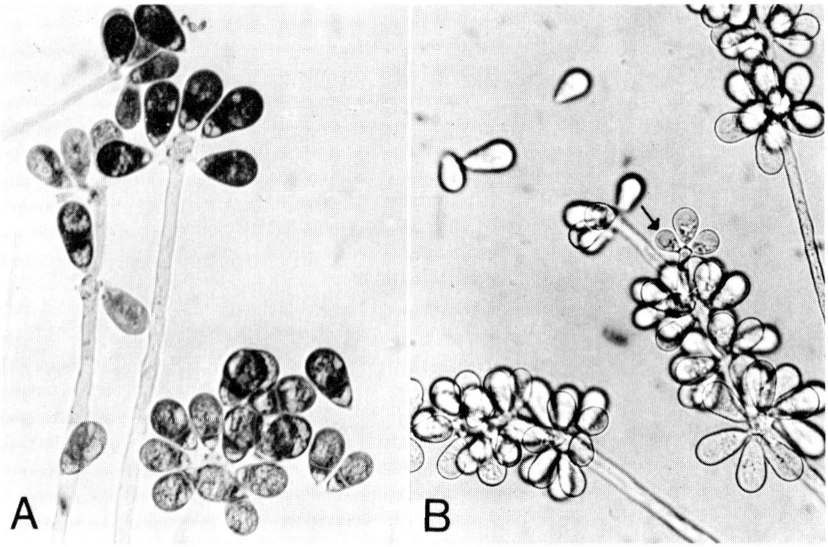


FIG. 24. *Arthrobotrys oligospora. A*, conidiophores with conidia partly dispersed to show peg-like outgrowths on which the spores are borne. *B*, conidia *in situ* with closely packed spore clusters at the "nodes." Note that the conidia are formed successively (*arrow*) at the apex of the conidiophore (OAC 10082).

Montégut (1957) and Agnihothrudu (1961). Chesters (1960) records *A. cylindrospora* as colonizing "excised roots" and buried root pieces. There is no monograph on this genus, but an excellent treatment is presented by Drechsler (1937). More recent evaluations of *Arthrobotrys* and related genera are given by Rifai and Cooke (1966) and Sidorova *et al.* (1964).

This genus belongs to the Sympodulosporae, but in the mature state it could be confused with the Botryoblastosporae. The key difference is seen at the apex of the conidiophore where several different sizes of spores indicate successive rather than simultaneous development (see *Candelabrella*).

## ARXIELLA Papendorf

Type Species: *Arxiella terrestris* Papendorf.

**Generic Description:** Hyphae septate, branched, thick-walled, hyaline to faintly olivaceous; conidiophores one- to many-celled, continuous with hypha or basally septate, simple or branched, hyaline or faintly olivaceous, all cells conidiiferous; conidia apical or lateral on cells of conidiophore, sessile or on short or elongate sterigma-like projections, single or in short chains or in small irregular groups of interconnected members, reniform with ends obliquely cornute, hyaline to faintly colored, smooth, medianly one-septate. (Description from Papendorf, 1967.)

**Diagnostic Features:** The most distinctive feature of *Arxiella* according to Papendorf is the shape of the spores (Fig. 25).

**Notes:** *Arxiella* was erected on a single species, *A. terrestris*, by Papendorf (1967), who considered the genus to be closely related to *Scolecobasidium* and *Diplorhinotrichum* but emphasized that the conidia in *Arxiella* may occur in short chains or irregular groups with interconnected members and are frequently sessile. I have not seen this species.

## ASPERGILLUS Link

Type Species: *Aspergillus candidus* or *Aspergillus flavus*?

**Generic Description:** Conidiophores erect or suberect, frequently originating from a well developed foot cell, hyaline to subhyaline, becoming dark brown in age in some species, swelling at the apex to produce a vesicle; vesicle spherical to ovoid or clavate, bearing sporogenous cells directly or on metulae; sporogenous cells phialides, flask-shaped, producing phialospores in long dry chains; phialospores continuous, smooth or roughened, hyaline to darkly pigmented, globose to ovoid; sclerotia or cleistothecia, or both, produced by some species.

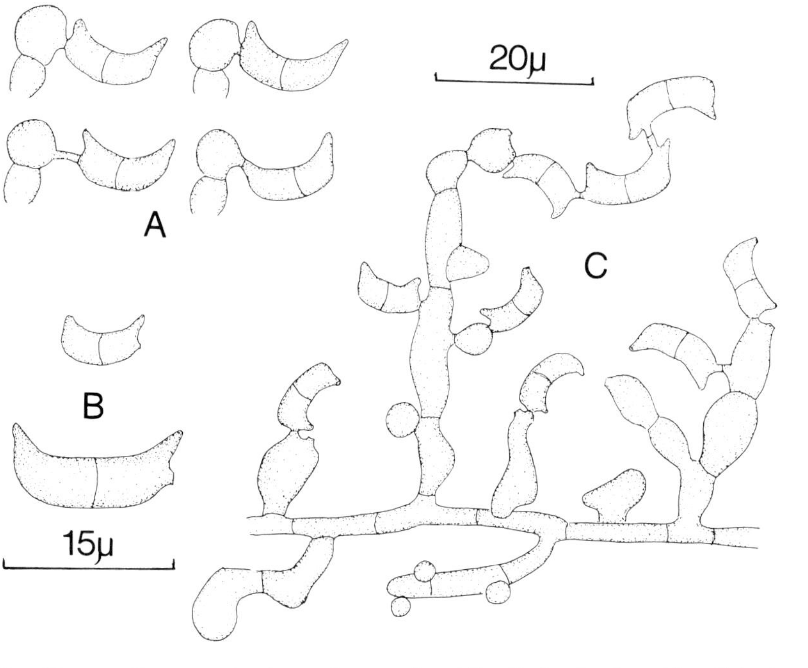


FIG. 25. *Arxiella terrestris*. *A*, attachment of conidia. *B*, details of conidia. *C*, hyphae conidiophores and conidia. (Figure redrawn from Papendorf, *Trans. Brit. Mycol. Soc.*, *50*: 69–75, 1967. Fig. 3).

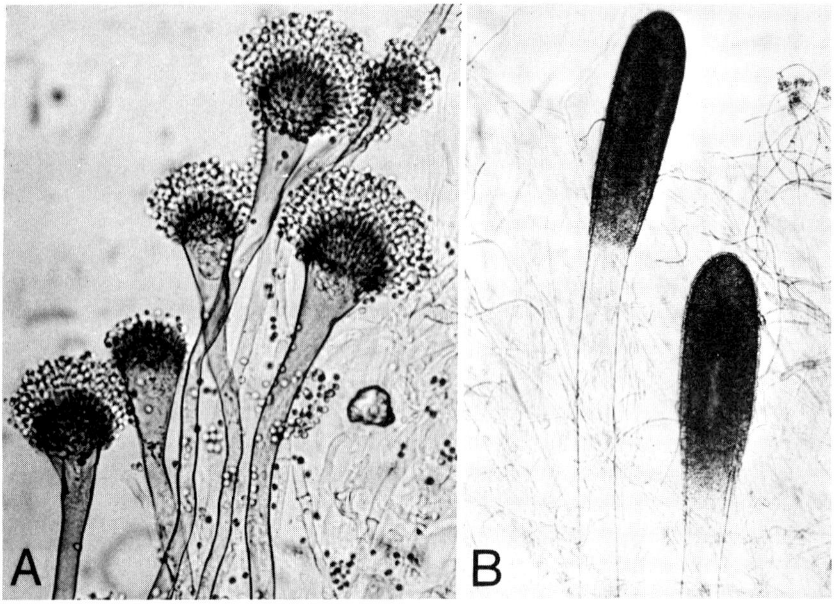


FIG. 26. *A*, *Aspergillus fumigatus*. *B*, *Aspergillus clavatus*

**Diagnostic Features:** The conidiophore with apical vesicle bearing numerous phialides and with spores in long, dry chains make *Aspergillus* one of the most readily recognized of fungi (Fig. 26).

**Notes:** The genus *Aspergillus* needs little introduction or elaboration and has a striking appearance both macroscopically and microscopically. Its taxonomy, morphology and biology are excellently covered in a revised monograph on the genus by Raper and Fennell (1965). *Aspergillus* is biologically one of the most successful of all fungi and shows a great physiological versatility. It is to be expected that it will occur on all sorts of organic debris. Members of the genus are readily recovered from soil by dilution plate techniques and are particularly common in tropical and subtropical soils, but less dominant in temperate regions. In our experience, *A. fumigatus* is far and away the most common species recovered in Canadian soils.

## ASTEROMYCES F. and Mme. Moreau

Type Species: *Asteromyces cruciatus* F. and Mme. Moreau.

**Generic Description:** Hyphae hyaline to brown, septate, branched and often funiculose; sporogenous cells arising singly and laterally, sessile or short-stalked, at first hyaline and narrow, later becoming fuscous and globose with successive conidium production, one or two subterminal cells of the stalk may become sporogenous in turn; conidia nonseptate,

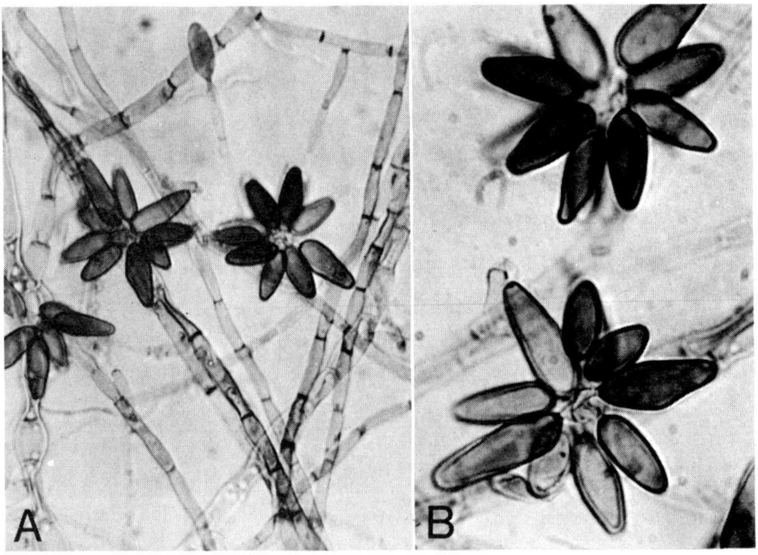


FIG. 27. *Asteromyces cruciatus*. *A*, typical spore clusters. *B*, higher magnification showing spores borne on long pedicels arising from the swollen apex of the sporogenous cell (IMI 53748, ex type).

obclavate to pyriform, dark brown, thin-walled, borne singly on long denticles; first conidium apical, successive conidia lateral and basipetal; conidia secede by rupture of the denticle and germinate through an irregular rupture of the wall. (Description based on Hennebert, 1962.)

**Diagnostic Features:** The large, dark, obclavate to pyriform conidia, produced on long, tubular denticles radiating from a vesiculose sporogenous cell, are diagnostic for *Asteromyces* (Figs. 20 and 27).

**Notes:** Moreau and Moreau (1941) erected *Asteromyces* to include a single species, *A. cruciatus*. No Latin description was published for either taxon, but the genus and species were later validated by Hennebert (1962). *Asteromyces* is monotypic, it was originally recorded from sand dunes under halophilic plants on the Normandy coast, and has subsequently been reported by Brown (1958) as occurring in open sand between low and high tides in Dorset, England, and by Nicot (1958a) from a similar habitat in France.

## **AUREOBASIDIUM** Viala and Boyer

Type Species: *Aureobasidium pullulans* (de Bary) Arnaud.

**Generic Description:** Conidiophores lacking, conidia arising individually, directly from the vegetative hyphae on short denticles; vegetative hyphae hyaline to darkly pigmented; conidia ovoid, continuous, hyaline to

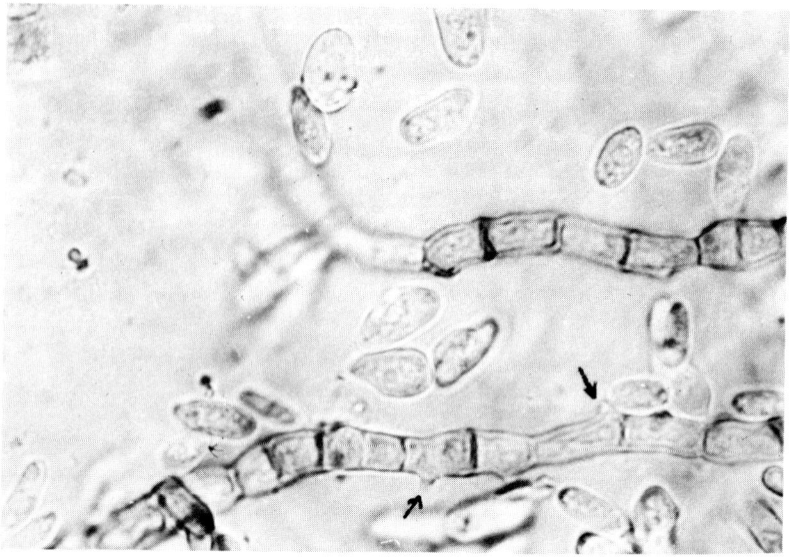

FIG. 28. *Aureobasidium pullulans*. Conidia arise on short denticles (*arrows*) directly from the vegetative hyphae (OAC 10300).

pigmented, produced in slime; primary spores derived from the vegetative hyphae frequently give rise to secondary and tertiary series of blastospores.

**Diagnostic Features:** The production of blastospores on denticles directly from pigmented, vegetative hyphae (Fig. 28) is characteristic of *Aureobasidium*. The colonies are dark brown in color and tend to be water-soaked or yeast-like in appearance.

**Notes:** An excellent treatment of *A. pullulans* has been presented by Cooke (1959) who noted that this fungus is ubiquitous and apparently omnivorous. It occurs on painted surfaces, on wood, in flower nectar and in soil. It is associated with the retting of flax and may be important in the deterioration of pears and oranges in storage or transit. It occurs in paper mill slimes and old books and many other likely and unlikely substrates. Cooke pointed out that this fungus is very variable in culture. Strains on different media or even on different batches of the same medium show considerable variation, and this explains in part the large number of names applied to this organism. Two of the better known synonyms under which this fungus is found in the soil literature are *Pullularia pullulans* and *Dematium pullulans*.

*A. pullulans* is reported by some workers to form arthrospores. This is regarded by Cooke as being due to the chlamydospore nature of the thick-walled mycelial cells rather than to any inherent arthrosporic characteristics.

### BACTRIDIOPSIS Hennings

The genus *Bactridiopsis* is one of uncertain standing. It was regarded by von Höhnel as congeneric with *Coccospora* (Ainsworth, 1963). In their review of *Coccospora*, however, White and Downing (1951) suggested that this latter genus should either be reduced to synonymy under *Sphaerosporium* or discarded. Hughes (1958) satisfactorily disposed of *Coccospora* when his study of type material of *C. aurantiaca*, the type species, proved it to be a *Mycelium Sterilium*.

The only record of *Bactridiopsis* from soil is by Sewell (1959). Sewell's isolate is disposed in the herbarium of the Commonwealth Mycological Institute and is shown in Figures 29 and 139*H*.

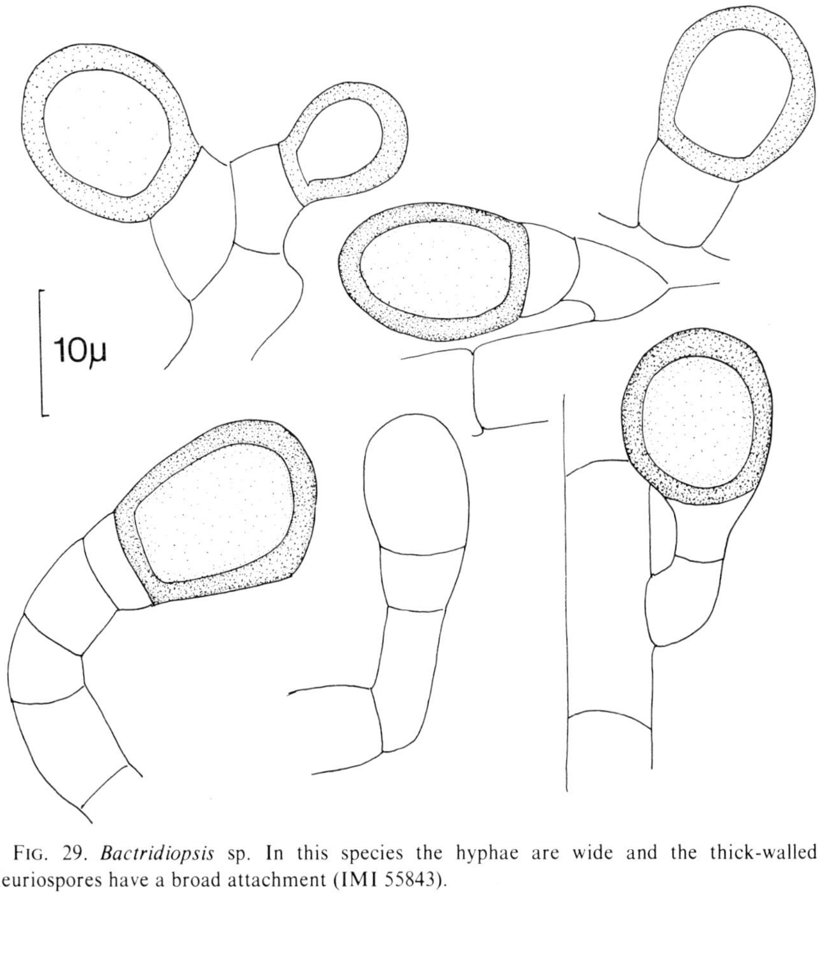


FIG. 29. *Bactridiopsis* sp. In this species the hyphae are wide and the thick-walled aleuriospores have a broad attachment (IMI 55843).

FIG. 30. *Beauveria bassiana*. Sporogenous cells arise in dense, radiate clusters on the aerial hyphae. The small, globose spores are borne in acropetal succession from the sporogenous cell which is typically swollen below in this species (OAC 1304).

## BEAUVERIA Vuill.

Type Species: *Beauveria bassiana* (Bals.) Vuill.

**Generic Description:** Conidiophores not obvious; sporogenous cells sympodulae, arising more or less directly from the vegetative hyphae, simple, hyaline to subhyaline, frequently produced in groups or dense clusters, consisting of a basal cell which is often inflated, and producing sympodulospores in acropetal succession; spore-bearing tip at maturity long, rachis-like, sometimes bent or tortuous; spores minute, spherical to ovoid, continuous, hyaline to subhyaline. (Description based on *B. bassiana.*)

**Diagnostic Features:** The clusters of inflated sporogenous cells each with a long, narrow rachis bearing minute, hyaline spores, are characteristic of *Beauveria* (Fig. 30).

**Notes:** The species most frequently isolated from soil in Ontario is *B. bassiana*, which appears on dilution plates as a white, hemispherical colony. Sporogenous cells are produced abundantly in culture in radiating masses and under the dissecting microscope appear as little, white, powdery balls in the aerial hyphae. The frequency of *Beauveria* in undisturbed soils is probably related to its normal ecological habit as a parasite of insects. Records of *Beauveria* from soil are not extensive, possibly due to difficulties in its identification. *B. bassiana* has been recorded from soil by Sewell (1959) and *B. tenella* by Pugh (1963), Rall (1965) and Christensen and Whittingham (1965). The genus has also been recorded by Williams and Schmitthenner (1956) in Ohio and by Mosca (1964) in Italy.

The morphology and taxonomy of *Beauveria* have been studied by Benham and Miranda (1953) and by MacLeod (1954) who also considered the closely related *Tritirachium*. According to Tubaki (1963), there is little to distinguish *Beauveria* from *Tritirachium*.

## BELTRANIA Penzig

Type Species: *Beltrania rhombica* Penzig.

**Generic Description:** Setae simple, erect, septate, dark brown, smooth or roughened, arising from flat, radially-lobed, basal cells; conidiophores mostly simple, erect, straight or flexuous, septate, pale brown, arising from the basal cells of the setae, paler above, often inflated, bearing one to several denticles; separating cells sometimes present, arising as blown out ends of the conidiophores, subhyaline to pale brown, oval to subglobose; conidia continuous, biconic, pale yellow-olive to dark red-brown, with a distinct, pale, transverse band just above the widest part of the spore, rounded or truncate at the base, free end spicate or apiculate. (Description from Pirozynski, 1963.)

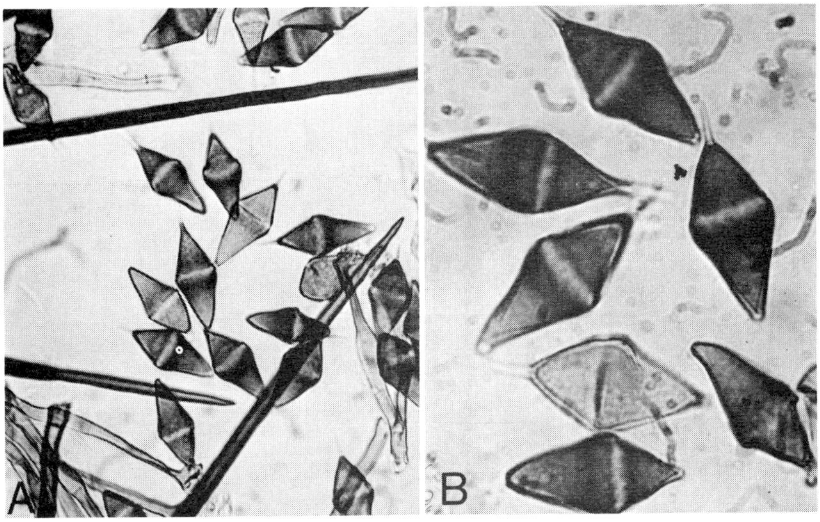


FIG. 31. *Beltrania rhombica*. *A*, conidia, sporogenous cells, and setae. *B*, conidia are nonseptate, apparent wall is an artifact and corresponds to the "shoulder" of the conidium (IMI 89066).

**Diagnostic Features:** The pigmented setae and conidiophores (Fig. 14*D*) arising from expanded basal cells are characteristic of *Beltrania*. The key character, however, is undoubtedly the distinctive biconic conidia (Fig. 31).

**Notes:** An excellent paper on *Beltrania* and related genera has been published by Pirozynski (1963) and is the basis of the generic description above. Species of the genus *Beltrania* are infrequently recorded from soil. *B. multispora* Swart, recorded by Swart (1958) from sand in Mozambique, has been shown by Pirozynski to be a synonym of *B. rhombica*. *B. rhombica* was also recorded several times from soil in the former Belgian Congo by Meyer (1959) and from soil in paddy fields in India by Das (1963). *Beltraniella humicola* described by Ramarao (1962) is disposed in Herbarium IMI under *B. rhombica*.

## BENIOWSKIA Raciborski

Type Species: *Beniowskia sphaeroidea* (Kalchbr. and Cooke) Mason.

**Generic Description:** Fructification a sporodochium; hyphae broad, hyaline, septate, branching more or less at right angles with the tips of the branches anastomosing to form a network of hyphae with few free ends; conidia sessile, spherical, hyaline, continuous, leaving small raised scars on the subtending hyphae; simple, spirally twisted hyphae may radiate from the sporodochium.

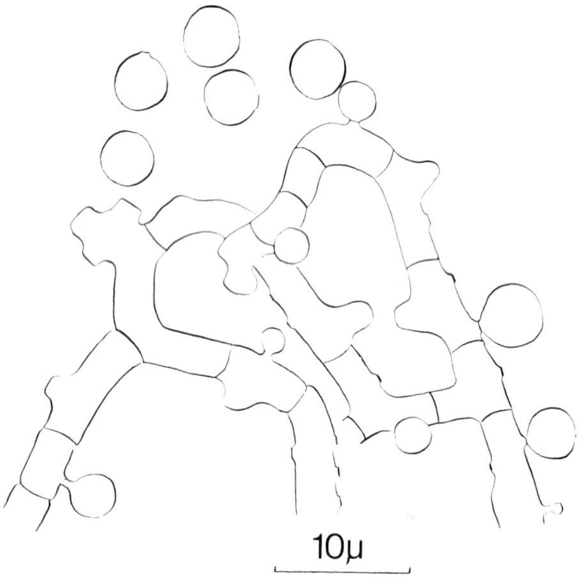


FIG. 32. *Beniowskia sphaeroidea*. The spores in this species are solitary and secede to leave small, raised scars on the sporogenous hyphae (IMI 96723).

**Diagnostic Features:** This genus is characterized by nonpigmented sporodochia with anastomosing hyphae bearing sessile, spherical, hyaline amerospores (Fig. 32).

**Notes:** *Beniowskia* is an unusual Hyphomycete on which little information is available. It is normally found as a foliar parasite of tropical plants (Mason, 1925). The Commonwealth Mycological Institute, Kew, has a single record of this genus from soil. Characteristics of this fungus in culture are not known, it may be that the sporodochia are not produced *in vitro*. Mehrotra (1963) has described a new species of this genus from soil in India.

## BIPOLARIS Shoemaker

Type Species: *Bipolaris maydis* (Nisikado) Shoemaker.

**Generic Description:** Conidiophores brown, producing conidia through an apical pore and forming a new apex by growth of the subterminal region; conidia fusoid, straight or curved, germinating by one germ tube from each end; exosporium smooth, rigid, brown; endosporium hyaline, amorphous, separating cells of mature phragmospores; parasitic chiefly on Gramineae; perfect state, where known, in *Cochliobolus* Drechs. (Description from Shoemaker, 1959.)

**Diagnostic Features:** *Bipolaris* has indeterminate conidiophores which extend sympodially producing a succession of dark, transversely septate, porospores (Fig. 33). These are basically fusoid in shape (Fig. 191) and germinate only from the ends.

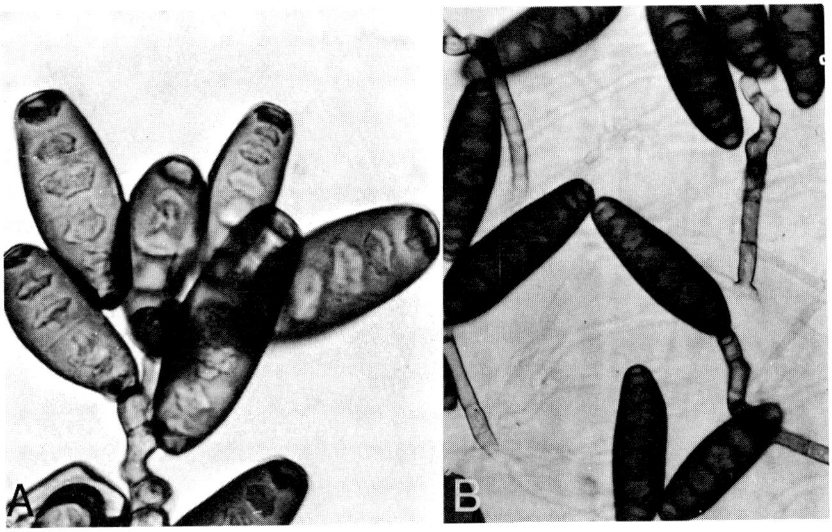


FIG. 33. *Bipolaris* sp. *A*, conidia produced in quick succession may appear in apical clusters. *B*, the conidiophores elongate sympodially and are markedly geniculate.

**Notes:** The genus *Bipolaris* was erected by Shoemaker (1959) as a segregate of *Helminthosporium*. Hughes (1953) studied the method of conidium formation in *H. velutinum*, the type species of *Helminthosporium*, and showed that the conidia in this species develop terminally or laterally in more or less regular verticils. The growth of the conidiophores in *Helminthosporium* ceases with the production of terminal conidia (Ellis, 1961). Hughes (1958) restricted *Helminthosporium* (= *Helmisporium*) to a comparatively few species and excluded amongst others all the graminicolous species. Ito (1930) proposed the name *Drechslera* for the species with cylindric spores that germinate from any cell. The species with fusoid conidia and which show bipolar germination were placed in *Bipolaris* by Shoemaker. (See *Helminthosporium*.)

*Helminthosporium* species are infrequently reported from soil. In many cases these isolates are not identified to species. There is no way of knowing which of these are legitimate *Helminthosporium* species and which belong to *Bipolaris* or *Drechslera*.

## BISPORA Corda

Type Species: *Bispora antennata* (Pers.) Mason.

**Generic Description:** Conidiophores inconspicuous, short, darkly pigmented, simple or sparingly branched; conidia produced in simple chains in acropetal succession, two-celled, dark brown with a very dark septum.

**Diagnostic Features:** The simple acropetal chains of two-celled, dark spores with a conspicuous septum make the common species distinctive (Fig. 34).

**Notes:** *Bispora*, commonly found on dead wood and bark, is rarely recorded from soil. We have isolated it on only two occasions from forest soils in Ontario. The only record I am aware of in the literature is that of *B. pusilla*, reported by Jensen (1912). Hughes (1958) regards this latter name as a synonym of *B. betulina*.

## BLODGETTIA Harvey

In 1858 W. H. Harvey described an alga-fungus association under the name *Blodgettia*, a new genus of algae. Later it was shown to be a species of *Cladophora* associated with a fungus which was named *Blodgettia borneti* by Wright (1881). Feldman (1939) proposed that Harvey's plant was a lichen and applied to it the quadrinomial *Cladophora fuliginosa-Blodgettiomyces borneti*. *Blodgettia indica* has been described by Subramanian (1954) from dead stubble submerged in water. In studies on fungi isolated from Antarctic materials, Tubaki (1961) isolated *Blodgettia borneti* Wright from 13 soil samples and also from some algal material. His identification was tentative in that no type material was available for comparison. Tubaki's description (1961) of this fungus is as follows:

"Conidia acrogenous or intercalary, on poorly differentiated conidiophores; conidia thick-walled, torulose, 4–7 septate, markedly constricted at the septa, the middle cells larger and broader, basal and apical cells progressively narrower, the basal cell mostly doliform, 25–40 μ long and 9–10 μ wide at the broadest parts, dark-colored."

## BOTRYOSPORIUM Corda

Type Species: *Botryosporium diffusum* (Alb. and Schw.) Corda.

**Generic Description:** Conidiophores erect or nearly so, main axis stout, simple or more often dichotomously branched, hyaline, septate, bearing numerous, small, fertile branches more or less at right angles to the main axis; fertile branches narrower at the base, somewhat swollen at the apex, bearing a cluster of swollen sporogenous cells (ampullae) at the apex; spores produced simultaneously over the surface of the ampullae, non-septate, hyaline, ellipsoid. (Description based on *B. longibrachiatum*.)

**Diagnostic Features:** *Botryosporium* is one of the most distinctive and beautiful members of the Hyphomycetes. The stout, main axis bears lateral, fertile branches. These in turn produce several ampullae on which the conidia develop synchronously (Fig. 35).

**Notes:** *Botryosporium* is recorded from soil infrequently. *B. longibrachiatum* has been recorded by Stenton (1953) and a *Botryosporium* species by Williams and Schmitthenner (1956). We have isolated *Botryosporium* only once in Ontario, although I have seen it on a number of occasions

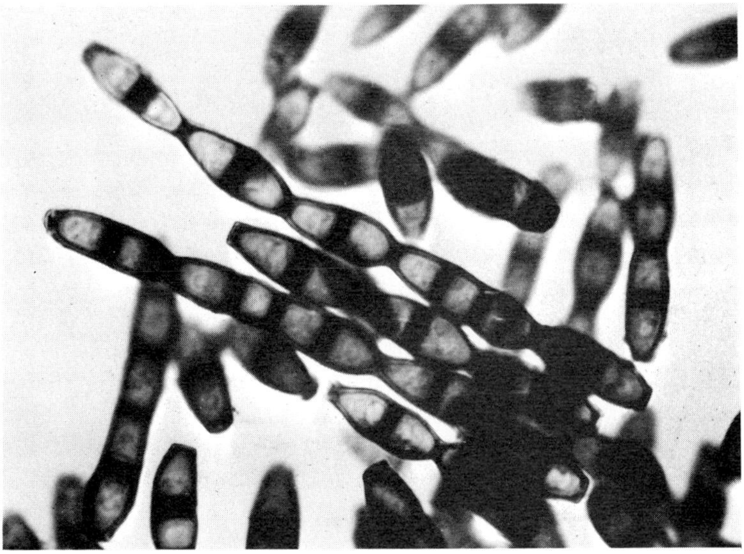

FIG. 34. *Bispora antennata*. The very dark, central septum is characteristic (IMI 9882)

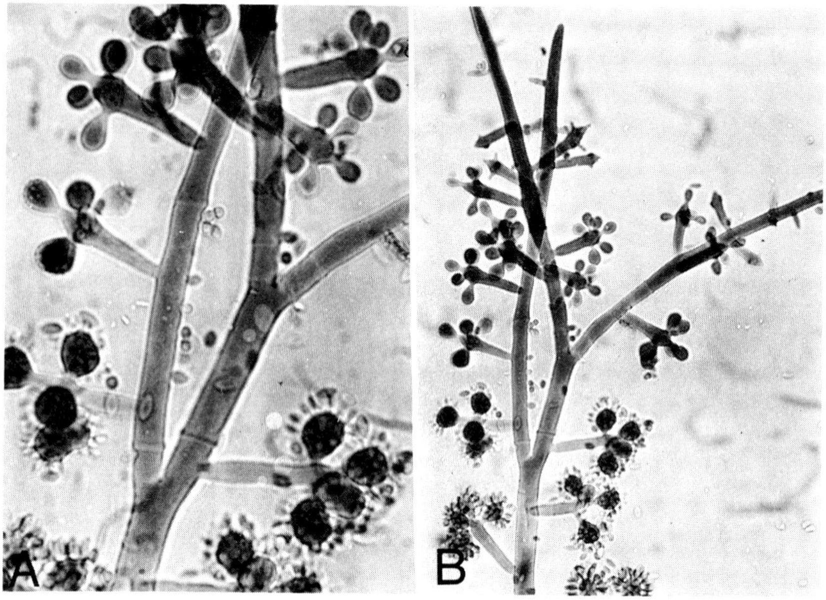


FIG. 35. *Botryosporium longibrachiatum*. *A* and *B*, typical dichotomous or subdichotomous branching of the main axis with acropetal series of lateral, fertile branches (IMI 89331).

growing from organic particles on the surface of potting soil in the greenhouse. It may well be more common in soil than the records indicate but is screened out by usual laboratory techniques.

## BOTRYOTRICHUM Sacc. and March.

Type Species: *Botryotrichum piluliferum* Sacc. and March.

**Generic Description:** Conidiophores inconspicuous; aleuriospores borne more or less directly on the vegetative hyphae or on short lateral branches, frequently produced in botryose clusters; spores borne singly on individual stalks, large, spherical, continuous, thick-walled, golden brown to dark brown; sterile setae intermixed with the vegetative hyphae; setae erect or suberect, tapering acutely at the apex to a narrow, sometimes flexuous, point, smooth or more often roughened, darker below, hyaline or subhyaline near the apex; phialospores sometimes found; phialides short, simple, arising directly from the vegetative hyphae; phialospores hyaline, continuous, in short chains. (Description based on *B. piluliferum*.)

**Diagnostic Features:** The clusters of golden brown to dark brown, thick-walled aleuriospores with intermixed setae are characteristic of *Botryotrichum* (Fig. 36).

**Notes:** The genus was reviewed by Downing (1953) who also considered the related genera *Peziotrichum* and *Coccospora*. Downing concluded that *Peziotrichum* was distinct and placed *Coccospora agricola* as a synonym of *B. piluliferum*. Downing regarded *Botryotrichum* as monotypic and exluded the only other described species, *B. atrogriseum* v. Beyma, on the basis that it was dematiaceous and lacked setae and phialides.

*B. piluliferum* is commonly found in soils, especially those high in organic matter. It is closely related to the genus *Staphylotrichum*, which bears the aleuriospores on well developed conidiophores and lacks a phialospore state. *B. piluliferum* has been connected to its ascus state, *Chaetomium piluliferum*, by Daniels (1961). There are possibly several *Chaetomium* species with *Botryotrichum*-like conidial states (see Ames, 1963). It is difficult to make a distinction between *Botryotrichum* and *Humicola* save that the latter genus lacks setae and has a much darker pigment in the spores. Despite the similarities to *Humicola*, the genus seems to be well founded on *B. piluliferum*.

## BOTRYTIS Pers. ex Fr.

Type Species: *Botrytis cinerea* Pers. ex Fr.

**Generic Description:** Conidiophores tall, erect or nearly so, branching irregularly or dichotomously, dark, septate, terminal cells swell to pro-

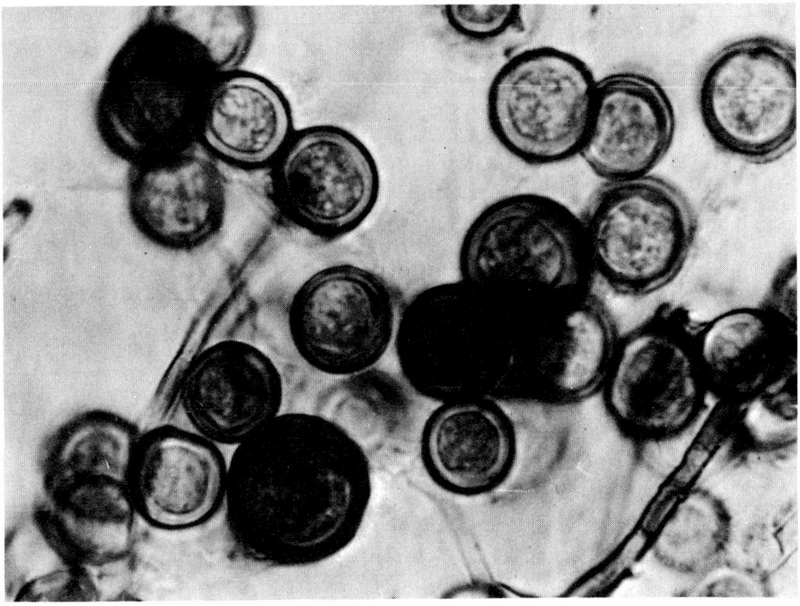

FIG. 36. *Botryotrichum piluliferum*. The aleuriospores are thick-walled and interspersed with sterile setae (OAC 9707).

duce sporogenous ampullae; numerous conidia arising simultaneously on each ampulla, produced on short denticles, hyaline or pigmented, continuous, globose to ovoid.

**Diagnostic Features:** In this genus the stout, dark, branching conidiophores bear clusters of paler conidia (grey in mass) on denticles from apical ampullae (Figs. 8*C* and 37).

**Notes:** The genus *Botrytis* is best known as a pathogen of ornamental and crop plants. Because of the symptoms produced, it is commonly referred to as "the grey mould." In culture, the colonies become grey with the powdery masses of spores. Numerous, fairly large, black sclerotia are frequently produced. In some species the ultimate cells of the conidiophore collapse, to give the branchlets a concertina-like appearance. *Botry-*

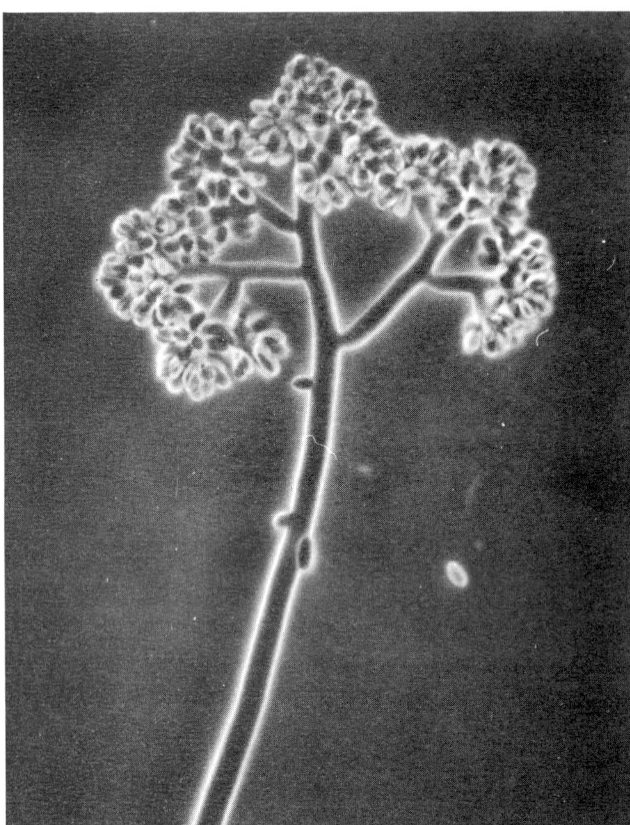

FIG. 37. *Botrytis cinerea.* The main axis of the conidiophore is dichotomously branched near the apex with the ultimate branchlets producing terminal ampullae on which spores develop simultaneously (OAC 10213).

*tis* is recorded infrequently from soil. Whenever I have found colonies on dilution or soil plates, they have been near the edge of the plate or arising directly from a soil particle. Possibly it is relatively common in soil, but it has a relatively low sporulating capacity. This would tend to screen it out in dilution plates and soil plate or washing techniques would be required for accurate estimations of occurrence. *Botrytis* is the conidial state of many members of the Sclerotiniaceae of the Discomycetes (Whetzel, 1945). Cain and Hastings (1956) described *Sphaerospora minuta*, recovered from sandy soil below *Pinus banksiana* in Ontario, as having a *Botrytis*-like conidial state. The *Botrytis* state of *Trichophaea abundans*, recovered from burnt ground in England, was figured and described by Webster *et al.* (1964).

## CALCARISPORIUM Preuss

Type Species: *Calcarisporium arbuscula* Preuss.

**Generic Description:** Conidiophores erect, septate, hyaline; sporogenous cells borne in whorls of three to seven along the main conidiophore axis; conidia borne in acropetal succession at the upper end of the sporogenous cells, which become inflated with repeated spore formation; inflated tip of the sporogenous cell bears conspicuous and irregularly disposed denticles; sometimes the apex elongates rachis-like, bearing conidia along its length; conidia hyaline, continuous, elongate-ovoid. (Description based on *C. arbuscula*.)

**Diagnostic Features:** The hyaline, long-ovoid spores borne on apically swollen sporogenous cells with prominent denticles and with the sporogenous cells arranged in whorls are distinctive for *Calcarisporium* (Fig. 38).

**Notes:** The common species, *C. arbuscula*, is associated with decaying fructifications of higher fungi, particularly Basidiomycetes of the families Boletaceae and Agaricaceae. An account of the morphology and taxonomy of this species is given by Hughes (1951). It is rarely reported from soil and, in our laboratory, we have recovered it only once. *C. pallidum* was described from soil by Tubaki (1955). While this species bears some resemblance to *C. arbuscula*, it is perhaps more closely related to *Sporothrix schenckii* and is not unlike some of the so-called "*Sporotrichum*" states of certain *Ceratocystis* species. *C. parasiticum* described by Barnett (1958) has not been reported from soil, but an unassigned *Calcarisporium* species was recovered by Hodges (1962) from forest nursery soils.

According to Watson (1955), *C. arbuscula* has the ability to live as an endophyte in apparently healthy sporophores of *Lactarius* and *Russula*. Watson (1965) has also shown that *C. arbuscula* is a poor competitor and is unable to survive more than short periods in soil.

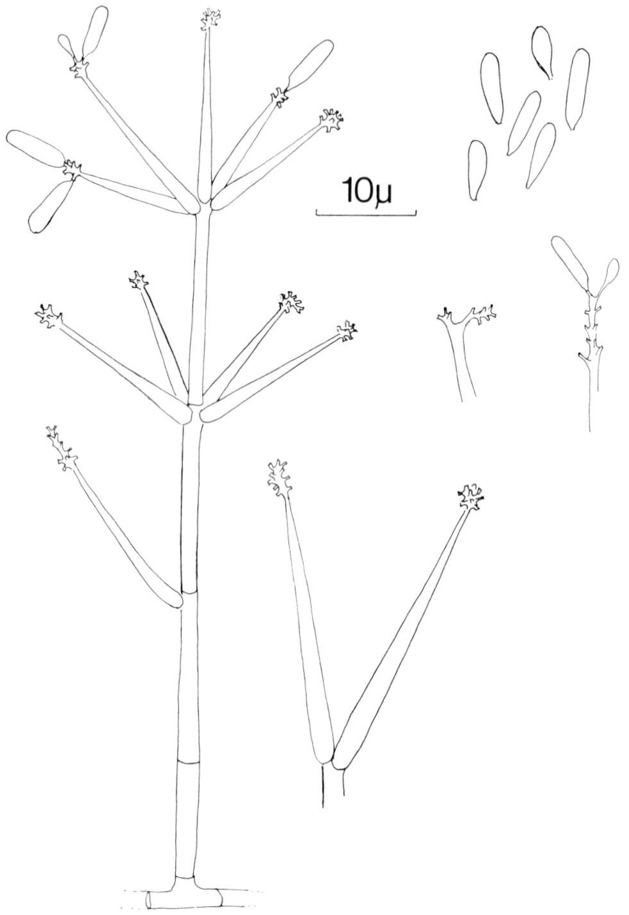


FIG. 38. *Calcarisporium arbuscula*. The sporogenous cells are in verticillate arrangement and usually enlarge at the apex with successive spore formation (OAC 10018).

## CANDELABRELLA Rifai and R. C. Cooke

Type Species: *Candelabrella javanica* Rifai and R. C. Cooke.

**Generic Description:** On cornmeal agar, colonies effused, at first translucent, becoming watery-white to very pale pinkish-white; mycelium composed of septate, branched, hyaline hyphae, sometimes aggregated to form creeping hyphal cords; hyphae sometimes bearing lateral or terminal thick-walled chlamydospores; conidiophores smooth, erect, straight, septate and hyaline, terminated by a small candelabrum-like branching system arising by subapical proliferations of the conidiophore apex; conidia arise singly as blown out ends of the successively produced new growing points to form a lax head at the conidiophore apex, hyaline, obpyriform, ellipsoidal or curved, smooth-walled and one-septate. (Description from Rifai and Cooke, 1966.)

**Diagnostic Features:** The long, distinctive, subcylindrical conidial pegs and the candelabrum-like branching of the conidiophore apex serve, according to the authorities, to distinguish this genus from *Arthrobotrys* (Fig. 39).

**Notes:** *Candelabrella* was erected by Rifai and Cooke (1966) to include two species. *C. javanica*, the type species, was recovered from small

FIG. 39. *Candelabrella javanica*. The conidia are borne on long denticles which are aggregated at the apex of the conidiophore in loose heads. (Figure redrawn from Rifai and Cooke, *Trans. Brit. Mycol. Soc.*, *49:* 147–168, 1966. Fig. 6.)

crumbs of soil from Bogor Botanic Garden, Java, and found to be predaceous on nematodes. *C. musiformis* was originally described by Drechsler (1937) in *Arthrobotrys*. An *Arthrobotrys* recovered by Meyer (1959, Fig. 30) is morphologically like *Candelabrella*.

Rifai and Cooke noted that *C. javanica* produces adhesive networks similar to *A. oligospora*, and regarded *A. entomopaga* Drechsler as congeneric with *C. javanica*. In the latter species, the conidiophores are capable of renewed growth after the first head has been formed so that one or more branching systems are produced at some distance from one another on the main conidiophore axis, and irregular whorls of branches are formed along the length of the conidiophore.

The only real difference between *Candelabrella* and *Arthrobotrys* is the length of the pegs bearing the conidia. This, in my opinion, can be quite a variable character, and I have seen *Arthrobotrys* species in which, under certain cultural conditions, the pegs are very short and accompanied by a slight terminal swelling in the spore-bearing region. At other times the pegs are much longer and approach *Candelabrella*. The segregation of *Candelabrella* from *Arthrobotrys*, is in my opinion, not warranted, the differences being of degree rather than kind.

## **CANDIDA** Berkhout

Type Species: *Candida albicans* (Robin) Berkhout.

**Generic Description:** Cells of varying shape; reproduction by multilateral budding; chlamydospores sometimes present; pseudomycelium more or less abundantly developed; true mycelium may also occur; blastospores attached to the pseudomycleium in a way typical of the species. In liquid media bottom growth, often ring formation and often a pellicle; oxidative assimilation present, and, in many species, a strong fermentative dissimilation. (Description from Lodder and van Rij, 1952.)

**Diagnostic Features:** Yeast-like organisms with a strong mycelial or pseudomycelial growth and multilateral budding are typical of *Candida* (Figs. 6*C*, 40, and 59*D*).

**Notes:** As pointed out by Lodder and van Rij (1952), *Candida* was erected by Berkhout for those yeast-like fungi which were erroneously incorporated into the genus *Monilia*. There is still some controversy surrounding the taxonomy of this genus. Members of this genus are common in soil.

## **CATENULARIA** Grove

Type Species: *Catenularia cuneiformis* (Richon) Mason.

**Generic Description:** Conidiophores solitary or in tufts, stout, dark brown, septate, erect, simple, straight or sinuous, cylindrical, sometimes with a basal swelling, terminating in a solitary phialide with a prominent

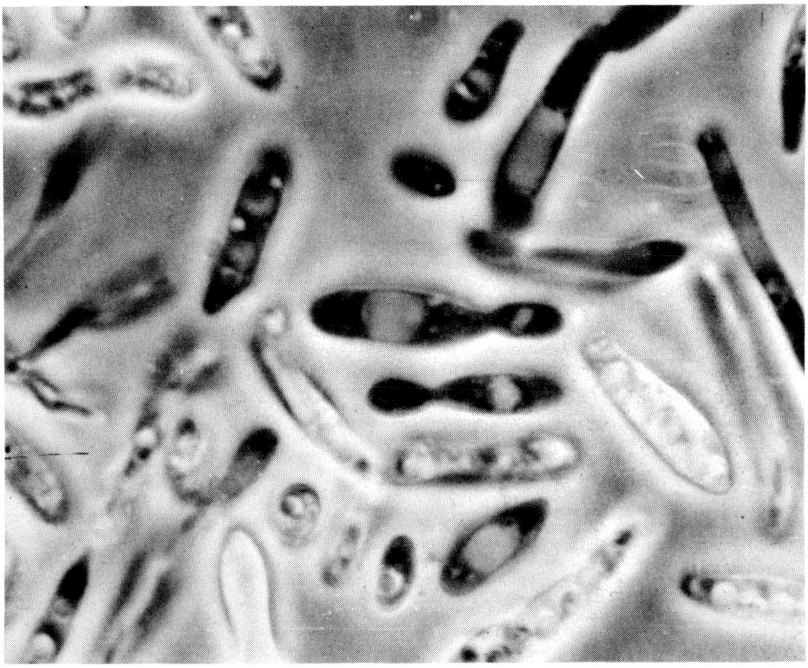

FIG. 40. *Candida humicola.* The lateral branches from the filamentous hyphae bud repeatedly to form a pseudomycelium which breaks up readily on mounting (ATCC 14438).

collarette, accompanied by sterile capitate hyphae in most species; capitate hyphae similar to young conidiophores but terminating in a hyaline, possibly mucilaginous, cap; conidiophores may elongate by repeated proliferations through the collarettes; phialospores develop singly and successively within collarette, forming short chains or sometimes gathering in clusters, nonseptate, smooth, thick-walled, dark brown, rounded-obconic or obovoid, truncate at the point of attachment, angular in section with a germ pore at each corner. (Description based on Hughes, 1965.)

**Diagnostic Features:** The stout, dark, cylindrical conidiophores bearing short chains of large, dark, angular amerospores from a well marked collarette are typical of *Catenularia* (Figs. 41 and 139*C*).

**Notes:** The genus *Catenularia* has been recently reappraised by Hughes (1965), who recognized four species. Hughes excluded 10 additional taxa which had been described under *Catenularia.* Among these were the conidial states of *Chaetosphaeria innumera* and *Chaetosphaeria myriocarpa. Catenularia heimii,* the conidial state of *C. myriocarpa,* has been reported on several occasions from soil and we have recovered it from forest soils in Ontario on at least a dozen occasions. Hughes suggested that this species be considered under the genus *Chloridium.* No

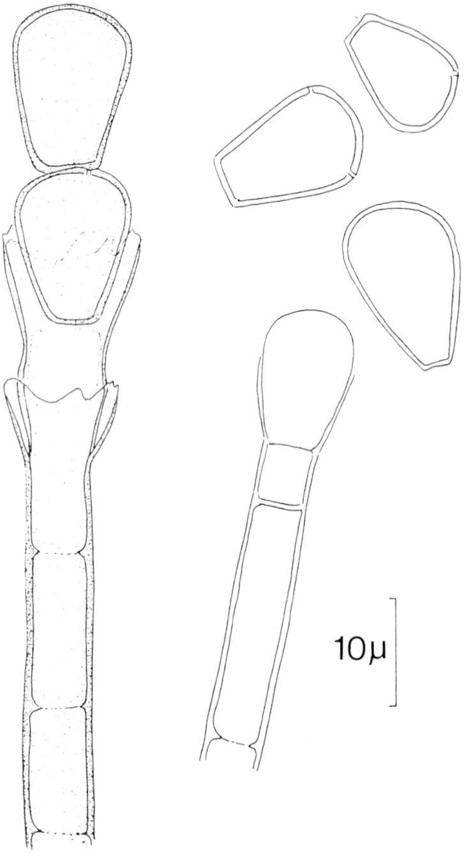


FIG. 41. *Catenularia cuneiformis*. The pronounced collarette, with short chains of large, dark, thick-walled phialospores, is characteristic of the genus (IMI 19069).

member of the genus *Catenularia* as described by Hughes has been recovered from soil, if we may presume that isolates referred to as "*Catenularia* sp" in the floristic listings are probably not *Catenularia* as described here (see *Chloridium*).

## CEPHALIOPHORA Thaxter

Type Species: *Cephaliophora tropica* Thaxter.

**Generic Description:** Conidiophores short or lacking; conidia arising from a swollen sporogenous cell (ampulla) which may be sessile or borne on a more or less short, slender pedicel; conidia didymospores or phragmospores, arising simultaneously as bulbous outgrowths blown out over the surface of the ampulla, becoming elongate and finally one- or several-septate; conidia large, thick-walled, lightly pigmented, salmon-colored to pale orange in the mass. (Description based on *C. tropica*.)

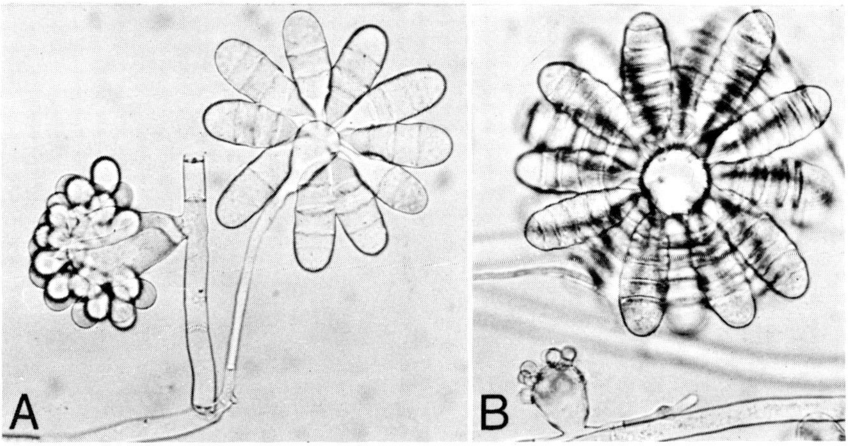


FIG. 42. *Cephaliophora tropica. A* and *B*, various stages of development of the phragmospores from the lateral or terminal ampullae. Note that spores develop synchronously (OAC 10016).

**Diagnostic Features:** The large, septate conidia radiating out from a bulbous ampulla are distinctive for *Cephaliophora.* (Figs. 7*C*, 42, and 117*F*).

**Notes:** The genus was erected by Thaxter (1903) on two species, *viz. C. tropica* and *C. irregularis.* Both of these species have been isolated or recorded from the dung of various animals in several countries. *C. tropica* has been recovered from soil in India by Agnihothrudu and Barua (1957) and in Barbados by Routien (1957). We have recorded it once from soil in Canada. *C. irregularis* is more rare but has been recorded from soil by Goos (1964). There is some question as to whether *C. irregularis* is a valid species or represents merely a variety of *C. tropica.* Goos considered the shape of the conidia and the one-septate nature of the majority of spores in *C. irregularis* as sufficient reason for maintaining it as distinct.

## CEPHALOSPORIOPSIS Peyronel

Type Species: *Cephalosporiopsis alpina* Peyronel.

**Generic Description:** Vegetative hyphae repent, hyaline, septate; conidiophores simple, broader below; conidia hyaline, oblong, one-septate, aggregated into spherical heads.

**Diagnostic Features:** Simple, slender conidiophores bearing two-celled conidia in heads (Fig. 43) are characteristic of *Cephalosporiopsis.*

**Notes:** This genus is essentially a *Cephalosporium* with two-celled conidia. Taxonomically it is a difficult genus because there are a number of nondescript moniliaceous Hyphomycetes which could be classified here. Certain microconidial species of *Fusarium* produce a number of

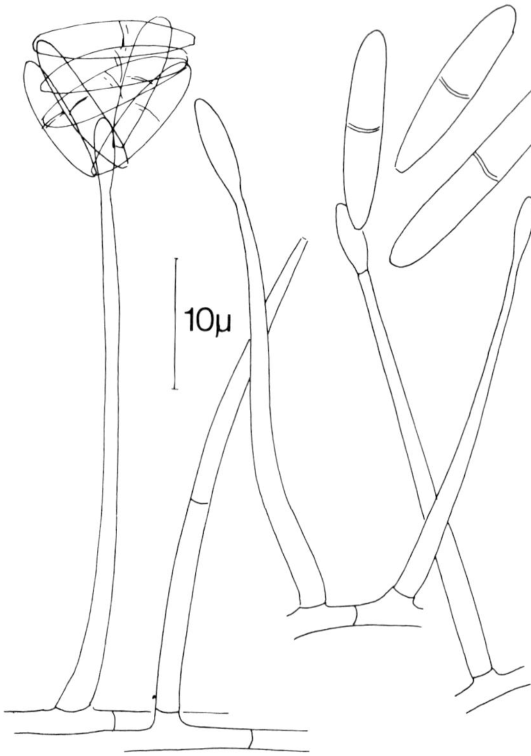


FIG. 43. *Cephalosporiopsis* sp. The two-celled phialospores gather in balls at the mouth of the simple sporogenous cells. The spores are large in this species and it approaches the microconidial states of *Fusarium* or *Cylindrocarpon* in its morphology (IMI 91373).

two-celled spores. There are two-celled *Cylindrocarpon* and *Cylindrocladium* species which under certain cultural conditions might be confused with *Cephalosporiopsis*. Inasmuch as septation of conidia in moniliaceous Hyphomycetes is an inconstant character, then *Cephalosporiopsis* will remain something of a catch-all.

Kamyschko (1961) erected a new genus, *Cephalodiplosporium*, based on *C. elegans* isolated from soil in Russia. There is apparently no type material of this species available, but, considering his generic description and figures, there seems no reason to regard *Cephalodiplosporium* as distinct from *Cephalosporiopsis*.

Records of the genus from soil are rare. The only records I am aware of are those of Guillemat and Montégut (1957), who recorded *C. imperfecta*, and Kamyschko noted above.

## CEPHALOSPORIUM Corda

Type Species: *Cephalosporium acremonium* Corda.

**Generic Description:** Conidiophores lacking; sporogenous cells phialides arising directly and singly from the vegetative hyphae or from funiculose strands of hyphae; sporogenous cells hyaline, tapering, producing phialospores in balls or rarely in fragile chains at the apex; phialospores hyaline, globose to ovoid or short-cylindric, nonseptate.

**Diagnostic Features:** The key character of *Cephalosporium* is the spores balls produced on top of solitary, tapering phialides (Fig. 44).

**Notes:** A casual identification will place many fungi in the genus *Cephalosporium.* Unfortunately, the genus can be readily confused with others such as *Gliomastix*, *Verticillium*, microconidial *Fusarium* or *Cylindrocarpon*, to mention a few. Nevertheless, it is perhaps one of the easiest fungi to identify to genus and one of the most difficult in which to make species determinations. Durrell (1963) studied isolates of 59 named species obtained from a number of sources. The great difficulties experienced in this study were summarized by Durrell as follows:

> From the above description and discussion the question arises. "Is the specific classification of the genus *Cephalosporium* possible where the perfect stage is unknown?" Isolates of the organism from many sources are so variable that it is doubtful if they can be permanently described. Are they merely cultivars? The variations of single spore strains of any "species" are as great as the variations of the genus. We should then return to the original species *Cephalosporium acremonium*, unless a perfect stage is associated.

Sukapure and Thirmulachar (1966) presented a treatment of *Cephalo-*

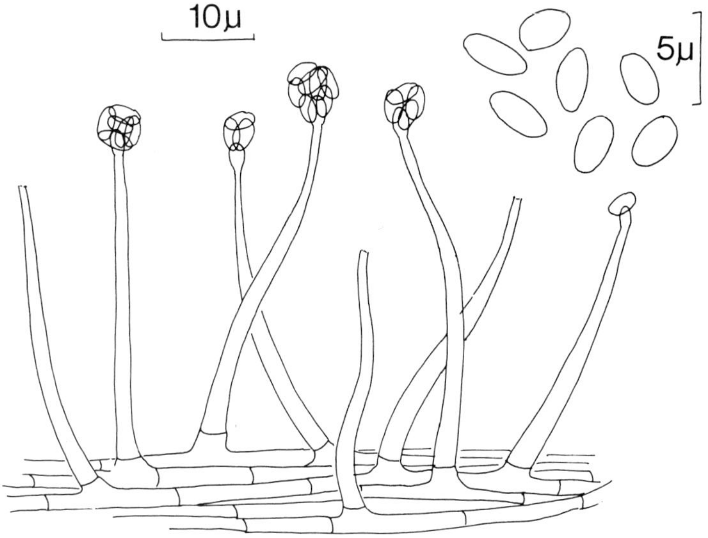


FIG. 44. *Cephalosporium* sp. The hyphae in this genus are typically funiculose with phialides arising directly from hyphae or hyphal strands.

*sporium* with particular reference to the Indian species. Their generic diagnosis differed from the one above in that they included species in which the phialides are opposite or in verticils. They did not indicate how "verticillate" *Cephalosporium* species could be distinguished from *Verticillium*.

## CERCOSPORA Fres.

Type Species: *Cercospora apii* Fres.

**Generic Description:** Conidiophores borne singly or in fascicles, arising from a stroma or directly from the vegetative hyphae, pale to darkly pigmented, septate or continuous, simple or branched, uniform to clavate or sometimes obclavate, straight or curved, sometimes tortuous, smooth or geniculate; spore scars invisible or prominent, considerable variation in length and width between species; conidia borne terminally and singly, becoming lateral by sympodial development of the conidiophore, acicular to obclavate or cylindrical, rarely clavate, hyaline to darkly pigmented, thin-walled, smooth, without appendages, usually multi-septate, straight or strongly curved, sometimes sinuous; base sharply obconic to truncate; tip acute to obtuse; conidia varying between species from 1 to 10 $\mu$ wide and from 30 to 600 $\mu$ long. (Description from Chupp, 1953.)

**Diagnostic Features:** *Cercospora* is characterised by long, hyaline or pigmented phragmospores borne in acropetal succession from a usually simple, sympodially extending, pigmented conidiophore (Fig. 45).

**Notes:** A comprehensive treatment of the genus *Cercospora* has been presented by Chupp (1953). Species of this genus are commonly found as parasites of higher plants, particularly on leaves, stems and fruits. According to Chupp, they are never wholly saprophytic and this perhaps accounts for their rarity in soil. The only record I am aware of is that of *Cercospora salina* which was described by Pugh (1962) from salt-marsh soil. More recently, however, this species has been reassessed and placed in *Dendryphiella* by Pugh and Nicot (1964). *Cercospora* is of some interest, however, in its close morphological resemblance to *Dactylaria* and *Diplorhinotrichum*.

## CERCOSPORELLA Sacc.

Type Species: *Cercosporella persicae* Sacc.

**Generic Description:** Conidiophores erect or suberect, hyaline, septate, simple or sparingly branched, extending by sympodial growth, bearing conidia apically and singly in acropetal succession; conidia hyaline, multiseptate, long ovoid, cylindric, or filiform, straight or curved, usually truncate at the point of attachment.

**Diagnostic Features:** *Cercosporella* is a moniliaceous fungus bearing phragmospores acrogenously from a sympodially extending conidiophore. It is essentially a hyaline *Cercospora*.

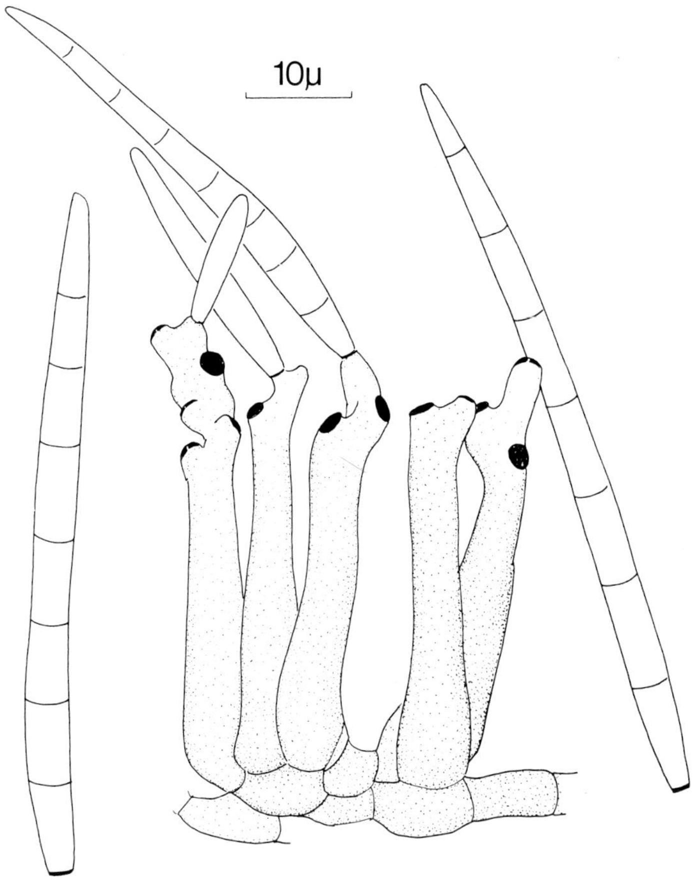


FIG. 45. *Cercospora* sp. The elongate phragmospores are borne in acropetal succession from conidiophores which are frequently aggregated into fascicles.

**Notes:** Members of the genus are commonly found as parasites of higher plants (Chupp, 1953). They are similar to *Cercospora* but differ in lacking pigmentation in both the conidiophore and conidia. There is some doubt as to the validity of pigmentation as the criterion for distinguishing *Cercospora* from *Cercosporella*; because of the large number of described species in these genera, however, the separation is a convenient one.

The genus is rarely recorded from soil. The Commonwealth Mycological Institute, Kew, has a single record of a *Cercosporella* species from this substratum.

## CHAETOPSINA Rambelli

Type Species: *Chaetospina fulva* Rambelli.

**Generic Description:** Conidiophores erect or suberect, stout, septate, brown, thick-walled, frequently swollen at the base, tapering to an acute point, apex sometimes with incrustations; fertile hyphae arising on the conidiophores from just below the septa in the lower half of the main axis, closely adpressed to the conidiophore and giving rise to a compact corticating layer; phialides arising in a dense stand, more or less at right angles to the corticating hyphae, ampulliform, inflated below, with a long neck, sometimes with a small apical collarette, hyaline when young, becoming lightly pigmented in age; phialospores long and narrow, gathering in a mucoid mass as a collar around the conidiophore. (Description based on *C. fulva*).

**Diagnostic Features:** The stout, seta-like, brown conidiophore with a collar of glistening phialospores makes *Chaetopsina* readily distinguishable (Figs. 46 and 47).

**Notes:** This genus is close to *Chaetopsis* but differs in having simple phialides rather than the polyphialide type of sporogenous cells described for *Chaetopsis*. In some ways it resembles *Chaetosperma*, but in this latter genus the corticating hyphae arise from the basal substratum and grow up around what is essentially a sterile seta. In *Chaetopsina* the corticating hyphae arise directly from the main axis of the seta, which is

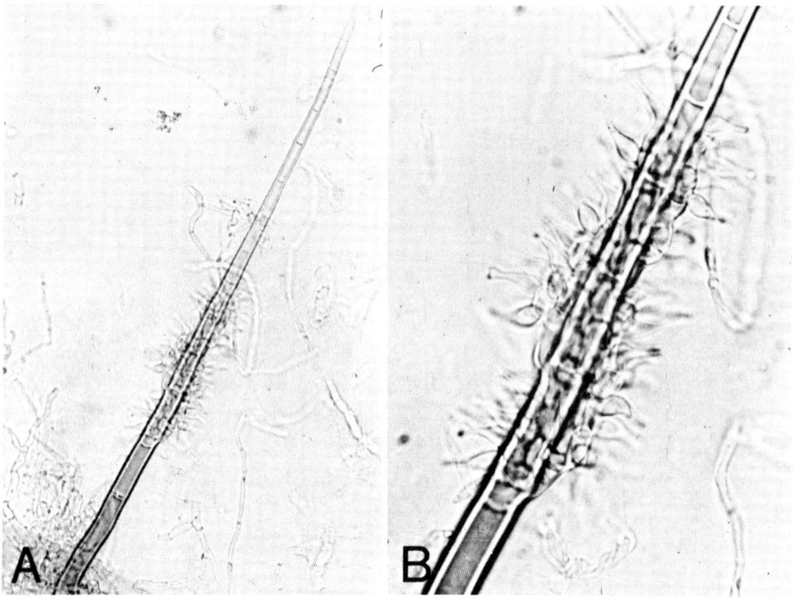


FIG. 46. *Chaetopsina fulva*. *A* and *B*, acutely-tapered, main axis with cortex of fertile hyphae giving sheath of phialides in the lower half. The flask-shaped phialides arise from corticating hyphae (OAC 10274).

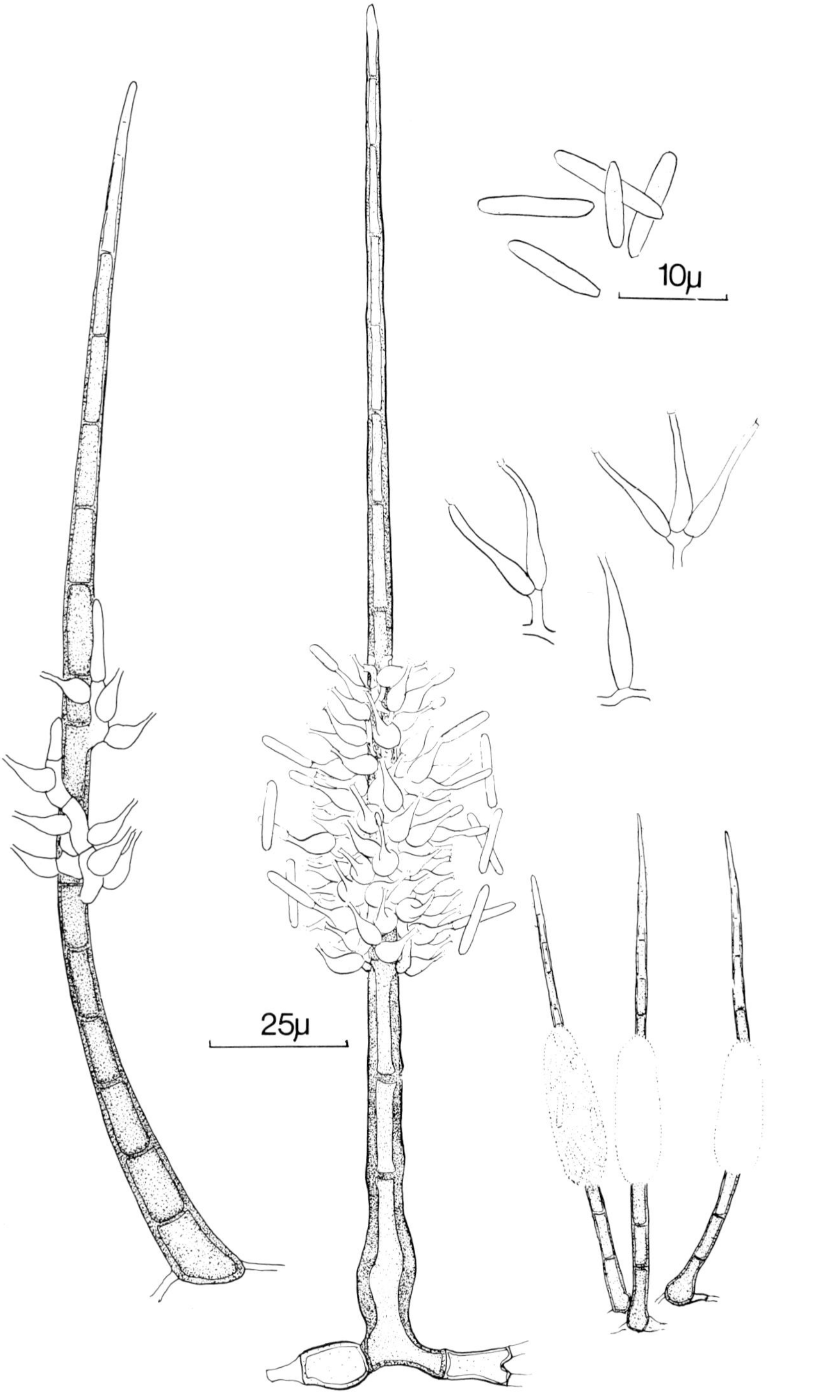


FIG. 47. *Chaetopsina fulva*. Conidiophores, phialides, and phialospores. Phialospores gather to form slimy sheaths around the main axis of the conidiophore.

therefore a conidiophore. *Chaetopsina* shows relationships to *Zanclospora* described by Hughes and Kendrick (1966) from New Zealand. In the latter genus, the phialides arise in whorls directly from the conidiophores. Each whorl originates just below a septum.

This genus was erected by Rambelli (1956), who recovered it from the dead leaves of *Cedrus*, *Quercus* and *Carpinus* in Italy. It has been recorded on several occasions in this laboratory from peat soil under *Thuja occidentalis*.

## CHAETOPSIS Grev. ex Corda

Type Species: *Chaetopsis wauchii* Grev. ex Corda.

**Generic Description:** Main stalk of the conidiophore unbranched, erect, with sterile apex, bearing short, single or paired primary lateral branches in the lower part; secondary lateral branches are also produced; sporogenous cells polyphialides, terminating the lateral branches, more or less cylindrical; phialospores more or less cylindrical, slimy, one-septate, hyaline. (Description based on Hughes, 1951c.)

**Diagnostic Features:** *Chaetopsis* produces stout, erect conidiophores with tapering sterile apices. The lateral branches terminate in polyphialides which produce long-cylindric, two-celled spores (Fig. 48).

**Notes:** *Chaetopsis* is commonly found as a saprophyte on bark or wood. A detailed account of the morphology and taxonomy of the genus was given by Hughes (1951c). As described by Hughes, the conidia are produced through a terminal collarette which becomes lateral by subsequent growth of the sporogenous cell. The new apex develops a new collarette, and, by repetition of the process, up to six collarettes are produced on each sporogenous cell. Hughes refers to this structure as a polyphialide.

Members of this genus are rarely reported from soil. The only record I am aware of is that of Sewell (1959).

## CHALARA (Corda) Rabenh.

Type Species: *Chalara fusidioides* (Corda) Rabenh.

**Generic Description:** Conidiophores short and undistinguished, consisting of a short series of cylindrical cells with the uppermost cell giving rise to an elongate sporogenous cell; sporogenous cells phialides, sometimes arising more or less directly from the vegetative hyphae, hyaline to pigmented, sometimes slightly inflated in the lower half and long and tubular in the upper half; phialospores differentiated internally in the sporogenous cell, short- or long-cylindric, truncate at both ends, hyaline or pigmented, variable in width and length, produced in long, fragile chains.

**Diagnostic Features:** The long chains of cylindrical spores produced endogenously are typical of *Chalara* (Fig. 49). It is distinguished from similar genera by the absence of chlamydospores.

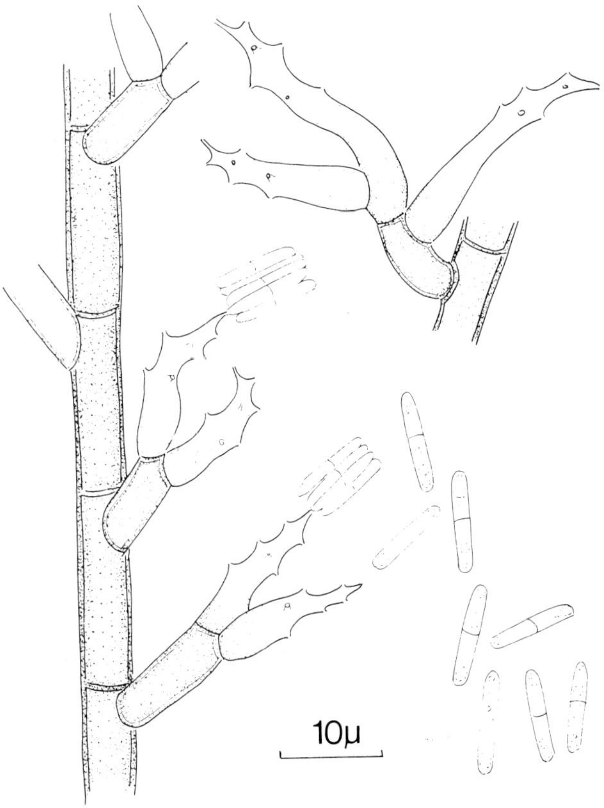


FIG. 48. *Chaetopsis wauchii*. The sporogenous cells have the appearance of sympodulae but the conidia are produced through collarettes and a succession of collarettes on the same sporogenous cell gives rise to a polyphialide (IMI 14024).

**Notes:** *Chalara* is reported from soil infrequently. No comprehensive treatment of the genus is yet available and it is difficult to identify soil isolates to species. There are species presently disposed in other genera which fall within the limits of the generic concept of *Chalara*. In *Thielaviopsis*, aleuriospores are produced which are large, very dark, and several-septate. In *Chalaropsis* aleuriospores are also present, but are nonseptate and produced individually or in short chains. The distinctions between these three genera become arbitrary and there seems no good reason why they should not be incorporated into a broader concept of *Chalara*.

## CHALAROPSIS Peyronel

The genus *Chalaropsis* was erected by Peyronel (1916) based on *C. thielavioides* recovered from *Lupinus albus*. The essential feature of this

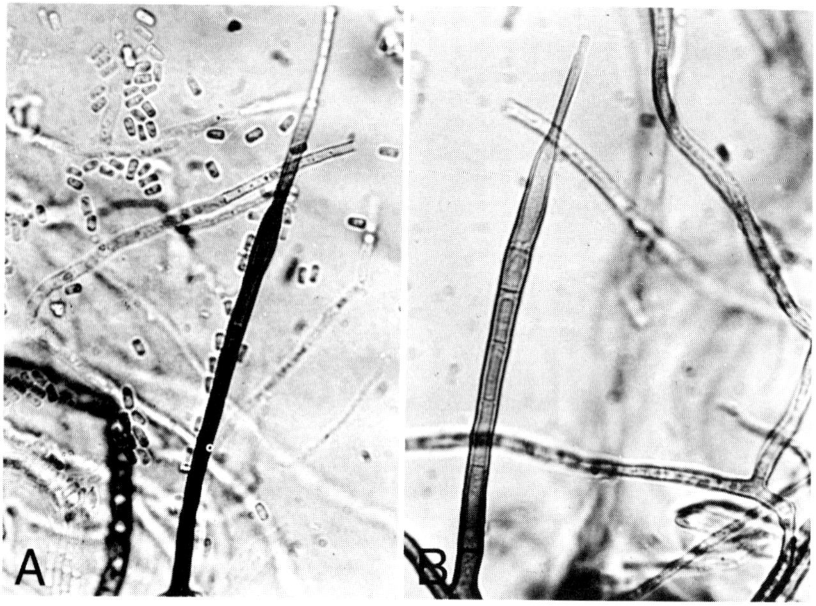

FIG. 49. *A*, *Chalaropsis* sp. Note typical truncate nature of endogenously produced phialospores (OAC 10340). *B*, *Chalara* sp. (OAC 10056).

genus is that it possesses both phialospore and aleuriospore conidial states. The phialospore state is of the endoconidial type described for *Chalara*. The aleuriospore state consists of nonseptate spores borne singly, in short chains, or in acropetal succession from a sympodially extending conidiophore. The generic description for *Chalaropsis* therefore is identical with that of *Chalara*, with the addition of the aleuriospore characteristic. (See *Chalara*.)

## CHLAMYDOMYCES Bain.

Type Species: *Chlamydomyces diffusus* Bain.

**Generic Description:** Conidiophores lacking; conidia aleuriospores, borne more or less directly on the vegetative hyphae or on short pedicels; conidia two-celled; apical cell large, ovoid, tuberculate; basal cell much smaller, wedge-shaped, hyaline, smooth; phialospore state also produced, asperigilliform, irregularly swollen at the apex; phialides short and broad; phialospores oblong, hyaline, abstricted in chains or sliming down to form mucilaginous heads.

**Diagnostic Features:** The large, unequally two-celled aleuriospores (Fig. 50), which are pigmented and roughened are characteristic of *Chlamydomyces*. The accessory spores state is *Aspergillus*-like.

**Notes:** The genus *Chlamydomyces* was founded by Bainier on a single

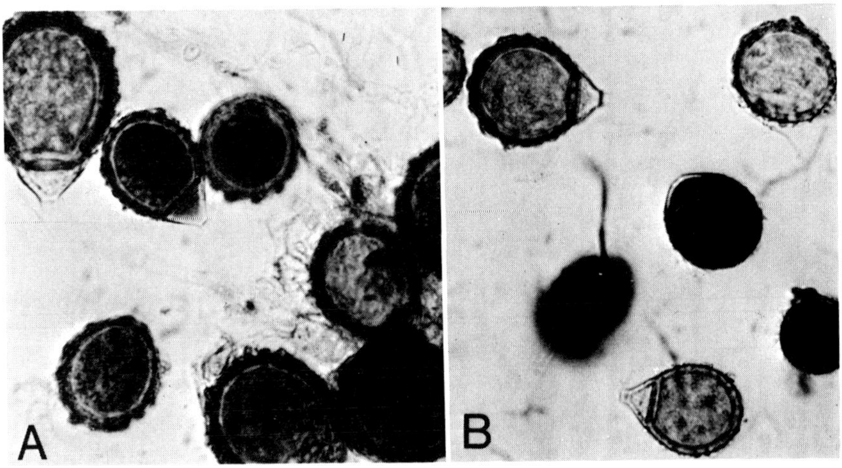


FIG. 50. *Chlamydomyces palmarum*. *A* and *B*, the upper cell is pigmented and roughened, the lower cell hyaline and smooth (IMI 104934). Compare with *Mycogone*.

species, *C. diffusus*, a fungus resembling *Mycogone*. The taxonomy, morphology, and relationships of *Chlamydomyces* were considered in detail by Howell (1939). (See *Mycogone*.)

This genus is rarely recorded from soil, the only published records being those of Pugh (1962) and Thornton (1956).

## CHLORIDIUM Link

Type Species: *Chloridium viride* Link.

**Generic Description:** Conidiophores simple, erect, septate, brown, terminating in a solitary phialide with a more or less prominent collarette; conidia arising successively in the collarette with one or several developing concurrently; conidia nonseptate, hyaline to pigmented, globose to ovoid, gathering in slimy balls or sometimes in loose columns; conidiophores frequently proliferating through the collarette to form successive growing points at higher levels and leaving the mucoid spore masses as "nodes"; chlamydospores found in some species.

**Diagnostic Features:** The simple, erect, pigmented conidiophore terminating in a distinct collarette with several conidia being produced concurrently (Fig. 51) is diagnostic for *Chloridium*.

**Notes:** In the earlier literature *Chloridium* is figured and described under what now would be considered as *Rhinocladiella* or *Rhinocladiella*-like. The revised concept of the genus is based on the analysis by Hughes (1958) of the type material of the type species, *C. viride*. Hughes also recognized that *Bisporomyces* van Beyma and *Cirrhomyces* v. Höhnel were congeneric with *Chloridium*.

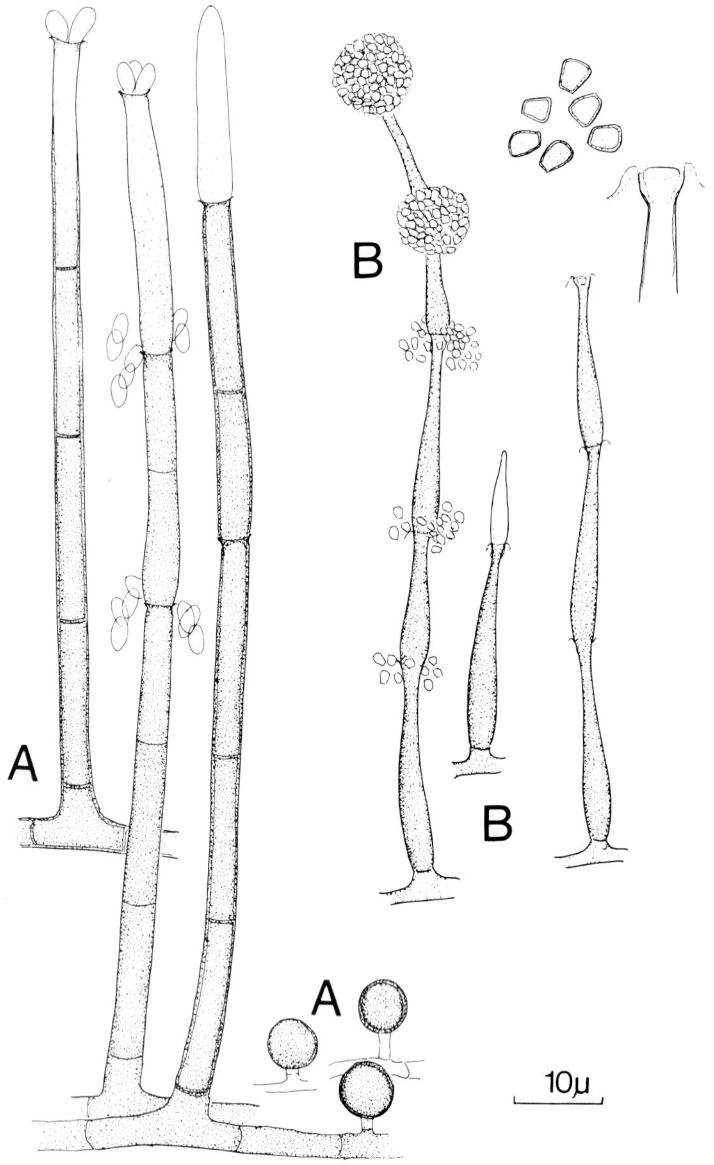


FIG. 51. *A*, *Chloridium chlamydosporis*. Several conidia may develop concurrently (but successively) at the mouth of the shallow collarette. *B*, conidial state of *Chaetosphaeria myriocarpa*. In this species the conidia are produced singly within the cup. Elongation of the conidiophore by proliferation through the collarette is characteristic of *Chloridium*.

The species most commonly isolated in our experience is *C. chlamydosporis* (van Beyma) Hughes which is frequently found in soils high in organic matter. It is very typical of the genus *Chloridium*, and mounts will frequently show at least two conidia originating at the mouth of the phialide. This species is characterized by the presence of spherical chlamydospores scattered throughout the vegetative hyphae.

*Chloridium* is very close to *Gonytrichum* in the method of spore production from the phialides. Several spores may also develop concurrently in the latter genus and short columns of spores may be produced as figured by Hughes (1951c). The relationship between *Chloridium* and *Gonytrichum* was pointed out by Swart (1959). Barron and Bhatt (1967) described *G. chlamydosporium* from soil. In culture this species, after several transfers, lost the ability to produce the *Gonytrichum* conidiophores and collar hyphae, and appeared to all intents and purposes a *Chloridium*. I have seen isolates of *G. macrocladum* in which the *Chloridium* state was dominant, with the *Gonytrichum* state being more or less suppressed. Hughes (1965), in his treatment of *Catenularia*, suggested that the conidial states of *Chaetosphearia myriocarpa* (Fig. 117*B*) and *Chaetosphaeria innumera* (see Booth, 1957) should be considered under *Chloridium*. While this may be the best disposition for these conidial states at the moment, I am not convinced that it is a natural association. In *Catenularia heimii* (conidial *C. myriocarpa*) the conidia are produced singly within a relatively deep collarette, whereas in *Chloridium chlamydosporis* several conidia may develop concurrently from the phialide tip which has a relatively shallow collarette (Fig. 125*D*).

## CHRYSOSPORIUM Corda

Type Species: *C. merdarium* (Link) Carmichael.

**Generic Description:** Conidiophores poorly differentiated, not readily distinguished from the vegetative hyphae, sometimes erect and branching in irregular or verticillate or subverticillate fashion, hyaline, septate; conidia nonseptate, hyaline or brightly colored, globose to subglobose or pyriform, sometimes clavate, usually with a broad basal scar, terminal or intercalary, solitary or in short simple or branching chains, released by dissolution or fracture of the parental hyphal walls; sometimes formed by basipetal septation and differentiation from sporogenous hyphae in the manner of *Oidiodendron*.

**Diagnostic Features:** *Chrysosporium* has a very distinctive spore form and the common species are recognized by the pyriform or clavate spores with the broadly truncate base bearing an annular frill (Fig. 52).

**Notes:** This genus has been monographed by Carmichael (1962), who also gave information on the structure and relationships of similar genera of Hyphomycetes. Carmichael's concept of the genus is a broad one and may not be acceptable to all students of the group.

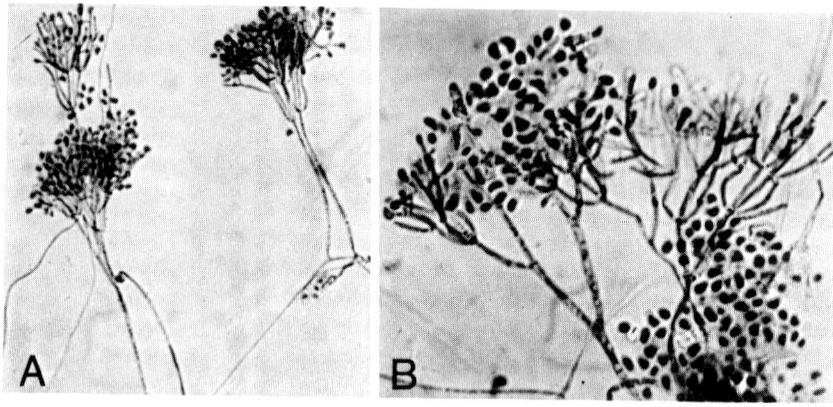


FIG. 52. *Chrysosporium pannorum*. *A*, mount from edge of old colony showing development of erect conidiophores bearing tree-like heads of sporogenous hyphae. *B*, higher magnification showing arthrospores developing in basipetal succession from sporogenous hyphae (OAC 10039).

The spore form in *Chrysosporium* is generally considered and described as an aleuriospore. In certain species of the genus, notably *C. pannorum*, the method of spore formation is more or less similar to that of *Oidiodendron* in origin and development. The spores might be referred to as arthrospores in *C. pannorum* and similar species (Figs. 4*D* and 59*C*).

Members of the genus *Chrysosporium* are common in soil. In our experience, *C. pannorum* is found most frequently in Ontario soils when using dilution plate techniques, although baiting techniques may eventually indicate that other species have an equally wide distribution. In soil studies, *Chrysosporium* species have also been listed under *Geomyces*, *Aleurisma* and *Myceliophthora*, generic names which, as pointed out by Carmichael, are to be regarded as synonyms of *Chrysosporium*.

The perfect states of *Chrysosporium*, where known, are usually found in the Gymnoascaceae (see Carmichael, 1962). *Pseudogymnoascus vinaceus* has been reported by Dal Vesco (1957) to have a *Geomyces* (=*Chrysosporium*) conidial state. Frey and Griffin (1961) have shown that the conidial state of *Ctenomyces serratus* is a *Chrysosporium*. Carmichael placed the conidial state of *Thielavia sepedonium* in *Chrysosporium*, but this could be disputed.

There is some confusion surrounding the relationships between *Chrysosporium*, *Sporotrichum* and *Sporothrix*. This has been clarified in part by Carmichael, who points out that the lectotype of *Sporotrichum*, *S. aureum* Link, has dilute, yellow-brown hyphae and golden brown aleuriospores. The type species of *Sporotrichum* is quite unlike the popular concept of this genus (see Barnett, 1960), which is apparently based on *Sporothrix*. Carmichael points out the confusion which surrounded

*Sporothrix shenckii*, the type species of *Sporothrix* and "*Sporotrichum shenckii*," and notes that this fungus does not in the least resemble *Sporotrichum aureum* (see *Sporothrix*). The differences between *Sporotrichum* and *Chrysosporium* are unfortunately still not resolved. Apinis (1963) described *Sporotrichum thermophile* from pasture soil. This species was reappraised by Semeniuk and Carmichael (1966), who noted that *S. thermophile* resembles *Chrysosporium luteum* in the manner in which the spores were borne, in being cellulolytic but not keratinolytic, and in growing at temperatures above 37°C. They noted that the definite, orange-brown color of the spores in *S. thermophile* is apparent in *C. pannorum* and *C. asperatum*. They pointed out, however, that *S. thermophile* also resembled *Sporotrichum aureum* in the color, size and shape of the conidia, although the hyphae were hyaline rather than pigmented, as in *S. aureum*. If a species could be placed equally well in either genus, then the relationships between *Chrysosporium* and *Sporotrichum* should be more critically defined.

## CILICIOPODIUM Corda

Type Species: *Ciliciopodium violaceum* Corda.

**Generic Description:** Fructification a coremium; coremia cylindric-clavate, large, light or bright-colored, sometimes rough or hairy, sterile below the fertile spore bearing apex; conidia nonseptate, hyaline, globose to ellipsoid, borne in dry masses.

**Diagnostic Features:** *Ciliciopodium* is a hyaline coremial form bearing amerospores in dry masses (Fig. 53).

**Notes:** This genus is rarely reported from soil. Gilman (1957) includes a single record in his manual.

## CLADOBOTRYUM Nees

Type Species: *Cladobotryum variospermum* (Link) Hughes.

**Generic Description:** Conidiophores erect or suberect, frequently arising from the substratum or the aerial hyphae in floccose masses, hyaline, septate, branching frequently and irregularly, or more commonly in verticillate arrangements, ultimately terminating in groups of phialides; phialides hyaline, flask-shaped or swollen basally and tapering gradually towards the apex; phialospores hyaline, one- to several-septate, with well marked basal scar, ovoid or cylindric, produced in short, fragile chains. (Description based on *C. variospermum*.)

**Diagnostic Features:** The short, fragile chains of septate, hyaline conidia borne from phialides on verticillately branched conidiophores (Fig. 54) are distinctive for *Cladobotryum*.

**Notes:** From his study of the type material of genera and species of Hyphomycetes in the classical European exsiccati, Hughes (1958) determined that *Diplocladium* Bon. and *Didymocladium* Sacc. were con-

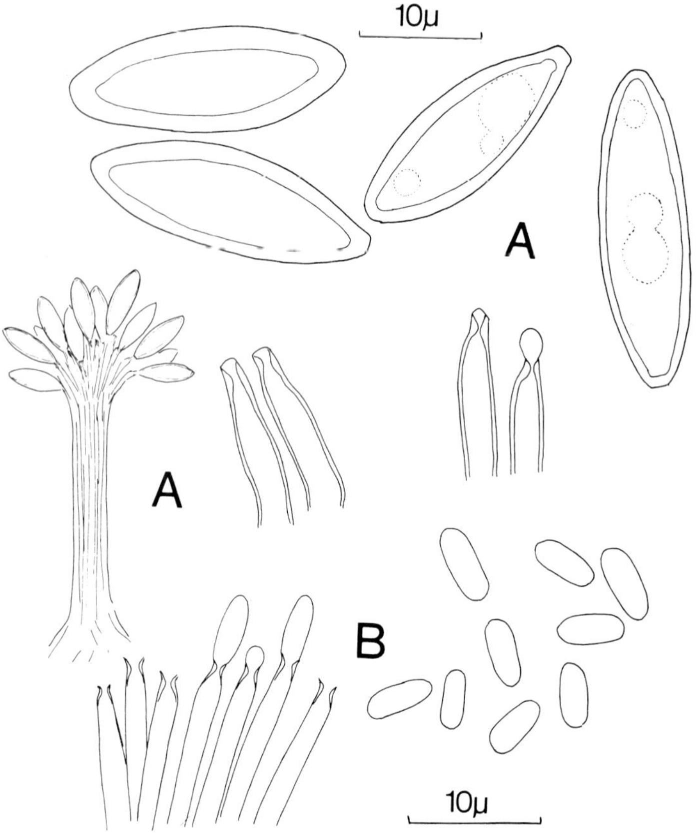


FIG. 53. *A*, *Ciliciopodium macrosporum* (IMI 34150). *B*, *C. sanguineum* (IMI 69912)

generic with *Cladobotryum*. *Cladobotryum* species are infrequently recorded from soil, but are commonly found as saprophytes or parasites on the fructifications of higher Basidiomycetes. We have recorded *C. variospermum*, the type species, only once from Ontario soils and the *Cladobotryum* state of *Hypomyces roseus* on several occasions. The conidia of *C. variospermum* are characteristically one-septate. Those of conidial *H. roseus* are several-septate at maturity. Nicot and Parguey (1963) have obtained *Hypomyces aurantius*, the perfect state of *Cladobotryum variospermum*, in culture.

## CLADOSPORIUM Link

Types Species: *Cladosporium herbarum* (Pers.) Link.

**Generic Diagnosis:** Conidiophores produced in dense stands from the substratum, erect, pigmented, branching irregularly at the apex, tree-like; branches producing conidia in acropetal succession by apical bud-

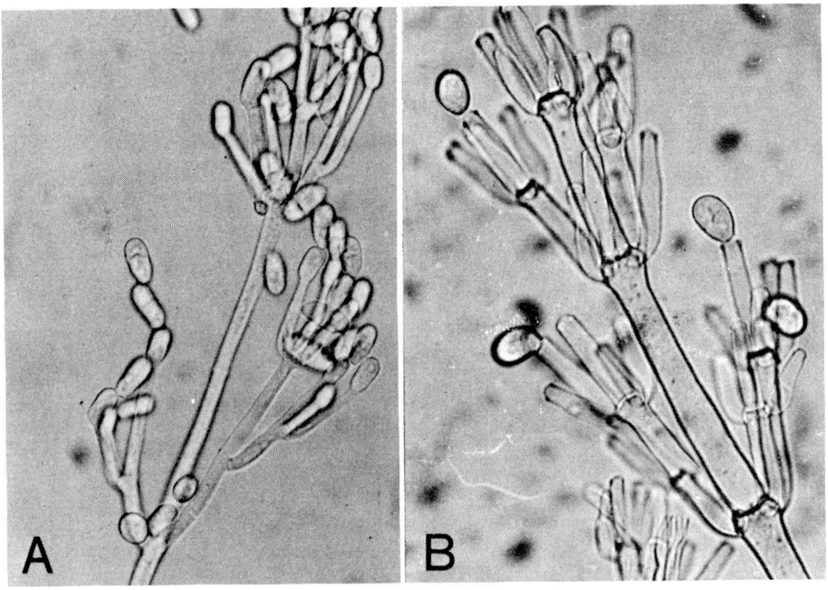


FIG. 54. *A*, *Cladobotryum variospermum*. Showing verticillate or subverticillate branching. Conidia are borne in short, fragile chains (OAC 10053). *B*, *Cladobotryum* state of *Hypomyces roseus*. The phialide nature of the sporogenous cells is seen clearly (OAC 10102).

ding; blastospores in branching chains, hyaline or pigmented, smooth or roughened, continuous or septate; conidial chains very fragile, breaking up readily into units; fragmentation at maturity frequently involving the branches, leaving only naked stumps of conidiophores entire.

**Diagnostic Features:** Erect, pigmented conidiophores with branching chains of blastospores in tree-like heads serve to characterize *Cladosporium* (Fig. 6*A*). This genus can frequently be identified by the conidia alone, which have well marked, dark attachment scars and show considerable variation in size and septation (Fig. 55).

**Notes:** A common saprophyte on organic debris, *Cladosporium* is a very common soil fungus (Gilman, 1957). The parasitic species of *Cladosporium* are rarely reported from soil.

Originally *Cladosporium*-like fungi with nonseptate spores were segregated into the genus *Hormodendrum*. There is, however, considerable variation within and between strains regarding number of septa, and this factor is also influenced by age and culture medium. *Hormodendrum* is now generally considered as a synonym of *Cladosporium*.

Regarding the status of the closely related *Heterosporium*, Jacques (1941) suggested that there was little distinction between this genus and *Cladosporium*. De Vries (1952) established that the lectotype of *Hetero-*

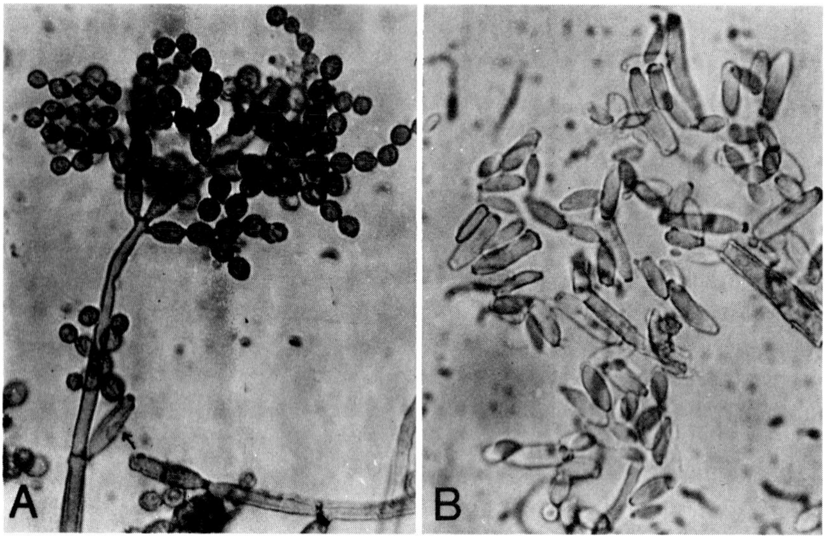


FIG. 55. *A*, *Cladosporium sphaerosporum*. Blastospores are produced in acropetal branching chains. Branches may break off (arrow) to function as conidia (ramo-conidia). *B*, *Cladosporium cladosporioides*. The conidial chains break up readily on mounting; the dark connectives between spores are characteristic.

*sporium* (*H. ornithogali* Klotzsch) belonged properly in *Cladosporium* and this disposition has been accepted by Hughes (1958).

The only species of *Heterosporium* reported from soil, *H. terrestre* Atkinson, was shown to be *Scolecobasidium constrictum* by Barron and Busch (1961).

## CLASTEROSPORIUM Schweinitz

Type Species: *Clasterosporium caricinum* Schweinitz.

**Generic Description:** Conidiophores short, arising singly and laterally on the hyphae, brown or dark brown, sometimes with a few successive terminal proliferations; setae present or absent; conidia formed singly as blown out ends at the apex of the conidiophore and of successive proliferations, obclavate or cylindrical, sometimes rostrate, medium to dark brown, transversely septate; small, darkly pigmented, irregularly angular hypho-podia also present. (Description based on Ellis, 1958.)

**Diagnostic Features:** *Clasterosporium* is typified by the formation of phragmospores singly as blown out ends from the tips of the conidio-phores. Hyphopodia (Fig. 56) are also a characteristic of the genus.

**Notes:** *Clasterosporium* and allied dematiaceous Hyphomycetes have been treated in some detail by Ellis (1958, 1959). Members of the genus are commonly found on the leaves of higher plants, and are rarely re-ported from soil. Bayliss Elliott (1930) reported *C. carpophilum* from soil

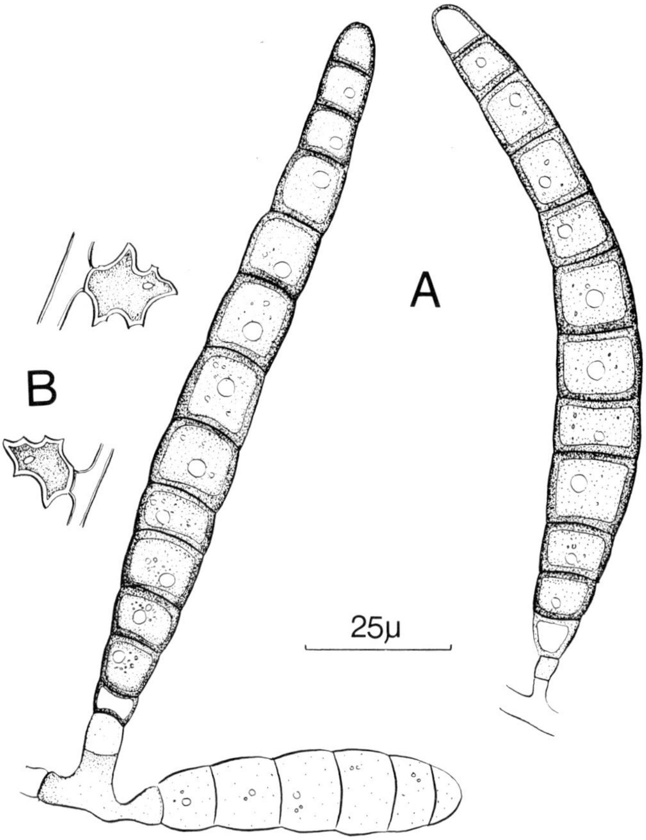


FIG. 56. *Clasterosporium caricinum*. *A*, the aleuriospores are typically very large with many cells. *B*, angular hyphopodia are a characteristic feature.

of the Dovey salt marshes in England, and Johnson and Osborne (1964) isolated a *Clasterosporium* species from soil in Tennessee.

## COCCOSPORIUM Corda

A *Coccosporium* species was recorded by Farrow (1954) from soils collected in Panama and Costa Rica. This species was isolated by Farrow on nine occasions from different geographic localities. Unfortunately, little is known about the genus. Hughes (1958), in his study of Corda's type materials, was unable to find any type or authenticated specimens of *C. maculiforme*, the type species.

## COLLETOTRICHUM Corda

Type Species: *Colletotrichum lineola* Corda.

**Generic Diagnosis:** Fruit body an acervulus; conidiophores produced in a dense, even stand on a thin or well developed stroma; conidiophores

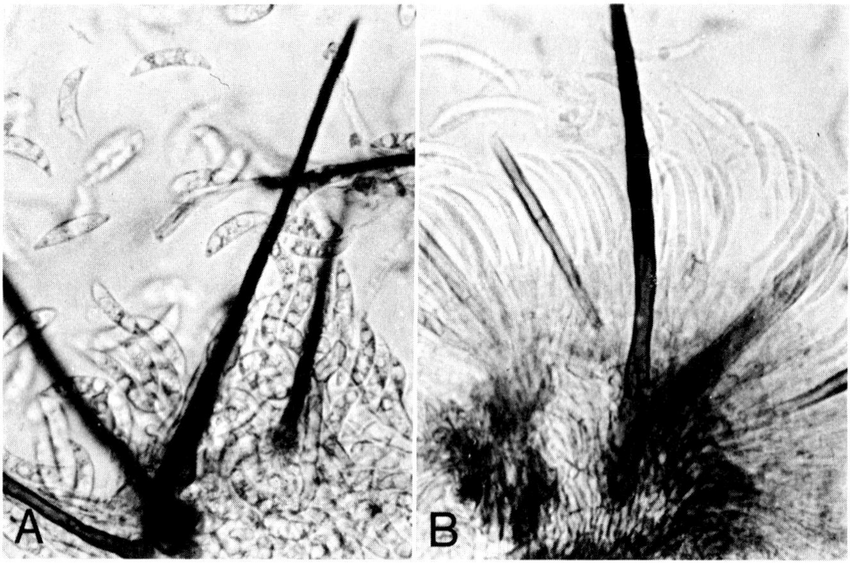


FIG. 57. *A*, *Colletotrichum graminicolum*. The conidia are falcate or crescent-shaped in this species. *B*, *Colletotrichum* sp.

simple, short, hyaline, producing abundant phialospores; phialospores produced in mucus, ovoid, nonseptate, short-cylindric, falcate or crescent-shaped, hyaline, pinkish in mass, frequently producing dark setae; setae stout, septate, darkly pigmented, acutely pointed at the apex.

**Diagnostic Features:** The pink masses of amerospores with the dark setae standing out in sharp contrast are diagnostic for *Colletotrichum* (Fig. 57).

**Notes:** Commonly a parasite of higher plants, *Colletotrichum* is frequently recovered from soil. In some cases the setae are lacking, and previously such forms would have been placed in *Gloeosporium*. Several workers have noted that the presence or absence of setae in *Colletotrichum* is a variable character, and there is little justification for separating out *Gloeosporium* on this basis.

In culture, colonies of *Colletotrichum* may have sparse setae and produce pinkish, water-soaked colonies. Because of the parasitic nature of this genus and the fine morphologic distinctions made between some species, it is difficult to identify a *Colletotrichum* isolate to species. A survey of the genus *Colletotrichum* and a concept of the species based on morphological characters has been presented by von Arx (1957).

## CORDANA Preuss

Type Species: *Cordana pauciseptata* Preuss.

**Generic Description:** Conidiophores erect or nearly so, simple, septate, darkly pigmented, bearing apical clusters of dark, two-celled spores; conidiophores sometimes slightly swollen in the spore-bearing region, sometimes proliferating and forming a second apical cluster; conidia two-celled, darkly pigmented, with prominent septum, proximal cell sometimes larger. (Description based on *C. pauciseptata.*)

**Diagnostic Features:** The apical cluster of dark, two-celled spores on a simple, dark conidiophore is diagnostic for *Cordana* (Fig. 58).

**Notes:** *Cordana* belongs to the Sympodulosporae. The conidia are produced successively at the tip of the sporogenous cell which is separated from the main part of the conidiophore by a slight constriction. In one of a series of papers dealing with mechanisms of spore discharge in fungi, Meredith (1962) showed that, in *Cordana musae*, a parasite of bananas, the spores are violently discharged under conditions of rapidly decreasing vapor pressure.

*Cordana* is infrequently recorded from soil and is more commonly found on dead wood or bark (Hughes, 1955). *C. pauciseptata* has been recorded from soil by Routien (1957) and we have recovered this same species in Ontario.

## CORDELLA Spegazzini

Type Species: *Cordella coniosporioides* Speg.

**Generic Description:** Colonies compact or effused, blackish-brown or black, each made up of a close carpet of setae mixed with small groups of

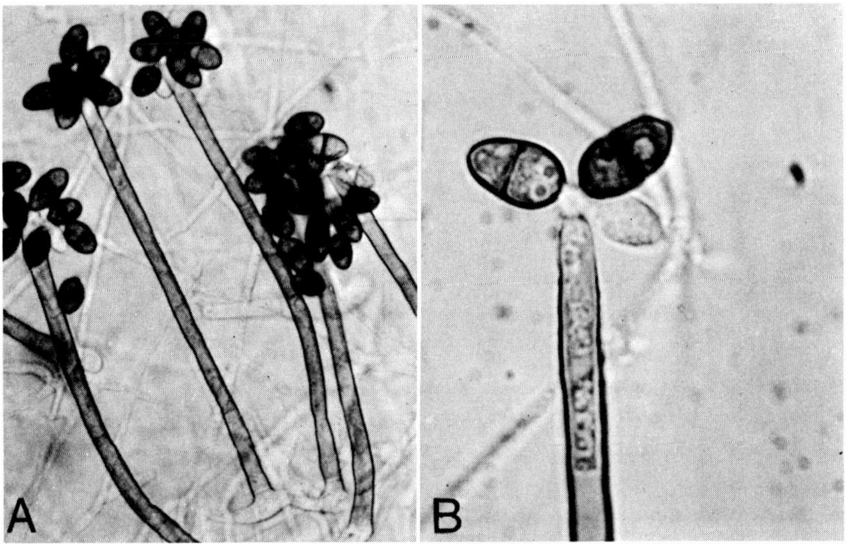


FIG. 58. *Cordana pauciseptata. A*, simple main axis bearing apical cluster of dark, two-celled spores. *B*, three conidia at different stages of development arising from the sporogenous cell (OAC 10079).

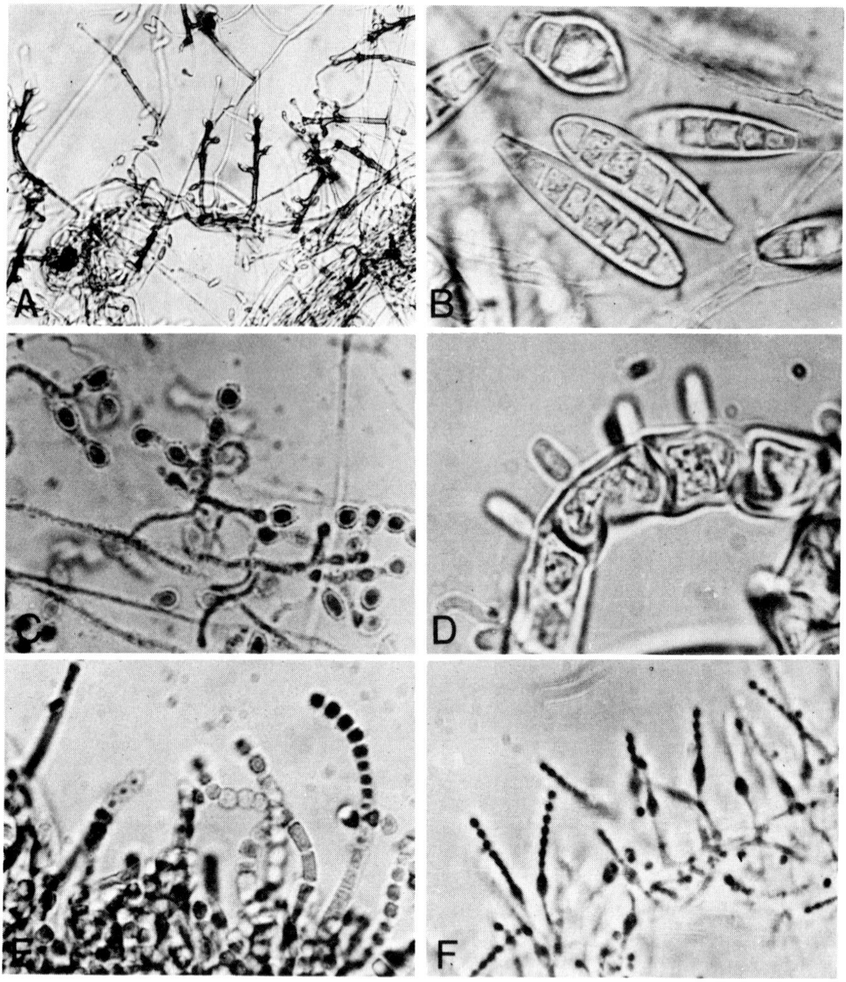


FIG. 59. Miscellaneous. *A*, *Zygosporium masonii*. *B*, *Microsporum* sp. (*Keratinomyces ajelloi*). *C*, *Chrysosporium* sp. *D*, *Candida* sp. *E*, *Wallemia ichthyophaga*. *F*, *Torulomyces lagena*.

conidiophores; mycelium partly superficial, partly immersed in the substratum; setae subulate, brown or black; conidiophore mother cells ampulliform or barrel-shaped, arising from cells of the superficial mycelial mat; conidiophores arise singly from conidiophore mother cells, simple, rather narrow, more or less cylindrical, colorless or pale-brown except for the thick, brown or dark brown, transverse septa; growth in length takes place only at the base; conidia terminal and lateral, usually formed on very short pegs, one-celled, lenticular, pale brown to brown with a hyaline band at the junction of the two sides. (Description from Ellis, 1965.)

**Diagnostic Features:** *Cordella* is a setose Hyphomycete in which the lenticular, darkly pigmented conidia with a hyaline rim are produced on simple conidiophores elongating from the base.

**Notes:** *Cordella* is rarely reported from soil. The only published record is that of England and Rice (1957). This genus is very close to *Arthrinium* differing principally in the presence of the dark setae.

## COSTANTINELLA Matruchot

Type Species: *Costantinella cristata* Matr.

**Generic Description:** Conidiophores erect or suberect, hyaline, septate, simple or branched; sporogenous cells borne singly or more often in pairs or verticils on the main axis or branches, bearing conidia singly and in acropetal succession at the apex; apex of sporogenous cells with prominent refractive denticles in the spore-bearing region, becoming strongly recurved with continuing unilateral spore production; conidia hyaline, continuous, globose to subglobose. (Description based on *C. micheneri.*)

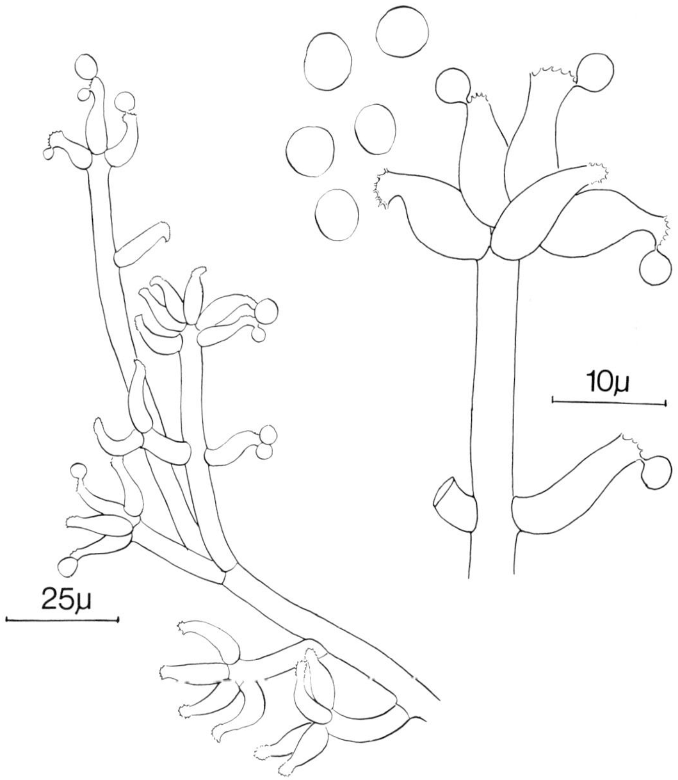


FIG. 60. *Costantinella micheneri.* A succession of conidia produces a reflexed "cockscomb" appearance at the apex of the sporogenous cell (IMI 78576).

**Diagnostic Features:** The whorls of sporogenous cells with strongly recurved, denticulate, spore-bearing regions make the genus quite distinctive (Fig. 60).

**Notes:** *Costantinella* is not found in the usual floristic listings. Records at the Commonwealth Mycological Institute, however, show that *C. terrestris* and *C. micheneri* have both been recorded from soil in England. The genus shows relationship to *Nodulisporium*.

## CRYPTOCOCCUS Kutzing

Type Species: *Cryptococcus neoformans* (Sanfelice) Vuill.

**Generic Description:** Cells round or oval, occasionally long-oval, amoeboid or polymorphic; reproduction by multilateral budding; cells surrounded by a capsule; under appropriate conditions, a starch-like compound is formed both in the capsule and in the medium; pseudomycelium absent or rudimentary; cultures on solid media have a mucoid appearance; in liquid media, bottom growth and ring formation, occasionally a pellicle, often entire contents of flask forming slimy mass; ability to ferment sugar lacking; red and yellow pigments of a carotinoid nature lacking. (Description from Lodder and van Rij, 1952.)

**Diagnostic Features:** The key features of *Cryptococcus* are the ability to form a capsule and "starch." Its dissimilation is strictly oxidative (Fig. 61).

**Notes:** *Cryptococcus* is a nonpigmented, nonfilamentous yeast characterized by the ability to form a capsule with starch. Di Menna (1965), in a comprehensive survey of yeasts from soils in New Zealand, showed that members of the genus *Cryptococcus* were among the dominant yeast species in almost all soils examined. She found that *C. albidus* and *C. terreus* were the most common species in soils in areas with rainfall of

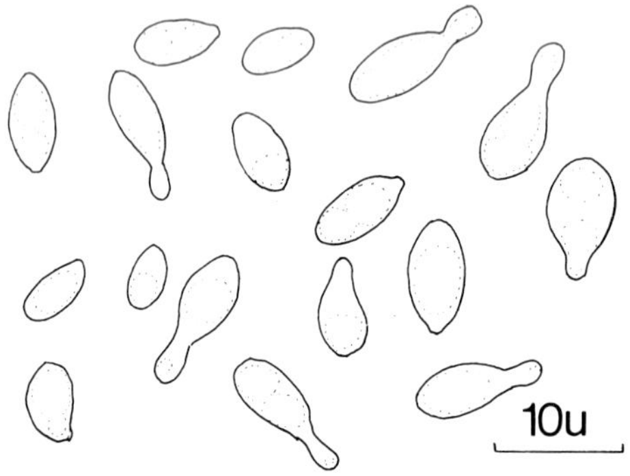


FIG. 61. *Cryptococcus* sp. Ellipsoid blastospores separate readily on mounting

30 inches or less. It is probable, from these and other studies, that *Cryptococcus* is common and of wide distribution in soil.

## CURVULARIA Boedijn

Type Species: *Curvularia lunata* (Wakker) Boedijn.

**Generic Description:** Colonies on natural substrata effused, brown to black, hairy; mycelium on natural substratum usually immersed; hyphae branched, septate, colorless or brown, smooth or verrucose; stromata often large, erect, black, cylindrical, sometimes branched, formed by many species in culture, especially on firm substrata such as rice grains; conidiophores arising singly or in groups, terminally and laterally on the hyphae, also on stromata when these are present, simple or (in culture) sometimes branched, often geniculate, sometimes nodose, septate, brown; conidia acropleurogenous, sometimes in whorls, arise through pores in the conidiophore wall, straight or curved, usually broadly fusiform, ellipsoidal, obovoid, clavate or pyriform, sometimes rounded at the base, sometimes with a distinctly protuberant hilum, septate, often with one or more cells larger and darker than the others, smooth or verrucose; in many species occasional triradiate staurospores are formed along with the normal conidia. (Description from Ellis, 1966.)

**Diagnostic Features:** The dark, curved porospores with paler end cells are typical of the common species of *Curvularia* (Fig. 62).

**Notes:** A well illustrated and detailed treatment of *Curvularia* has been

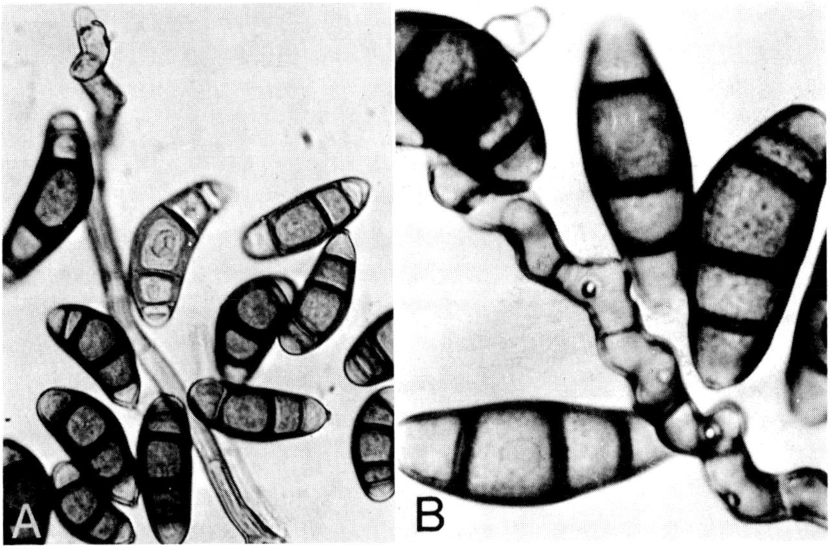


FIG. 62. *A*, *Curvularia* sp. showing curved, septate conidia with paler end cells. *B*, *C. geniculata*. Portion of older conidiophore with spores displaced to reveal pores (OAC 10278).

written by Ellis (1966), who gave descriptions and illustrations of 31 varieties and species of the genus, with keys to their identification. This genus is most frequently encountered as a parasite or saprophyte of graminaceous hosts. Although reported from soil from time to time, it is seldom recorded in high frequencies. The most commonly recovered species are *Curvularia lunata* (conidial state of *Cochliobolus lunatus*; see Nelson and Haasis, 1964) and *C. geniculata* (the conidial state of *Cochliobolus geniculatus*; see Nelson, 1964). *Acrothecium arenarium*, described from sand dunes by Moreau and Moreau (1941), is regarded by Ellis as being *C. inaequalis*.

## CYLINDROCARPON Wollen.

Type Species: *Cylindrocarpon cylindroides* Wollen.

**Generic Description:** Conidia slimy phialospores, formed from phialides in basipetal succession, generally not adhering in chains; microconidia, if present, hyaline, oval to ellipsoid, none- or one-septate; macroconidia always present, hyaline, straight or curved, cylindrical to fusoid but with rounded ends and without a *Fusarium*-type footcell, with 1 to 10 transverse septa; phialides simple, with a single apical pore bearing a collar, formed laterally on hyphae, terminally on simple lateral branches or singly or in groups as termination to branches of penicillately branched conidiophores; no continuation of conidiophore axis into a sterile appendage; chlamydospores present or absent, hyaline to brown, globose, formed singly, in chains or clumps, intercalarily or terminally or on lateral branches, or singly or in chains in cells of the macroconidia; cultures white-beige, orange-brown to purple, floccose to felted; sterile stromatic pustules or sporodochia present or absent. (Description from Booth, 1966.)

**Diagnostic Features:** The large, hyaline, multiseptate macrospores with rounded ends (Fig. 63) are characteristic of the phialosporous genus *Cylindrocarpon*.

**Notes:** *Cylindrocarpon* species are relatively common in soil but, because of confusion or misinterpretation in the literature, the records of this Hyphomycete are not as extensive as would be expected. It has also been recorded from soil under the genus *Moeszia*, a synonym of *Cylindrocarpon*. The genus merges with *Fusarium*, on the one hand, and *Cylindrocladium* on the other. It is distinguished from *Fusarium* by the absence of the characteristic "foot-cell" found in the macrospore of the latter genus and from *Cylindrocladium* by the absence of the sterile prolongation which terminates the conidiophore of that genus. Transitional species between these three related genera are sometimes difficult to place but a middle of the road *Cylindrocarpon* is relatively distinctive. The confusion and uncertainty surrounding this genus has been dispelled by the publication of an excellent and comprehensive treatment by Booth (1966). As pointed out

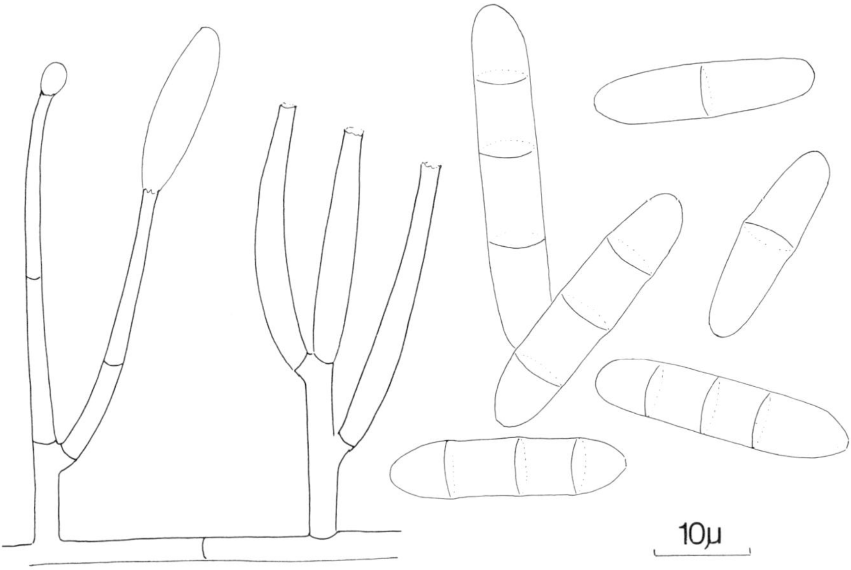


FIG. 63. *Cylindrocarpon destructans* (=*C. radicicola*). The septate phialospores have rounded ends (OAC 10482).

by Booth, *Cylindrocarpon* is associated with ascigerous states in *Nectria* and *Calonectria* of the family Hypocreaceae.

The distribution and significance of *Cylindrocarpon* species in soil have been studied by Matturi and Stenton (1964) who found them more prevalent in alkaline soils. *C. radicicola* (now *C. destructans*) was the most common species recorded.

### CYLINDROCLADIUM Morgan

Type Species: *Cylindrocladium scoparium* Morgan.

**Generic Description:** Vegetative mycelium cottony, white at first, becoming brown in age; conidiophores well differentiated, erect or suberect, branching repeatedly near the apex to give a penicillate appearance; ultimate sporogenous cells phialides; conidiophore axis prolonged in most cases to form a sterile appendage with a swollen end; conidia cylindrical, one- to multi-septate, produced in a palisade-like cluster, held together in mucus; chlamydospores commonly produced, singly or in chains, sometimes in compact sclerotium-like masses.

**Diagnostic Features:** Cylindrical, septate, hyaline phialospores produced in mucus from penicillately arranged sporogenous cells (Fig. 64), with the conidiophore terminating in a sterile appendage, are distinctive for *Cylindrocladium*.

**Notes:** *Cylindrocladium* is recorded infrequently from soil. *C. ilicicola*, *C. macrosporum*, *C. parvum*, and *C. scoparium* have all been recorded

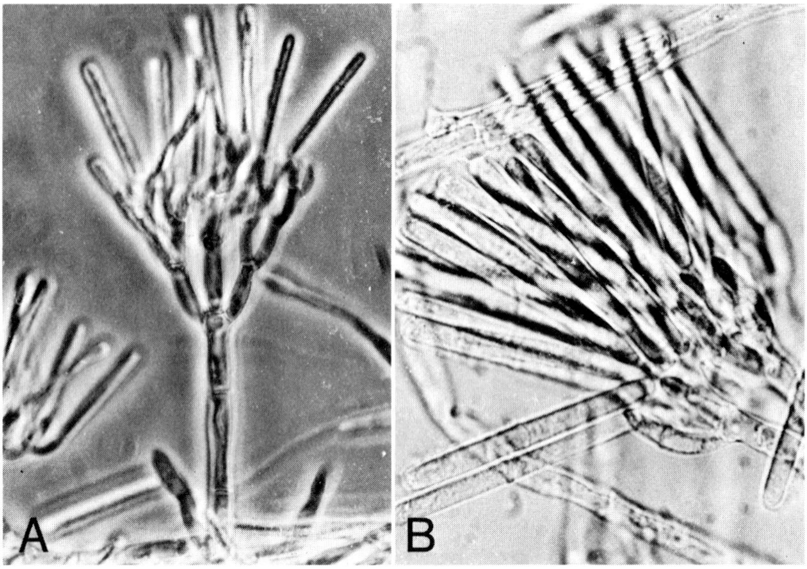


FIG. 64. *Cylindrocladium scoparium*. *A*, conidia are produced from penicillate conidiophores. *B*, conidia gather in slimy fascicles (OAC 10342).

from soil by Meyer (1959). We have isolated the common *C. scoparium* on several occasions from forest soils in Ontario. This species causes leaf blight and damping off of seedlings and cuttings of certain ornamental plants (Westcott, 1960). It can be a serious problem as the cause of crown canker of roses in greenhouses.

The genus has been treated taxonomically by Boedijn and Reitsma (1950), who gave descriptions and keys to the common species. *Cylindrocladium* is associated with the Ascomycete genus *Calonectria* (Boedijn and Reitsma, 1950; Booth and Murray, 1960).

## CYLINDROPHORA Bon.

*Cylindrophora* has been recorded from soil only once when Daszewska (1912) described *C. hoffmanni*. I have not seen any authenticated or type material of any member of this genus nor any specimen tentatively classified as *Cylindrophora*. The generic description given by Gilman (1957) serves to characterize the genus.

> Hyphae forming a turf, prostrate. Conidiophores erect, with or without septa, with simple or forked branches occurring on one or both sides, carrying single conidia at their tips; conidia cylindric, with rounded ends, hyaline.

## CYLINDROTRICHUM Bon.

Type Species: *C. oligospermum* (Corda) Bon.

**Generic Description:** Sterile setae produced, setae stout, darkly pigmented, thick-walled, septate; conidiophores produced on hyphae or frequently adjacent to sterile setae; conidiophores simple, more or less erect, terminating in a single sporogenous cell; sporogenous cell produces an apical collarette through which phialospores are produced in basipetal succession in mucus; sporogenous cell may extend by sympodial growth; successive sympodial extensions of the conidiophore produce a polyphialide; phialospores hyaline, cylindric, one-septate, gathering in slimy fascicles.

**Diagnostic Features:** This genus is characterized by simple conidiophores terminating in polyphialides, producing two-celled spores, and by sterile setae (Fig. 65).

**Notes:** *Cylindrotrichum* is rarely reported from soil. The only record is that of Meyer (1959), who recovered it from soil in the former Belgian Congo.

## DACTYLARIA Sacc.

Type Species: *Dactylaria purpurella* Sacc.

**Generic Description:** Conidiophores more or less erect, distinctive or little differentiated from the vegetative hyphae, simple, short, hyaline to lightly pigmented, septate, denticulate and sometimes swollen at the spore-bearing apex; conidia hyaline or lightly pigmented, multiseptate, cylindric or clavate, sometimes filiform, borne singly and acrogenously at the apex of the conidiophore on more or less prominent denticles.

**Diagnostic Features:** This genus is characterized by short, simple conidiophores, each with a denticulate apex bearing hyaline or lightly pigmented phragmospores in acropetal succession (Figs. 66 and 67).

**Notes:** Found frequently in soil, *Dactylaria* has been recorded by Williams and Schmitthenner (1956) and Routien (1957) described *D. lutea* from soil in Argentina. More recently Roy and Gujarati (1965) have described *D. fulva* from decayed roots in soil in India. In our own studies we have only recorded a *Dactylaria*-like species once, and this had scolecospores. Drechsler (1937) has described a number of *Dactylaria* species as parasitic on free-living terricolous nematodes but these are not *Dactylaria* as considered here. Taxonomically, the genus is close to *Dactylella* and *Diplorhinotrichum*. Meyer (1959) figured a *Mirandina* species from soil in the former Belgian Congo. *Mirandina* is an invalid genus erected by Arnaud (1953). From Meyer's figures and descriptions, this isolate could be considered under *Dactylaria*.

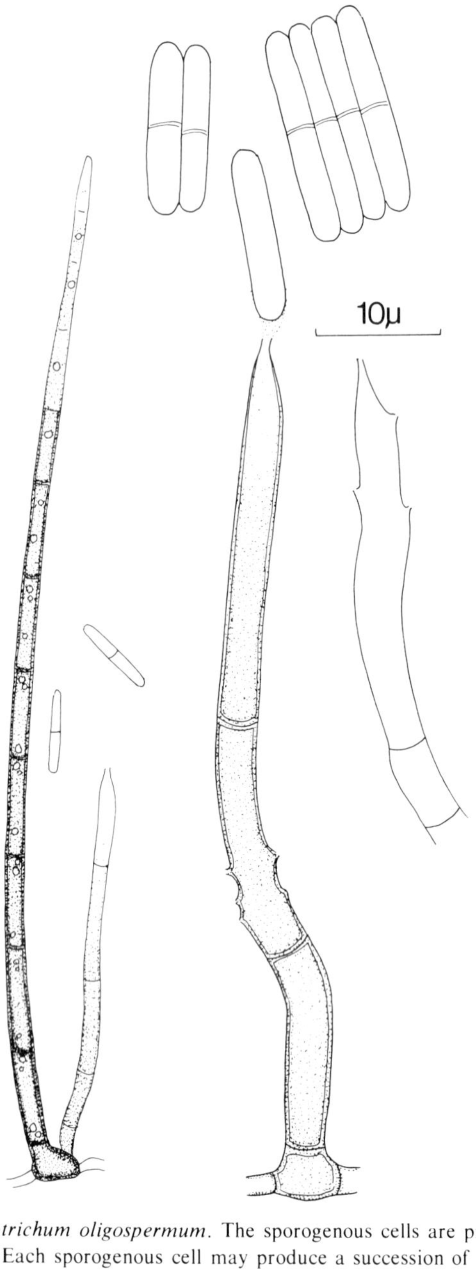


FIG. 65. *Cylindrotrichum oligospermum*. The sporogenous cells are produced in proximity to a sterile seta. Each sporogenous cell may produce a succession of collarettes with the conidia gathering in slimy fascicles (IMI 19208).

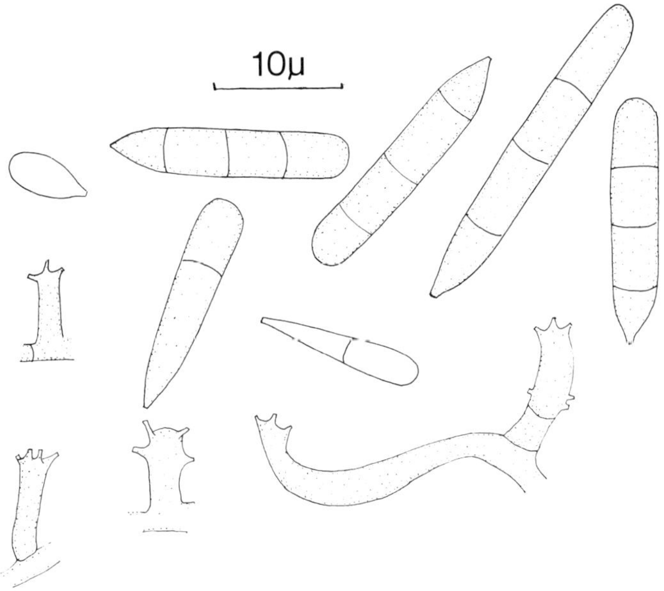


FIG. 66. *Dactylaria purpurella*. The cylindrical phragmospores are borne on truncate denticles on short conidiophores (IMI 104760, ex type).

## DACTYLELLA Grove

Type Species: *Dactylella minuta* Grove.

**Generic Description:** Conidiophores erect, simple, hyaline, septate, bearing spores singly and terminally; several spores may be produced in acropetal succession by sympodial growth of the conidiophore; conidia ellipsoid to fusoid, sometimes cylindrical, several- to multi-septate, hyaline, frequently broader in the center and tapering towards the ends.

**Diagnostic Features:** This genus is characterized by the large, hyaline phragmospores borne terminally on a simple conidiophore; the spores are usually solitary or few in number (Fig. 68).

**Notes:** The genera *Dactylella* and *Monacrosporium* have recently been reviewed by Subramanian (1963). Both of these genera are characterized by hyalophragmospores produced singly and acrogenously at the tips of simple conidiophores. Subramanian selected *M. elegans* as the lectotype for *Monacrosporium*, noting that in this species the conidia are fusiform, many-septate and have one of the medial cells much larger and wider than the others, a feature not seen in *D. minuta* (the type of *Dactylella*). On this basis, Subramanian disagreed with Yadav (1960), who suggested that *Monacrosporium* was a synonym of *Dactylella*. Subramanian presented an emended diagnosis of *Monacrosporium* as follows:

Conidiophores erect, usually simple, hyaline; conidia produced singly at the tips of the conidiophores, hyaline, usually fusoid, with two or more transverse septa, with one of the cells (usually intermediate) wider and longer than the others.

Thus, according to Subramanian, species with enlarged central cells belong in *Monacrosporium* and those in which no cell is particularly enlarged belong in *Dactylella*. Subramanian's interpretation is followed by Cooke and Dickinson (1965).

As pointed out by Drechsler (1943, 1950), Duddington (1951), Subramanian (1963) and Cooke and Dickinson (1965) members of the genus *Dactylella* are predaceous on nematodes. It is not surprising, therefore, to recover them from soil.

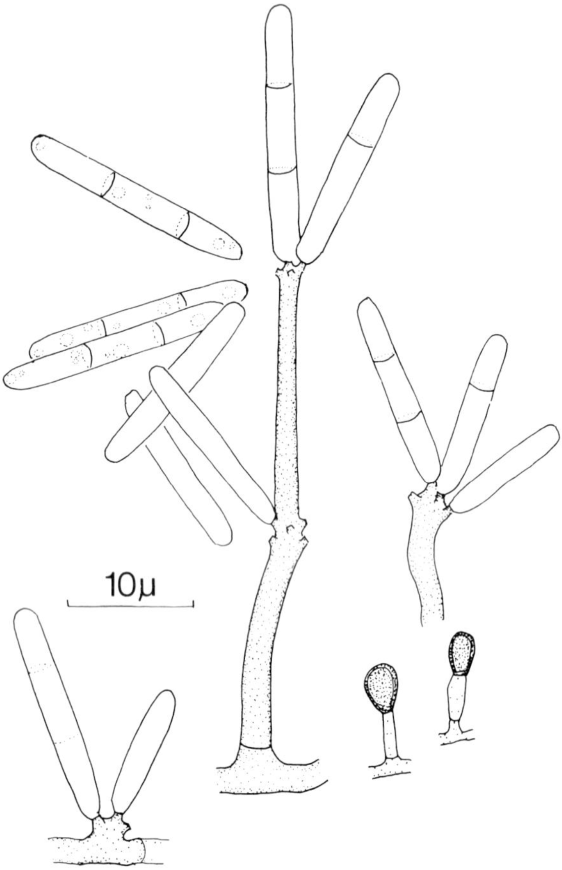


FIG. 67. *Dactylaria fulva* (IMI 104480, ex type)

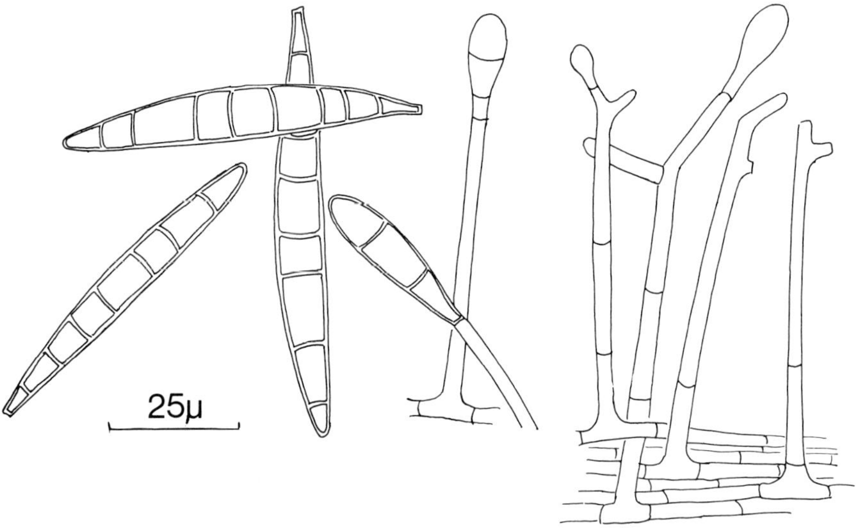


FIG. 68. *Dactylella* sp. (=*Monacrosporium oxysporum*). The large hyaline phragmospores are borne apically and are solitary or few in number (IMI 78728).

## DACTYLIUM Nees

De Vries (1962) checked the type material of *Cladobotryum variospermum* (Link) Hughes (=*Cladobotryum varium* Nees) and concluded that *Cladobotryum* Nees and *Dactylium* Nees had to be united. The type species of *Dactylium* is *D. candidum* Nees (Hughes, 1958). In his study of the classical exsiccati in European herbaria, Hughes apparently did not find type or authenticated material of this species. In view of the above observations, it would seem appropriate to consider *Dactylium* as a *nomen dubium* and dispose of the known *Dactylium* species in other genera. Lentz (1966) proposed conserving *Dactylium* Sacc. against *Dactylium* Nees. According to Lentz, this would effectively eliminate *Dactylium* Nees from competition with *Dactylaria* Sacc. as the name for fungi of this complex with simple conidiophores.

The popular concept of *Dactylium* is somewhat vague and the genus includes a heterogeneous assemblage of species. The *Dactylium* state of *Hypomyces roseus* (Fig. 54*B*) is best classified under *Cladobotryum*. *Dactylium dendroides* (Fig. 69) the conidial state of *Hypomyces rosellus* is not a *Cladobotryum* and is not readily classified into any form genus known to me. *Dactylium fusarioides* Frag. and Cif. was originally described from the dried pods of leguminous plants in the Dominican Republic. This species has been redescribed by De Vries (1962) on the basis of an isolate recovered from air in South Africa. De Vries showed that, as well as having *Fusarium*-like macrospores, this fungus also pro-

FIG. 69. *Dactylium dendroides*. In this species a succession of conidia is accompanied by a delicate extension of the sporogenous cell suggestive of an annellophore (IMI 104426).

duces a distinctive microspore state. This microspore state (Fig. 16*C*) may occur without the macrospore state, and as such it was described by Nicot (1956) as *Nodulisporium didymosporum*. According to Booth (private communication) this species should be considered as *Fusarium chlamydosporum*.

## DACTYLOSPORIUM Harz

Type Species: *Dactylosporium macropus* (Corda) Harz.

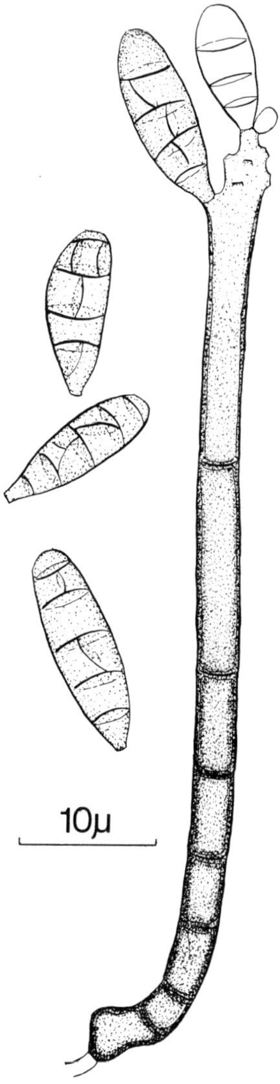


FIG. 70. *Dactylosporium macropus*. The muriform conidia are borne in acropetal succession on simple conidiophores (IMI 69717).

**Generic Description:** Conidiophores sparse or crowded, arising singly or in groups, simple, erect, stout, dark brown, thick-walled, septate; conidia developing singly and in acropetal succession to produce a bunch of dry spores at the apex of the conidiophore; conidia subhyaline to brown, smooth, oval or sometimes asymmetric, with flat basal scar, muriform. (Description based on *D. macropus*.)

**Diagnostic Features:** The dry clusters of muriform conidia produced on tall, stout, dark conidiophores make *Dactylosporium* distinctive (Figs. 70 and 117*D*).

**Notes:** *Dactylosporium* is commonly found on wood. It is rare from soil and the only record I am aware of is that of Ghosh and Dutta (1960). The most common species of the genus, *D. macropus*, has been redescribed by Hughes (1952) in some detail.

## DENDRODOCHIUM Bon.

Type Species: *Dendrodochium aurantiacum* and *D. flavum* described.

**Generic Description:** Fructification a sporodochium, white or lightly colored; conidiophores tightly packed, more or less erect, hyaline, septate, branched in verticillate fashion near the apex; sporogenous cells phialides; phialospores nonseptate, hyaline, subglobose to ellipsoid.

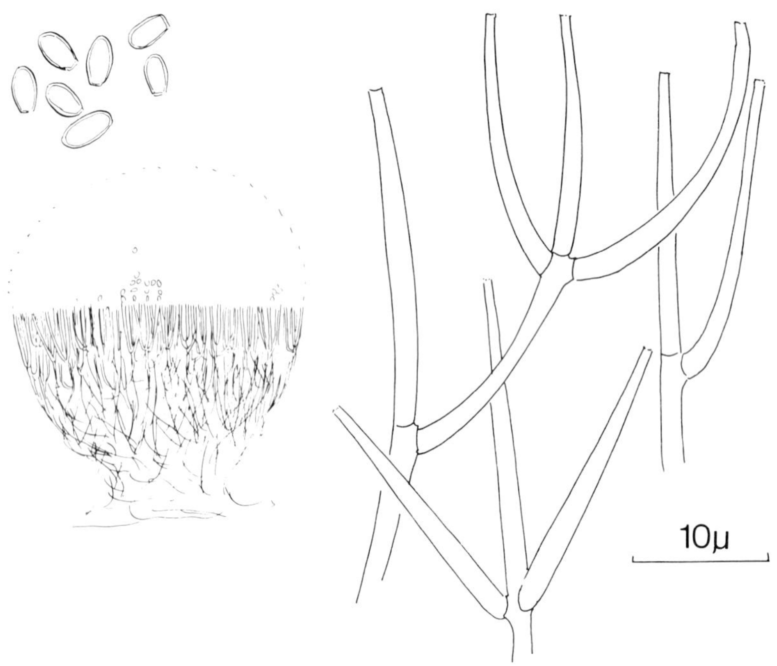


FIG. 71. *Dendrodochium* sp. (IMI 59467)

**Diagnostic Features:** *Dendrodochium* is a hyaline, sporodochial fungus with the sporogenous cells in verticils (Fig. 71).

**Notes:** This genus is rarely recorded from soil (Gilman, 1957). In the more recent literature Guillemat and Montégut (1956) recorded *Dendrodochium* in France and Nicholls (1956) recorded it from chalk soil in England. In the herbarium of the Commonwealth Mycological Institute, there is a single record of a *Dendrodochium* species isolated from chalk soil by V. Sankey.

## DENDROSTILBELLA Höhn.

Type Species: *Dendrostilbella prasinula* Höhn.

**Generic Description:** Hyphae aggregated to form more or less erect synnemata; synnema with sterile stalk and fertile head, lightly colored, composed of parallel, closely packed, brightly colored to hyaline hyphae; conidiophores branching at the apex, verticillate; conidia hyaline, oblong to elliptical, nonseptate, produced singly and successively, adhering in a mucoid mass.

**Diagnostic Features:** *Dendrostilbella* is a moniliaceous, coremial genus in which the conidiophores branch profusely in the head and produce numerous, small, hyaline spores in mucus (Fig. 72).

**Notes:** In its general appearance, the genus is very close to *Stilbum* (*Stilbella*). In the latter genus, the conidiophores are reputedly simple,

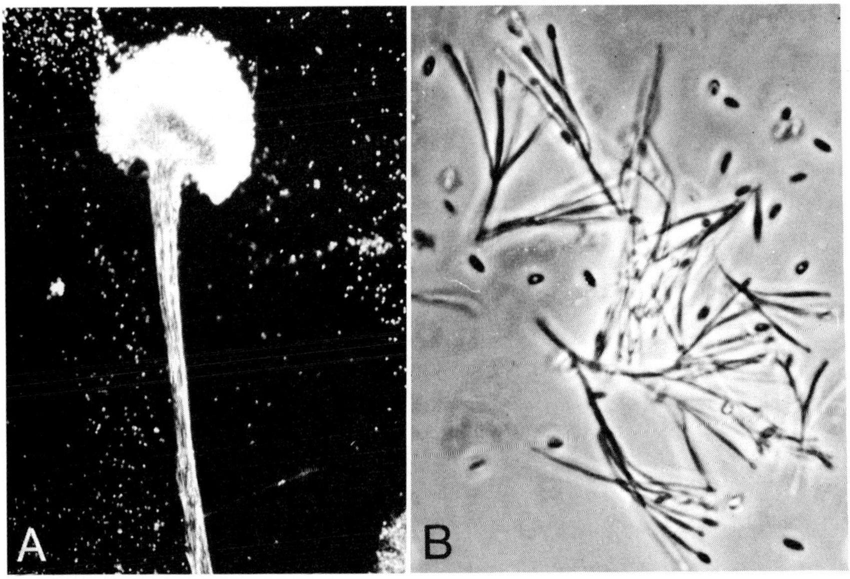


FIG. 72. *Dendrostilbella* sp. *A*, conidiophores are aggregated to form a synnema. *B*, sporogenous cells showing verticillate arrangement of phialides (OAC 10332).

and, in *Dendrostilbella*, branch in dendritic or verticillate fashion. Whether *Dendrostilbella* is related to *Verticillium* or *Gliocladium* is not clear. From a study of published figures, the sporogenous cells of *Dendrostilbella* appear to be phialides (see Morris, 1963). To the best of my knowledge, the type species of the genus has not been re-examined and the true nature of the sporogenous cells is still in doubt.

## DENDRYPHION Wallr.

Type Species: *Dendryphion comosum* Wallr.

**Generic Description:** Conidiophores more or less erect, main axis simple, branching irregularly or verticillately at the apex, septate, brown; branches terminating in sporogenous tips which produce porospores singly and in acropetal succession or in simple or branching chains; conidiophores increasing in length by sympodial extension of the sporogenous cells; sporogenous cells geniculate, bearing prominent scars; conidia porospores, darkly pigmented, usually multiseptate, smooth or roughened.

**Diagnostic Features:** Septate porospores on simple or branching conidiophores with the sporogenous cells extending by sympodial growth are characteristic of *Dendryphion* (Fig. 73).

**Notes:** The method of development of the conidiophores and the nature of the sporogenous cells in *Dendryphion* were discussed in detail by

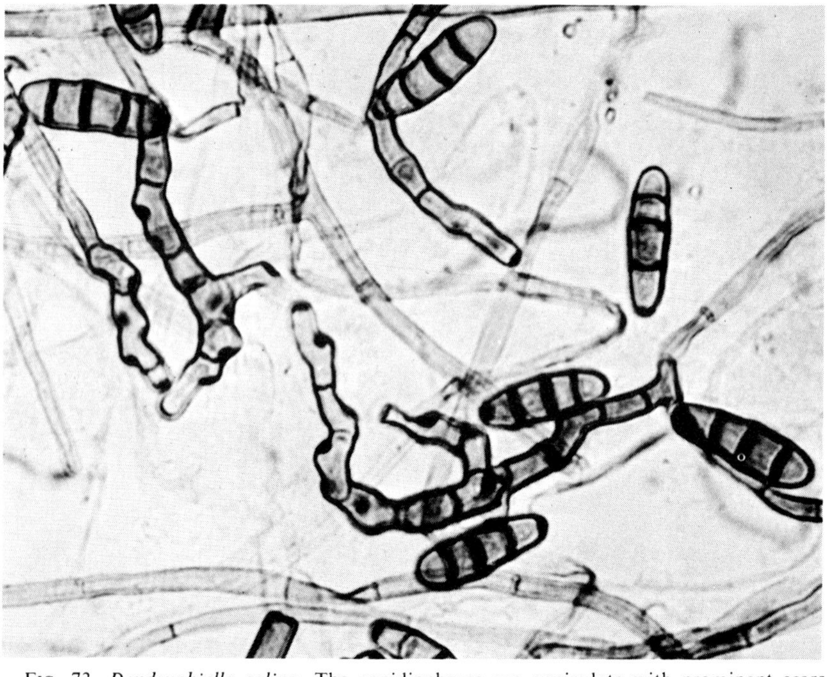

FIG. 73. *Dendryphiella salina*. The conidiophores are geniculate with prominent scars. The conidia are multiseptate porospores (IMI 81623).

Hughes (1953). Hughes noted that in *Dendryphiella interseminata* the branching of the conidiophores is longer and looser. He did not consider this distinction sufficient to warrant the separation of *Dendryphiella* from *Dendryphion*. Nicot (1958) disagreed with this viewpoint and described *Dendryphiella arenaria* as a new sand-inhabiting Hyphomycete. Pugh and Nicot (1964) supported this stand when transferring *Cercospora salina* to *Dendryphiella*.

Inasmuch as the difference between *Dendryphion* and *Dendryphiella* is one of degree rather than kind, I agree with the conclusions of Hughes. In *D. arenaria* and *D. salina* the conidia are frequently borne singly on the conidiophore and might be confused with *Dendryphiopsis*.

## DICHOTOMOPHTHORA Merlich and Fitzpatrick

Type Species: *Dichotomophthora portulacae* Mehrlich and Fitzpatrick.

**Generic Description:** Conidiophores brown, simple or regularly dichotomously branched; terminal cells of the main axis or branches four-, eight-, or multilobed, with each lobe bearing a single terminal conidium; conidia porospores, smooth, one- to several-septate, brown, ovoid to elongate-ovoid, or cylindrical, straight or rarely curved; sclerotia abundant, variable in shape, black.

**Diagnostic Features:** This genus is characterized by the unusual dichotomously branching habit of the main axis or sporogenous cells. Each lobe terminates in a solitary, septate porospore (Fig. 74).

FIG. 74. *Dichotomophthora indica*. *A*, simple main axis with spore bearing apex. *B*, higher magnification of apex showing dichotomous branching of sporogenous cells with each lobe bearing a solitary spore (OAC 10462).

**Notes:** *Dichotomophthora*, a dematiaceous Hyphomycete, was erected by Mehrlich and Fitzpatrick (1935) to include a single species, *D. portulacae*, isolated from common purslane (*Portulaca oleracea*) in Hawaii where the fungus caused epiphytotics on the weed. Neither genus nor species was validated by the original authors.

Rao (1966) described a second species, *D. indica*, from *Portulaca oleracea* in India and provided a Latin description for *D. portulacae*. According to the Commonwealth Mycological Institute (see *Index of Fungi, Vol. 3*, page 381) the genus is still not validly published.

I have isolated *D. indica* on one occasion from soil and repeatedly from *P. oleracea* in Ontario where it is apparently a common pathogen. *D. indica* (Figs. 13*A* and 74) is a striking species. The stout, erect, main axis is slightly swollen at the apex and bears the sporogenous cells as tightly packed dichotomous lobes.

## DICTYOSPORIUM Corda

Type Species: *Dictyosporium elegans* Corda.

**Generic Description:** Hyphae hyaline to pigmented, conidia borne on the vegetative hyphae or aggregated in sporodochium-like fructifications; conidiophores reduced to very short branches on the assimilative hyphae; conidium branches formed by the division of the terminal cells of the conidiophore in a cell-by-cell manner; conidium branches multicellular, arising from a single basal cell (although sometimes obscurely so) and

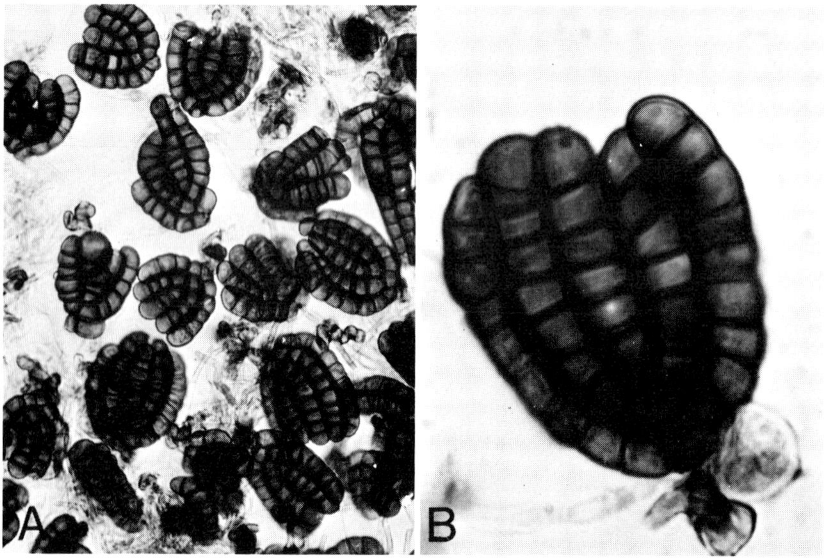


FIG. 75. *Dictyosporium toruloides*. *A*, conidia have a distinctly hand-like appearance. *B*, each conidium has usually five closely adpressed, finger like extensions.

fusing laterally or remaining free at maturity, more or less parallel, usually constricted at the septa, rarely slightly incurved at the tip; at maturity flat, dark-colored, usually U-shaped, morphology irregular in one species (*D. toruloides*), sterile setae may be present in one species (*D. chilensis*). (Description based on Damon, 1952.)

**Diagnostic Features:** *D. toruloides*, the common species and the only one recovered from soil, is recognized by its distinctive spores (Fig. 75).

**Notes:** *Dictyosporium* is rarely reported from soil. The Commonwealth Mycological Institute has a record of *D. toruloides* recovered from soil, Woburn Experiment Station, Rothamsted, by E. A. Peterson.

Studies on the genera *Dictyosporium* and *Speira* were carried out by Damon (1952) using the type material of the type species of both these genera. He reviewed the controversy surrounding *Dictyosporium* and *Speira* and regarded *Speira*, based on *S. toruloides* Corda, as congeneric with *Dictyosporium* based on *D. elegans* Corda. Damon gave descriptions and keys to the seven species he recognized in the genus.

## DIHETEROSPORA Kamyschko

Type Species: *Diheterospora chlamydosporia* (Kamyschko) Barron and Onions.

**Generic Description:** Aleuriospore state conspicuous; aleuriospores large, muriform, at first thin-walled and hyaline, later becoming yellowish or golden brown and very thick-walled, smooth, borne on short pedicels arising more or less at right angles to the vegetative hyphae; accessory phialospore state of the *Verticillium* or *Paecilomyces* types also produced; phialospore state sometimes *Cephalosporium*-like.

**Diagnostic Features:** The large, golden brown, muriform aleuriospores of *Diheterospora* are quite striking in appearance and make this genus readily distinguishable. The accompanying phialospore states aid diagnosis (Fig. 76).

**Notes:** This genus was erected by Kamyschko (1962) on two species; *D. heterospora*, the type species, and *D. catenulata*. It was pointed out by Barron and Onions (1966) that *D. heterospora* had been previously described under *Verticillium chlamydosporium* by Goddard (1913). The aleuriospore state is strikingly similar in both of the described species. One species, *D. catenulata*, has a *Paecilomyces* as the accessory phialospore state and the other, *D. chlamydosporia*, has a *Verticillium* as the accessory phialospore state (Fig. 77). Both species are apparently not uncommon in soil and are known to be parasitic on snail eggs (Barron and Onions, 1966).

Batista and Fonseca (1965) described *Pochonia*, based on *P. humicola*, as a new genus recovered from soil in several states of Northeastern Brazil. From the figures and descriptions of *P. humicola* it is clear that

Fig. 76.

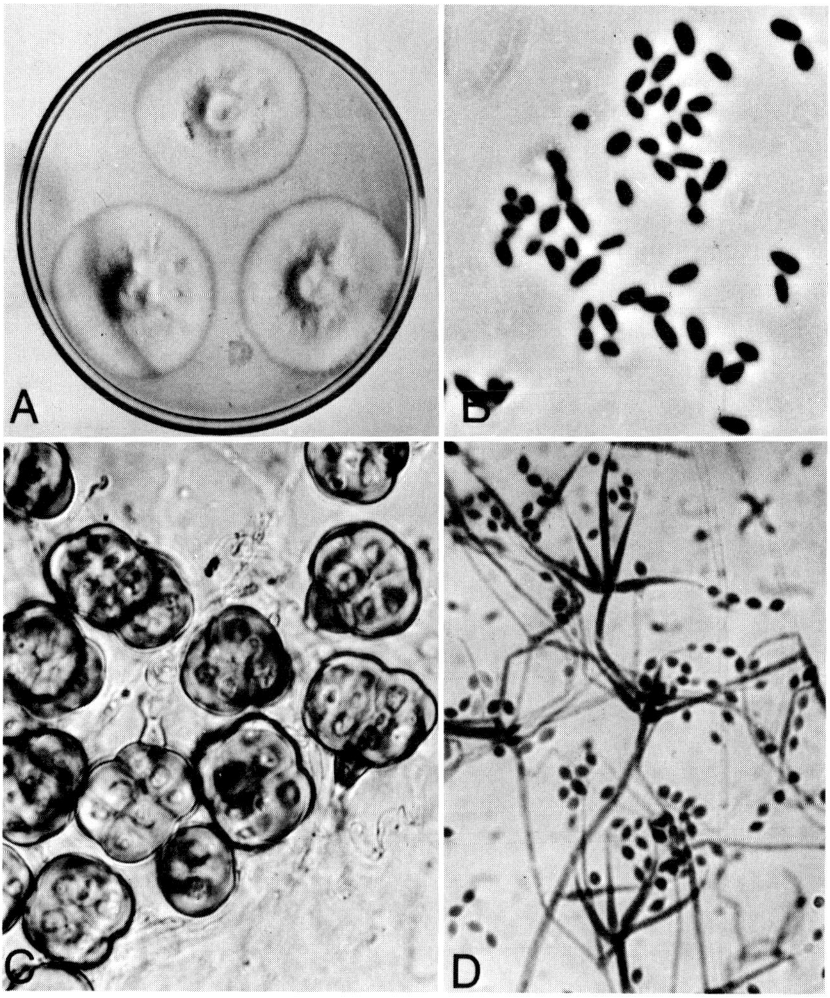


FIG. 77. *A*, *Diheterospora catenulata*, 2-week culture at 25°C on Czapek's agar. *B*, *D. chlamydosporia*, phialospores. *C*, *D. catenulata*, mature aleuriospores. *D*, *D. catenulata*, phialophores, phialides, and phialospores. (Reproduced by permission of the National Research Council of Canada, *Can. J. Bot.*, *44:* 861, 1966.)

this species is *D. chlamydosporia*. *Dictyoarthrinopsis kelleyi* described by Dominik and Majchrowicz (1966) is probably *Diheterospora*.

## DIPLOCOCCIUM Grove

Type Species: *Diplococcium spicatum* Grove.

**Generic Description:** Conidiophores more or less inconspicuous and little

---

FIG. 76. *Diheterospora catenulata*. Showing stages in the development of the aleuriospores (OAC 10250). (Reproduced by permission of the National Research Council of Canada, *Can. J. Bot. 44:* 861, 1966).

differentiated from the vegetative hyphae, erect or procumbent; conidia porospores, produced indiscriminately at any point from the supporting sporogenous cell, although frequently found in clusters at the apex of a branch or just below a septum; sporogenous cells sometimes slightly swollen in the spore bearing regions; conidia two-celled, constricted at the septum, darkly pigmented, more or less uniform in size and shape, produced singly or in chains.

**Diagnostic Features:** Dark, two-celled porospores borne indiscriminately on the sporogenous hyphae are diagnostic for *Diplococcium* (Fig. 78).

**Notes:** *Diplococcium* is commonly found on dead wood or bark (Ellis, 1963) and is rarely reported from soil. Bayliss Elliot (1930) recorded *D. resinae* from soil in England, and Thrower (1954) recorded a *Diplococcium* species from soil in Australia. The culture collection of the Commonwealth Mycological Institute lists an isolate of *D. avellaneum* from soil, Somalia, by J. Meyer.

The genus might be confused with *Scolecobasidium*, but in the latter genus the spores are borne in acropetal succession on slender pedicels and are not porospores.

## DIPLORHINOTRICHUM Höhn.

Type Species: *Diplorhinotrichum candidulum* Höhn.

**Generic Description:** Conidiophores short, hyaline, solitary, arising as sporogenous cells directly from the vegetative hyphae, straight or bent, more or less cylindrical, frequently bent sharply in the apical region, bearing numerous conspicuous denticles which stand out at a sharp angle;

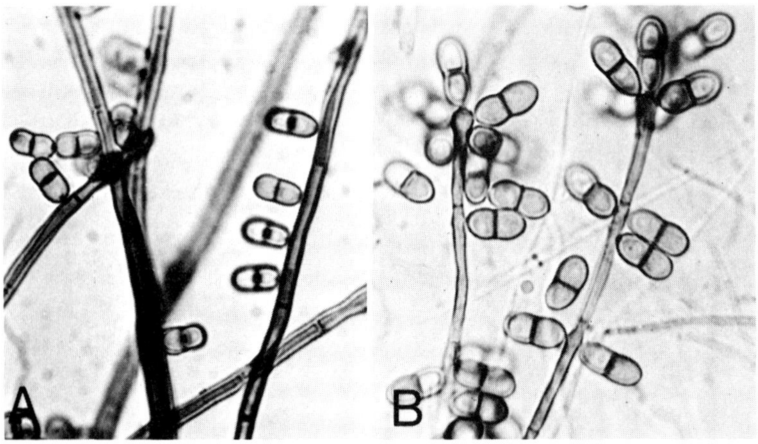


FIG. 78. *A*, *Diplococcium spicatum*. The two-celled porospores are borne more or less at random over the "conidiophores." *B*, *D. avellaneum*. In this species the conidia seem to be aggregated just below the septa (IMI 67827).

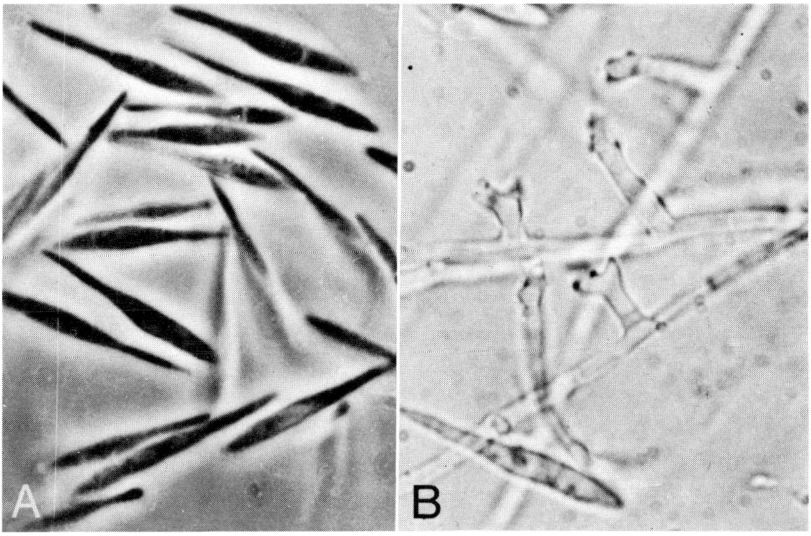


FIG. 79. *Diplorhinotrichum* sp. *A*, conidia. *B*, conidiophores (OAC 10075)

conidia cylindrical or fusoid, with a long apical prolongation in some species, borne singly and apically as blown out ends of the conidiophore apex from a succession of new growing points, one- or two-septate, hyaline.

**Diagnostic Features:** The large, hyaline, septate conidia borne in acropetal succession on conspicuous denticles from the sporogenous cells are characteristic of *Diplorhinotrichum* (Figs. 79 and 80).

**Notes:** *Diplorhinotrichum* is rarely reported from soil. We have isolated it only once, from greenhouse soil in Ontario. This species resembles *D. ampulliforme* described by Tubaki (1958), but the constriction between the apical prolongation and the main body of the conidium is not as marked in the Ontario isolate. Whether this is a species difference or a strain difference is not known. In the majority of the described *Diplorhinotrichum* species the conidia are two-celled and long-ovoid, lacking the prolongation of the apical cell shown in the illustrated species.

*Diplorhinotrichum* is close to *Dactylaria. Dactylaria arnaudii* described by Yadav (1960) has an apical prolongation composed of several cells. The relationships between *Diplorhinotrichum* and *Dactylaria* need clarification.

## DORATOMYCES Corda

Type Species: *Doratomyces stemonitis* Corda.

**Generic Description:** Conidiophores aggregated to form synnemata; synnemata more or less erect, each with a sterile stalk and fertile spore-

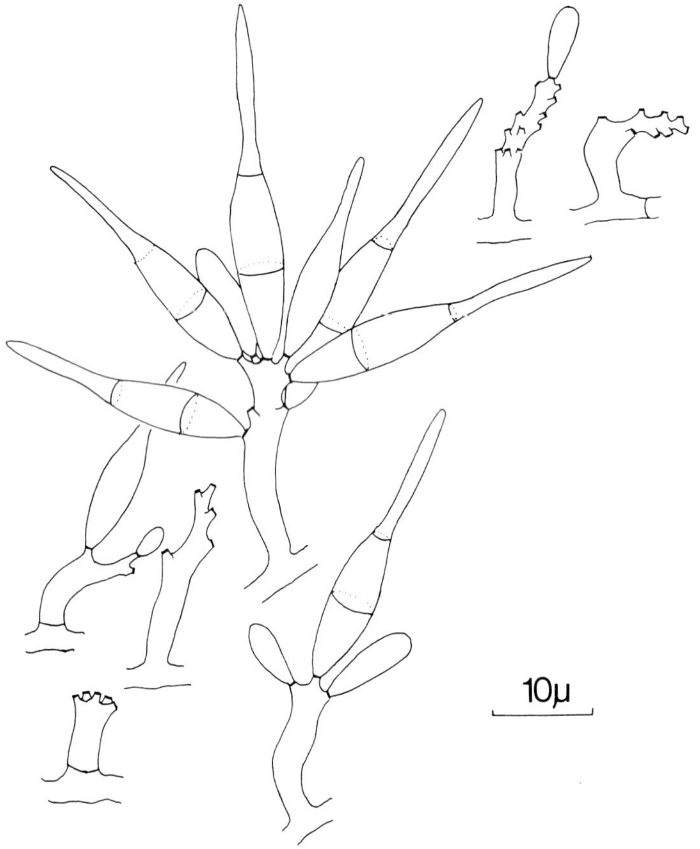


FIG. 80. *Diplorhinotrichum* sp. The hyaline phragmospores are borne in acropetal succession. They secede to leave well marked, truncate, highly refractive scars. The apical prolongation in the spore of this species is not typical of the genus. Note the similarity to *Dactylaria*.

bearing head; head usually elongate, feathery, composed of a central axis of anastomosing hyphae which branch towards the outside and bear numerous annellophores at the ultimate branchlets; annellophores short, inflated, producing long chains of spores; spores smooth or roughened, dark, nonseptate, usually truncate at the base, round or apiculate at the distal end, globose to ovate; sometimes associated with an *Echinobotryum* state.

**Diagnostic Features:** The stout, dark, synnemata with dry heads bearing chains of annellospores make *Doratomyces* readily recognizable (Fig. 81).

**Notes:** In the earlier literature this genus is to be found under the name

*Stysanus*. In his publication based on a study of the classical collections, however, Hughes (1958) proposed the name *Cephalotrichum*. Morton and Smith (1963), on the other hand, regarded this as a bibliographic transfer and preferred the name *Doratomyces*. The matter is not yet resolved and the writer has followed the most recent monographic treatment.

*Trichurus* was erected by Clements and Shear to include those forms which were *Stysanus*-like but in which sterile hairs were interspersed with the fertile hyphae in the sporiferous head. There is little need for a separate genus based only on the presence or absence of sterile hairs. *Trichurus* must eventually be considered under *Doratomyces*.

*Doratomyces* is essentially a coremial form of *Scopulariopsis*. It bears the same relationship to *Scopulariopsis* as *Penicillium claviforme* bears to the other *Penicillium* species. It seems possible that *Doratomyces* and *Scopulariopsis* could be lumped together under one generic name. This is especially worthy of consideration since most of the *Doratomyces* species produce a *Scopulariopsis* state arising from the "ground hyphae." Also, in some isolates from soil in our laboratory the coremial phase became suppressed and the cultures appeared as *Scopulariopsis*. *Dorato-*

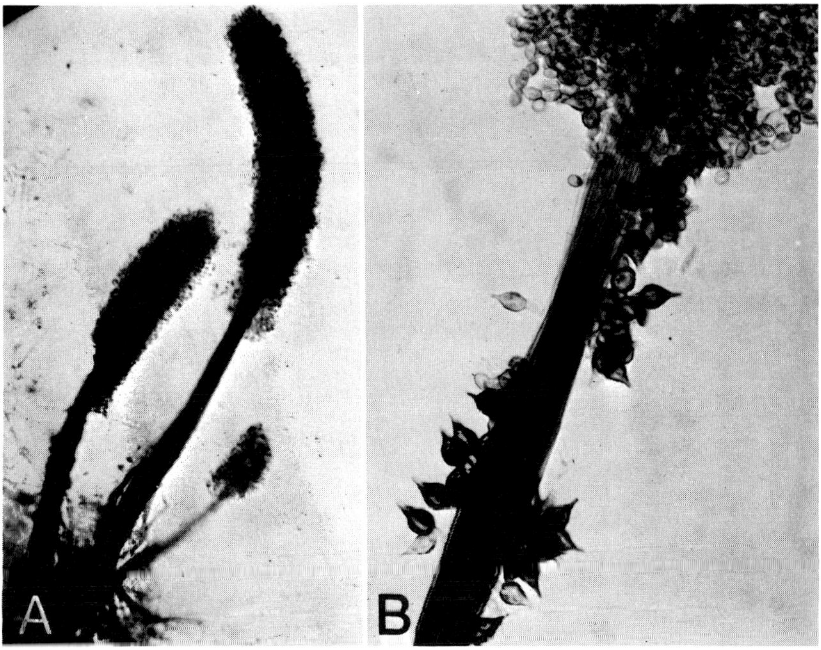


FIG. 81. *Doratomyces stemonitis*. *A*, young synnemata with sterile stalks and feathery spore heads. *B*, *Echinobotryum* state developing from stalk (OAC 10206).

*myces* species are relatively common in soils, particularly those with a high organic content.

A monograph on *Scopulariopsis* and *Doratomyces* has been published by Morton and Smith (1963). Unfortunately, these workers did not consider the closely related *Trichurus*.

## DRECHSLERA Ito

Type Species: *Drechslera tritici-vulgaris* (Nisikado) Ito.

**Generic Description:** Conidiophores brown, simple or less often sparingly branched, indeterminate; producing conidia singly at the apex of the conidiophore through small pores; conidiophore continuing growth sympodially from a point just below and to one side of the apex from which a second spore is then produced; a succession of spores produced in similar fashion acrogenously from the sympodially extending conidiophore; conidia multiseptate, darkly pigmented, cylindrical, germinating from any or all cells.

**Diagnostic Features:** In *Drechslera* the sympodially extending conidiophore produces an acropetal succession of multiseptate porospores which are cylindric in shape (Fig. 82), and germinate from any or all cells (see *Bipolaris*).

**Notes:** *Drechslera* was erected by Ito (1930) and reviewed by Shoemaker (1959, 1962). *Drechslera* differs from *Bipolaris* in having cylindrical porospores which germinate from any or all cells. In both *Drechslera*

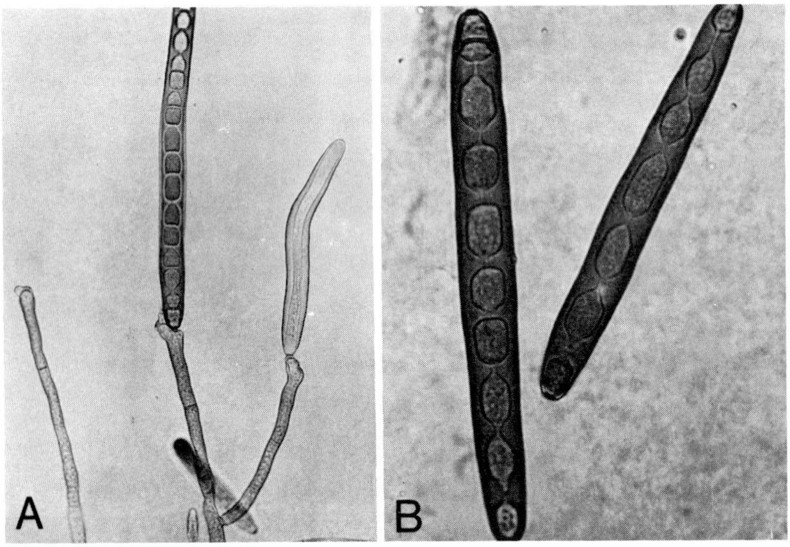


FIG. 82. *Drechslera* sp. *A* and *B*, the cylindrical porospores develop acropetally from a sympodially extending conidiophore (OAC 10485).

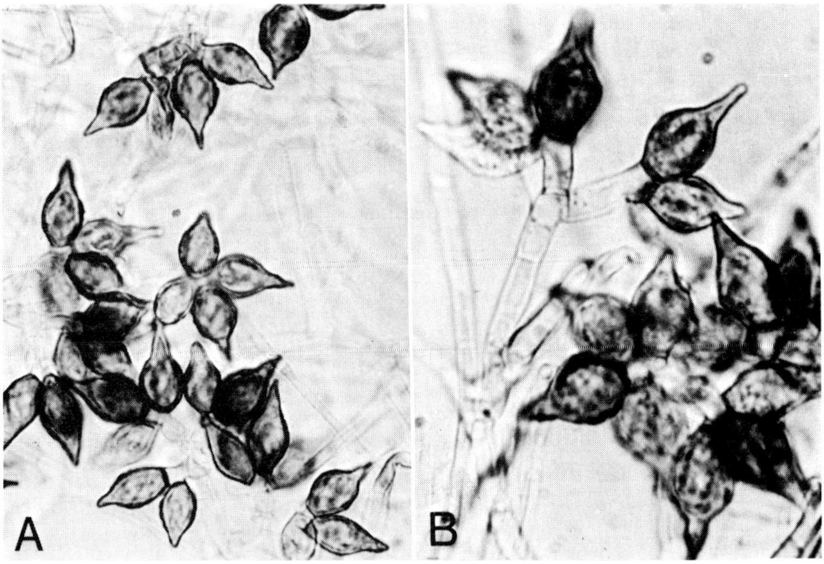


FIG. 83. *Echinobotryum atrum*. *A* and *B*, the aleuriospores are borne in groups or dense clusters. The roughened spore with apical beak is distinctive for this species (OAC 10206).

and *Bipolaris* the conidiophores are indeterminate, extending by sympodial growth and are thus distinct from *Helminthosporium*, in which the production of the apical conidia terminates the growth of the conidiophores. (See *Helminthosporium*.)

## ECHINOBOTRYUM Corda

Type Species: *Echinobotryum atrum* Corda.

**Generic Description:** Conidiophores not well developed, conidia borne directly on the vegetative hyphae or on short pedicels; conidia aleuriospores, borne singly or more often in groups or dense clusters, nonseptate, darkly pigmented, truncate at the base, usually tapering or papillate at the distal end, smooth or roughened.

**Diagnostic Features:** Lateral or terminal clusters of papillate or flame-shaped, dark aleuriospores are distinctive for *Echinobotryum* (Figs. 2*C* and 83).

**Notes:** *Echinobotryum* is reported infrequently from soil (Gilman, 1957). *E. laeve* was recorded in England by Bayliss Elliott (1930). *E. pulvinatum* was recorded in France by Guillemat and Montégut. This species, however, was shown by Dickinson (1966) to be a *Wardomyces*. The most commonly isolated species in my experience is *E. atrum*, the *Echinobotryum* state of *Doratomyces stemonitis*. *Echinobotryum* is very close morphologically to *Mammaria*, and there seems little to justify the separation of these two taxa.

## ELADIA Smith

Type Species: *Eladia saccula* (Dale) Smith.

**Generic Description:** Vegetative hyphae broad, hyaline, septate; conidiophores short, irregularly branched, ultimate branches bearing phialides; phialides clustered, short, ovoid; conidia nonseptate, solitary or in short chains, roughened.

**Diagnostic Features:** The type species, *E. saccula*, is distinguished by short chains of pigmented, spherical, rough-walled phialospores produced in very short chains from monoverticillate clusters of phialides on poorly differentiated conidiophores (Fig. 84).

**Notes:** This is a monotypic genus based on *Penicillium sacculum* described by Dale (1926) from soil. As reported by Smith (1961), *P. sacculum* was assigned to *Scopulariopsis* by Raper and Thom (1949). An isolate of *P. sacculum* was apparently submitted by Smith to Raper, who indicated that it was not a *Penicillium* in the sense of the manual, suggesting *Scopulariopsis* as a possible disposition. On the basis of the fact that the sporogenous cells were phialides, Smith created the new genus *Eladia*. *E. saccula*, while a distinctive species, is very close to the monoverticillate *Penicillium* species of the *nigricans* series. This species was listed by Chesters and Thornton (1956) as *Scopulariopsis* (*sacculum* Dale),

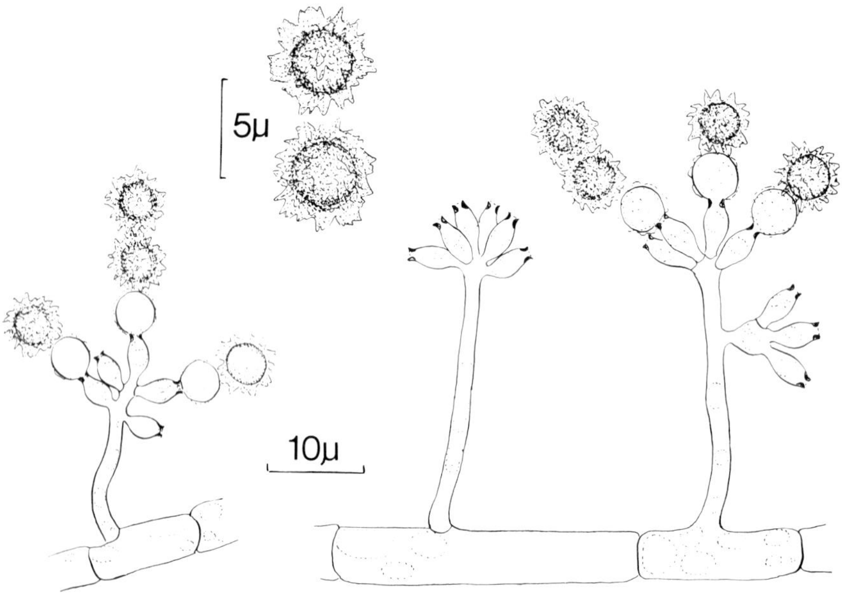


FIG. 84. *Eladia saccula*. The spherical conidia with membranous "spines" are borne in short chains. Phialides are frequently in a monoverticillate arrangement and have a highly refractive collar at the mouth (IMI 68319).

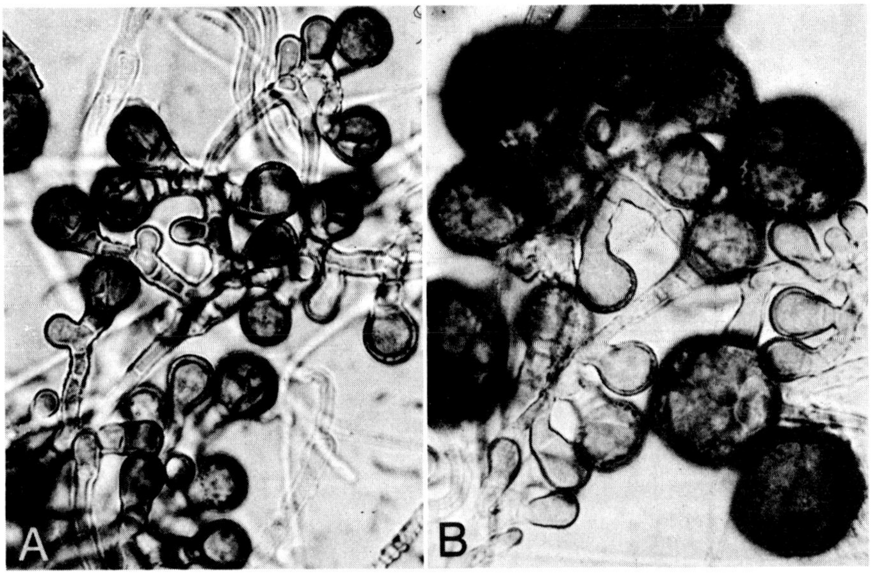


FIG. 85. *Epicoccum nigrum*. *A*, young aleuriospores develop in dense clusters from short lateral branches. *B*, mature spores are muriform and irregularly roughened (OAC 10333).

and the Commonwealth Mycological Institute record an isolate of this species from Ireland by A. Mangan. We have isolated *Eladia* only once, from organic soil in a cedar swamp in Ontario.

## EPICOCCUM Link

Type Species: *Epicoccum nigrum* Link.

**Generic Description:** Conidiophores very short, produced on sporodochia or in tight clusters from the vegetative hyphae, hyaline or lightly pigmented or lacking, with the conidia borne sessile on the hyphae; conidia aleuriospores, at first smooth and nonseptate, becoming muriform and roughened with warty incrustations in age, dark brown, globose to subglobose. (Description based on *E. nigrum.*)

**Diagnostic Features:** The clusters of dark, globose to subglobose, muriform aleuriospores serve to distinguish the common *E. nigrum* (Fig. 85). This species can frequently be identified by the yellowish or orange hyphae.

**Notes:** An excellent critical review of *Epicoccum* was presented by Schol-Schwarz (1959), who analyzed 70 isolates, 18 type specimens and 96 herbarium specimens of *Epicoccum* species. On the basis of a critical study of cell size, cell number and wall thickness in the spores, Schol-Schwarz concluded that all strains and herbarium specimens examined could be considered as *E. nigrum*. Schol-Schwarz further recommended that the genus *Cerebella* should be discarded and its only species included in *Epicoccum* as *E. andropogonis*.

*E. nigrum* is, in our experience, not uncommon in soil despite the scarcity of reports in the literature (Gilman, 1957).

## FUSARIELLA Sacc.

Type Species: *Fusariella atrovirens* Sacc.

**Generic Description:** Conidiophores more or less erect, short, simple or more often sparingly or irregularly branched, hyaline to pale brown, frequently darker at the apex, smooth or slightly roughened, terminating in solitary sporogenous cells; sporogenous cells phialides, frequently with a thickened apical ring, usually somewhat curved; phialospores in short chains, septate, darkly pigmented, fusoid.

**Diagnostic Features:** Chains of dark phragmospores produced from phialides serve to distinguish *Fusariella* (Fig. 86). The fusoid conidia, deflected laterally near the apex, are also diagnostic.

**Notes:** A detailed study of the morphology and taxonomy of *Fusariella* has been presented by Hughes (1949). The fungus is usually found as a saprophyte on the leaves of higher plants and is reported only rarely from soil. *F. bizzozeriana* and *F. obstipa* have been recorded by Nicot (see the Catalogue of the culture collection of the Commonwealth Mycological Institute).

## FUSARIUM Link

Type Species: *Fusarium roseum* Link.

**Generic Description:** Fruit body, when present, a sporodochium; sporodochium sometimes lacking and sporogenous cells arise directly from the vegetative hyphae or from conidiophores; conidiophores solitary and simple or aggregated and with complex branching, ultimate branches terminating in sporogenous cells; sporogenous cells phialides; phialides taper distally, sometimes with an apical collarette; phialospores frequently of two types; large macrospores, one- to several-septate, forming elongate phragmospores, hyaline, cylindric or curved, frequently boat-shaped, *with well marked foot cell at the attachment end* of the spore, produced in mucus and sliming down to form gloeoid heads or spore masses; microspores smaller, nonseptate or one-septate, ovoid to short-cylindric, gathering in short chains or more commonly in spore balls.

**Diagnostic Features:** The boat-shaped, hyaline, phragmospore with well marked foot cell is characteristic of most species of the genus (Fig. 87).

**Notes:** *Fusarium* is by definition a sporodochial fungus, classified in the Tuberculariaceae of the Moniliales under Saccardo's system. Unfortu-

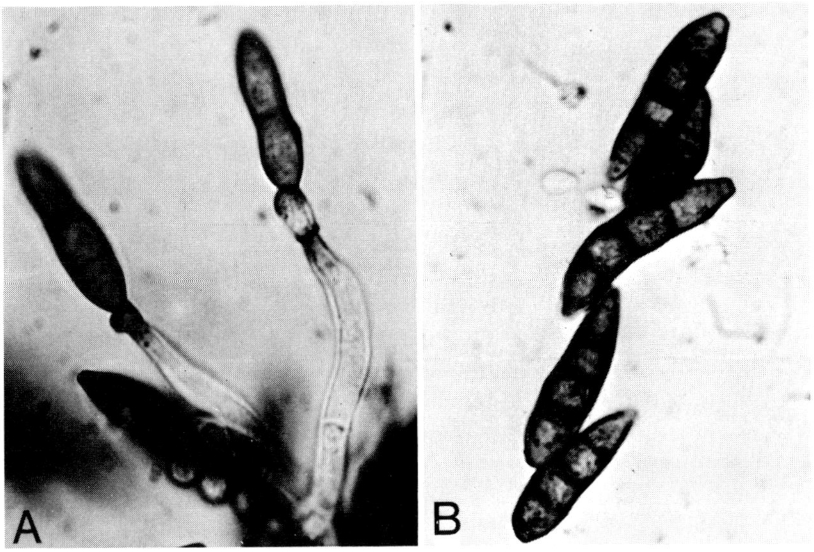


FIG. 86. *Fusariella bizzozeriana*. *A*, the phialides have a darkened apical ring. *B*, septate phialospores adhere in short chains (IMI 67732).

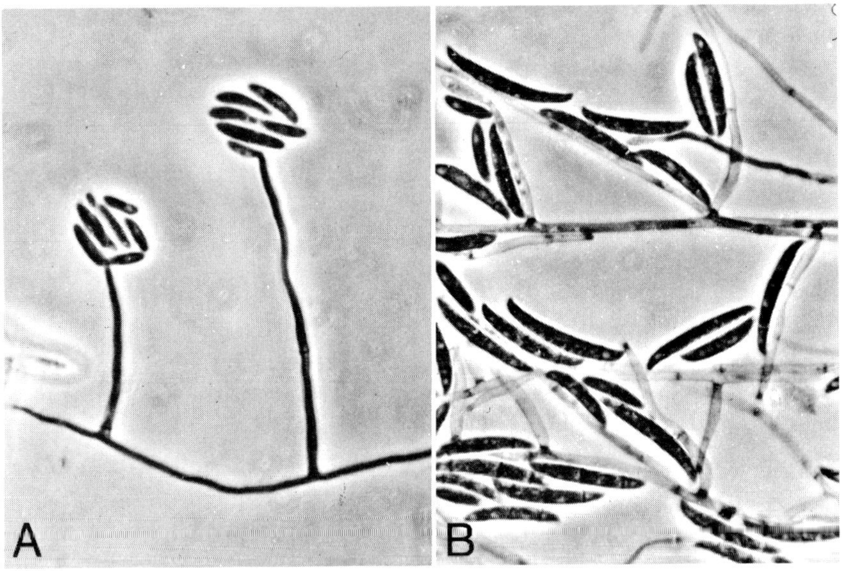


FIG. 87. *Fusarium solani*. *A*, microconidial state. *B*, macroconidial state. Note prominent collarettes on phialides (OAC 10564).

nately, the production of sporodochia is an inconstant character and many species fail to produce this fructification in culture.

Great variation is found in spore shape, size, septation and in growth habit. In culture, a number of *Fusarium* species lose the ability to produce macrospores after a few transfers and appear to all intents and purposes like a *Cephalosporium*. As pointed out by Booth (1959) and by Onions and Barron (1967) some microconidial states of *Fusarium* simulate the genus *Paecilomyces*; this is particularly true of *F. moniliforme* and *F. decemcellulare* (the conidial state of *Calonectria rigidiuscula*).

*Fusarium* species are frequently the conidial states of members of the Hypocreales (Booth, 1959). *F. moniliforme* and *F. graminearum* are the conidial state of *Gibberella fujikoroi* and *G. zeae*, respectively. *Nectria galligena*, the causal agent of cankers on apple trees, has a *Fusarium* conidial state, as have other *Nectria* species. The taxonomy of *Fusarium* is still controversial and there is no recent monograph of the genus. The treatment of Wollenweber and Reinking (see Gilman, 1957) is still perhaps the best source for identification of *Fusarium* species.

## FUSICLADIUM Bon.

Type Species: *Fusicladium virescens* Bon.

**Generic Description:** Conidiophores more or less erect, short, pigmented, continuous or septate, conspicuously denticulate in the spore bearing region, producing conidia singly and successively as blown out ends of successively produced growing points; conidia darkly pigmented, continuous or one-septate, truncate at the attachment point, variable in shape, ovoid or long-ellipsoid to flame-shaped. (Description based on *F. virescens*.)

**Diagnostic Features:** The short, dark, denticulate conidiophores, bearing pigmented conidia in acropetal succession, are characteristic of the common *Fusicladium* species (Fig. 88).

**Notes:** The genus *Fusicladium* is commonly found as a parasite of higher plants. *F. virescens*, the type species, is the conidial state of *Venturia pirina*, the causal agent of pear scab (Hughes, 1953a). As pointed out by Hughes, there has been in the past some confusion regarding *Fusicladium* on pear and *Spilocaea pomi* (*Fusicladium dendriticum*) on apple. In the latter, the sporogenous cells are annellophores and are annellated rather than denticulate, as in *Fusicladium*.

*Fusicladium* is not generally regarded as having great saprophytic potential and is rarely recorded from soil. The only such record of which I am aware is at the Commonwealth Mycological Institute, which lists a single isolate of this genus from soil.

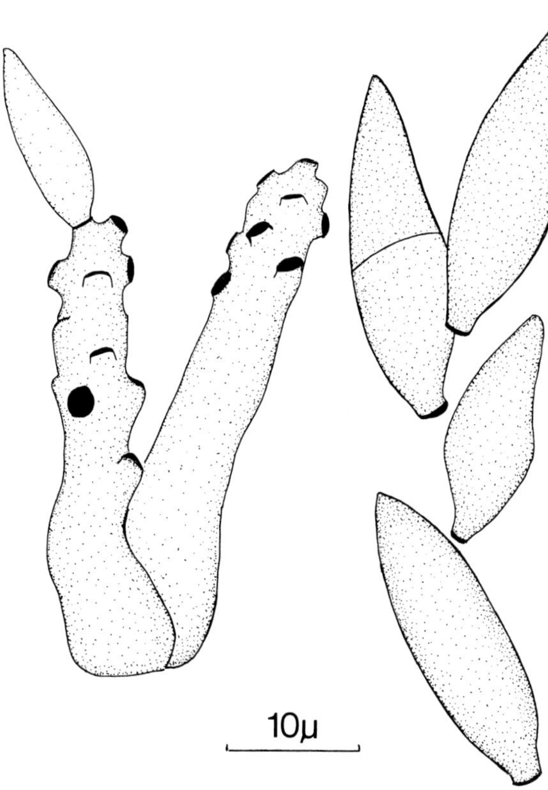


FIG. 88. *Fusicladium pirina.* Conidiophores are short and have conspicuous raised scars. Conidia are flame-shaped with one or two cells.

## FUSIDIUM Link

Type Species: *Fusidium griseum* Link.

**Generic Description:** Conidiophores not readily distinguished from the vegetative hyphae, short, simple or branched, hyaline to lightly pigmented, septate, arising as short lateral branches more or less at right angles to the substratum; conidia blastospores, produced in acropetal succession in simple or branching chains, nonseptate, hyaline to lightly pigmented, fusiform to long-cylindric, truncate or rounded at the ends.

**Diagnostic Features:** Simple or branching chains of elongate, hyaline blastospores arising from poorly developed conidiophores are typical of *Fusidium* (Figs. 89 and 191*B*).

**Notes:** The popular concept of *Fusidium* differs substantially from the generic description given above. Many authors consider *Fusidium* as producing chains of spores in basipetal succession from short phialides.

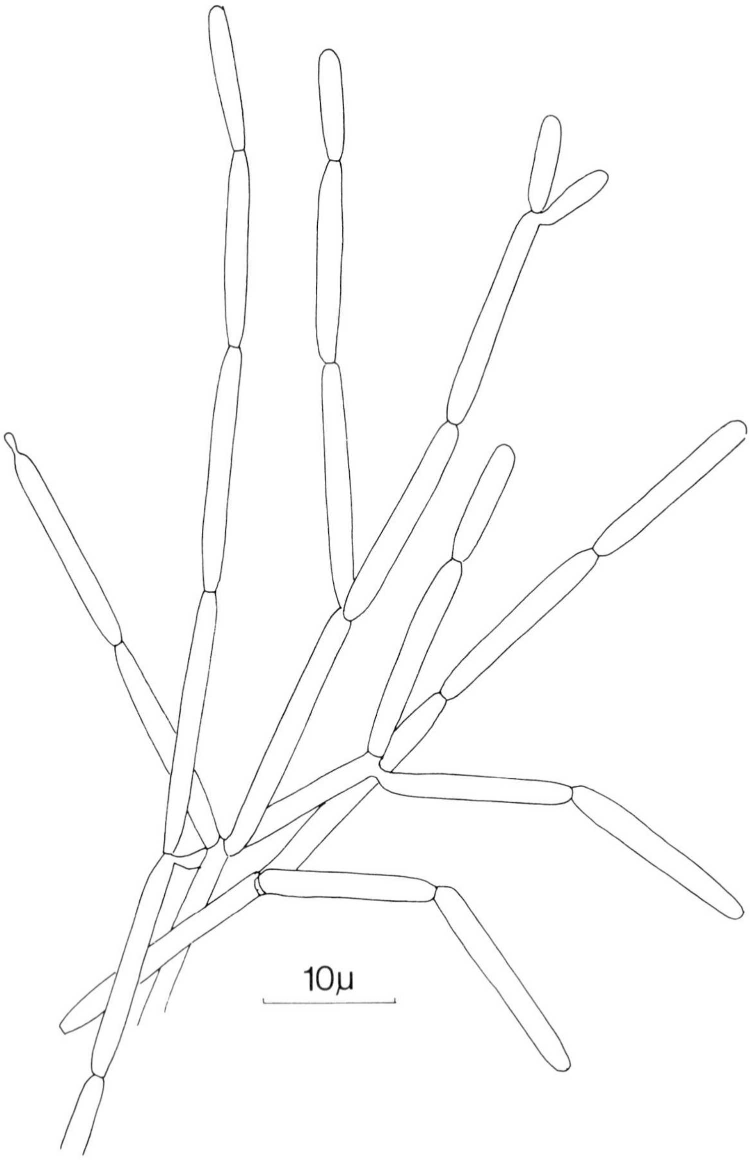


FIG. 89. *Fusidium griseum*. The conidia are produced in acropetal succession in branching chains (IMI 78926).

In some cases the illustrations in the literature are *Fusidium*-like in producing simple chains of conidia, but it is not clear whether the spores are produced in acropetal succession or basipetal succession. There are a number of records of *Fusidium* from soil, but, because of the confusion surrounding this genus, these records have little meaning. *Fusidium terricola*, described by Miller *et al.* (1957), produces its spores from phialides and was placed in *Paecilomyces* by Onions and Barron (1967).

As defined here, the genus *Fusidium* shows relationships to *Polyscytalum* and *Hormiactis*.

### GENICULARIA Rifai and R. C. Cooke

Type Species: *Genicularia cystosporia* (Duddington) Rifai and R. C. Cooke.

**Generic Description:** In pure culture on cornmeal agar, colonies effused, pale-pink to whitish; mycelium composed of septate, hyaline, branched hyphae with scanty aerial growth; conidiophores septate, erect or ascending, at first straight, becoming geniculate or flexuous, elongating (sometimes considerably) by repeated subapical renewal of growth, hyaline and smooth-walled; conidia arise singly as blown out ends of conidiophores; after the first conidium has been formed, a new growing point appears at one side of it and the second conidium is formed at the new apex, displacing the first conidium to a lateral position, this process being repeated several times; conidia obpyriform, smooth-walled, one-septate, the proximal end obconical and much smaller than the distal one, hyaline when viewed singly, appearing pale pinkish-white in mass. (Description from Rifai and Cooke, 1966.)

**Diagnostic Features:** The members of this genus are distinguished by producing two-celled, obpyriform conidia on geniculate conidiophores which elongate sympodially (Fig. 90).

**Notes:** Rifai and Cooke (1966), in their study of the genus *Trichothecium*, noted that *T. cystosporium* Duddington was not congeneric with *T. roseum*, the type species of this genus, because it produced a panicle of spores rather than the chain-like cluster produced by *T. roseum*. They therefore erected the new genus *Genicularia* to accommodate *T. cystosporium* and similar forms. Their description of *G. cystosporia* was based on an isolate from soil. This genus, although segregated from *Trichothecium*, is actually much closer to *Arthrobotrys* and *Candelabrella* and is predaceous on nematodes.

### GENICULISPORIUM Chesters and Greenhalgh

Type Species: *Geniculisporium serpens* Chesters and Greenhalgh.

**Generic Description:** Conidiophores erect, septate, smooth, regular, once or frequently branched; branches originating low down on the main axis and dichotomous or subdichotomous in appearance, such that the main axis becomes indistinct; apical region of the conidiophore appears irregularly geniculate; conidia produced in acropetal succession, hyaline to subhyaline, smooth, continuous, subspherical to ovoid-ellipsoid, with truncate base. (Description based on Chesters and Greenhalgh, 1964.)

**Diagnostic Features:** The uniform, branching conidiophores bearing

FIG. 90. *Genicularia cystosporia.* The large, two-celled conidia are borne acropetally from a sympodially extending conidiophore. (Redrawn from Rifai and Cooke, *Trans. Brit. Mycol. Soc.*, *49:* 147–168, 1966. Fig. 2.)

conidia in acropetal succession and having prominent scars on a somewhat rachis-like sporogenous cell, are typical of *Geniculisporium* (Figs. 91 and 117*C*). The spores in the described species have a slightly concave base.

**Notes:** The genus *Geniculisporium* was erected by Chesters and Greenhalgh (1964) to include the conidial state of certain *Hypoxylon* species. The conidial state of *H. repens* had been erroneously called *Haplaria grisea.* The type material of *H. grisea* has been examined by Hughes (1958), who found it to be identical with *Botrytis cinerea.*

I have isolated *Geniculisporium serpens* frequently from soil in Canada. One such culture consistently produced the stroma and perithecia of *Hypoxylon serpens* in culture. Until the perfect state was produced, I

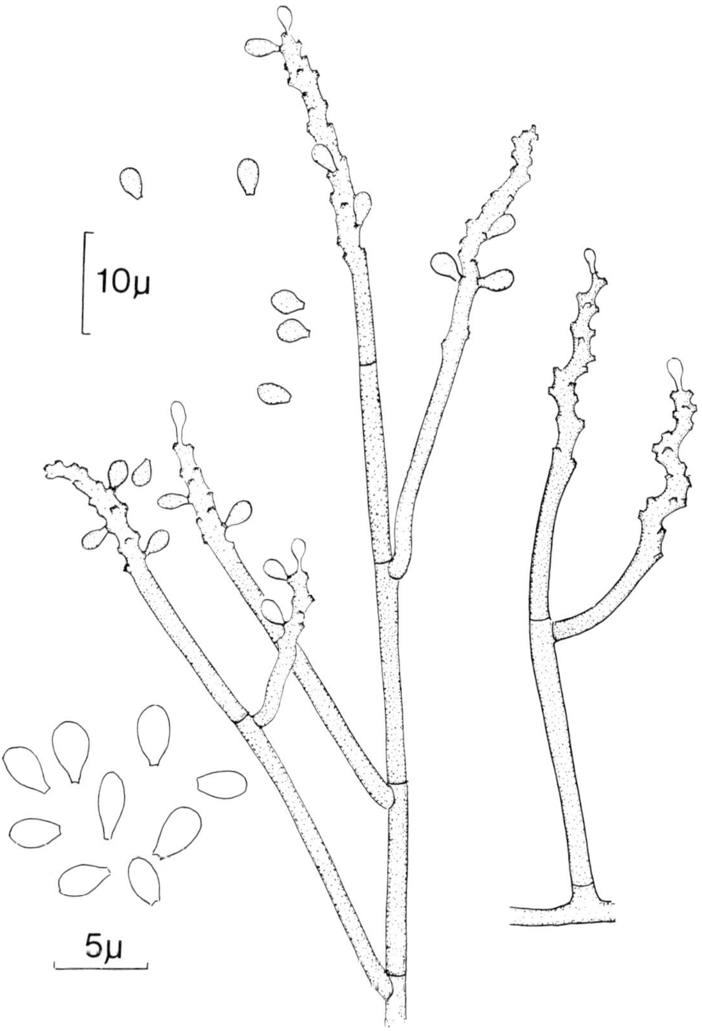


FIG. 91. *Geniculisporium serpens.* The conidiophores are irregularly branched and markedly geniculate. The conidia have a concave base (OAC 10202).

disposed the fungus as *Nodulisporium corticioides*. *Geniculisporium* shows affinities to *Nodulisporium* and *Rhinocladiella*.

## GEOTRICHUM Link

Type Species: *Geotrichum candidum* Link.

**Generic Diagnosis:** Conidiophores lacking, vegetative hyphae hyaline or lightly pigmented; conidia arthrospores, produced by basipetal septation and fragmentation of the vegetative hyphae; arthrospores very variable in width and length, nonseptate, hyaline or subhyaline, cylindric in shape with truncate ends, end walls sometimes slightly convex; in some species arthrospores "round off," to form chains of globose to subglobose spores.

**Diagnostic Features:** The key character of *Geotrichum* is the habit of the vegetative hyphae to fragment into unicellular units (Fig. 92). In older cultures of *G. candidum* almost all of the vegetative hyphae is converted to arthrospores and active vegetative hyphae are difficult to find.

**Notes:** *G. candidum* is the most common species. It is commonly associated with milk products and was formerly known as *Oospora lactis*. The many synonyms of this very common species are reviewed in some detail by Carmichael (1957). *G. candidum* is frequently isolated from soil. It grows vigorously but thinly over the agar and is often associated with a chlamydospore stage. Because of the dry, powdery nature of the spore in the mature colony, *G. candidum* can be a nuisance as a laboratory contaminant.

In another species of *Geotrichum*, isolated from organic soil in Ontario, spore formation is less typical, and delicate aerial hyphae grow up from the substratum. These hyphae are mostly simple and differentiate basipetally into globose or ovoid units (Fig. 117*H*).

Windisch (1951) connected *Geotrichum candidum* to its perfect stage *Endomyces lactis* (Fres.) Windisch. Carmichael (1957) reported that this may have been in error and that the question of sexual reproduction in *G. candidum* required further study. Butler *et al.* (1965) established *G. candidum* as the correct name for the causal agent of the "sour rot" disease of citrus. They noted that isolates of *G. candidum* attacking citrus occurred in soils in citrus groves along with nonpathogenic isolates. *Geotrichum* states are reported for certain of the Gymnoascaceae (Orr *et al.*, 1963; Apinis, 1964).

Arthrospore formation is a fairly primitive type of spore formation. It is not surprising, therefore, to find *Geotrichum*-like forms in quite unrelated fungi. This genus remains truly a form genus; a catch-all for fungi in which asexual states are more or less hyaline arthrospores produced by fragmentation of the vegetative hyphae.

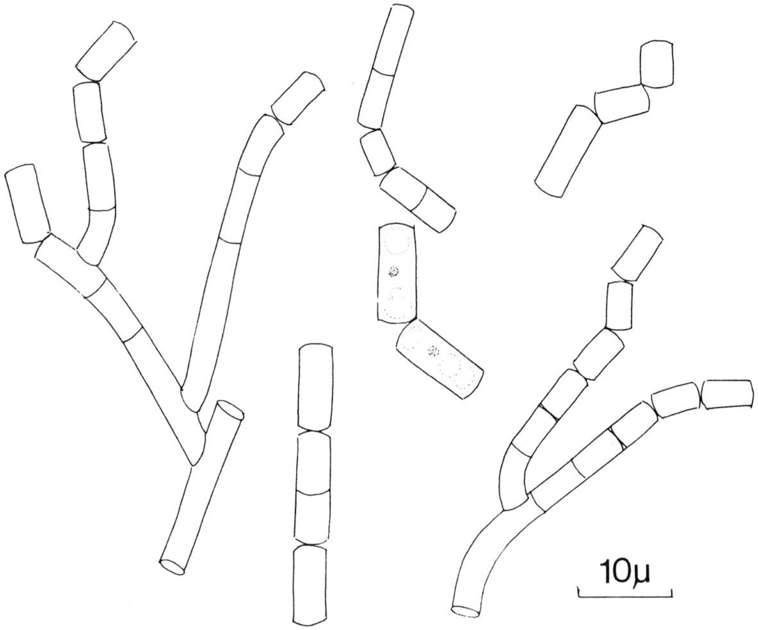


FIG. 92. *Geotrichum candidum*. The vegetative hyphae septate and eventually break up into arthrospores.

## GILMANIELLA Barron

Type Species: *Gilmaniella humicola* Barron.

**Generic Description:** Vegetative hyphae hyaline becoming brown, with prominent septa; conidiophores arising singly from the vegetative hyphae, short, simple or branched, stalk and branches frequently inflated, hyaline, septate; conidia apical, borne singly or in clusters, dark brown, with conspicuous apical germ pore.

**Diagnostic Features:** This genus is distinguished by the production of clusters of dark aleuriospores on short, inflated conidiophores (Fig. 93). The type species is distinctive in its very dark, spherical conidia with apical germ pores.

**Notes:** *Gilmaniella* was erected by Barron (1964a) on a single species *G. humicola* recovered from forest soil. Concurrently, Subramanian and Lodha (1964) described this same fungus as a new coprophilous genus, *Adhogamina*.

Records at the Commonwealth Mycological Institute indicate that *Gilmaniella* is not uncommon in soil. It has been recorded from soils in Cambridgeshire and Yorkshire in England and also from soil in Egypt. Miss A. Mangan (personal communication) isolated *G. humicola* from

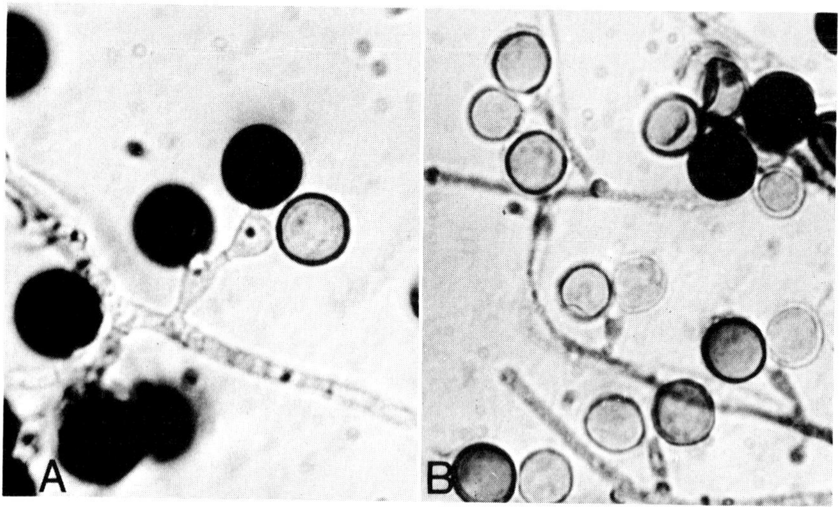


FIG. 93. *Gilmaniella humicola*. *A* and *B*, various stages of development of aleuriospores from short, inflated conidiophores (OAC 10076). (Reproduced by permission of the Mycological Society of America, *Mycologia 56:* 514, 1964.)

beet seedlings in sterile soil. The genus may well have been listed under other generic names in the earlier literature, possibly as *Humicola*.

On casual inspection, *Gilmaniella* appears as a mass of spherical aleuriospores with the general appearance of a *Humicola* or *Botryotrichum*. In these and similar genera, however, the spores are borne singly, either directly on the vegetative hyphae or on short lateral branches little differentiated from the vegetative hyphae. The conidia of *Gilmaniella* are frequently borne in groups at the ends of individual conidiophores, and the conidiophores themselves show a much greater degree of differentiation and, at their best development, are similar to the lower levels found in *Wardomyces*. The prominent apical pore of *Gilmaniella* is not found in *Botryotrichum*, but I have seen pores in some *Humicola*-like isolates from soil. In *Wardomyces* there is an elongate germ slit running the length of the spore. The vegetative hyphae of *Gilmaniella* have highly refractive septa, with the hyphae frequently swollen and node-like, in the vicinity of the septum.

## GLIOCEPHALIS Matruchot

Type Species: *Gliocephalis hyalina* Matruchot.

**Generic Description:** Mycelium immersed, sparse, irregularly septate, swollen at the base of the conidiophores; conidiophores erect, stout, hyaline, nonseptate, swollen apically to form a vesicle; vesicle bearing primary series of prophialides, frequently in the upper half only; each prophialide bearing two to five sporogenous cells; sporogenous cells phialides, producing spores successively in mucus; spores hyaline, non-septate, sometimes slightly truncate at the attachment point, gathering in gloeoid masses. (Description based on *G. hyalina*.)

**Diagnostic Features:** *Gliocephalis* is an *Aspergillus*-like fungus in which the spores slime down to form glistening masses. (Figs. 94 and 117*A*).

**Notes:** The type species, *G. hyalina*, was described from the roots of beetroot by Matruchot (1899), who gave a careful and detailed description. Matruchot was initially unable to grow the fungus in pure culture, but he eventually succeeded in growing *Gliocephalis* on a peptone-maltose agar along with an unknown bacterium.

The only record from soil is that of Embree (1963), who recorded a *Gliocephalis* species from potting soil in San Francisco. According to Embree, the conidiophores of this species were darkly pigmented. It is possible that the species recorded by Embree is closer to *Goidanichiella*, if this genus is to be regarded as distinct from *Gliocephalis*.

The figures and descriptions of *G. hyalina* are based on an isolate found on a potato which had undergone considerable deterioration in soil. Attempts to recover this fungus in pure culture were unsuccessful, as the conidia germinated poorly or not at all on all of the media tried. The fungus was maintained for a limited period in association with a *Fusarium* species and bacterial contaminants from the same potato. All colonies eventually petered out or were overgrown by the associated organisms. The ecology of *G. hyalina* should prove interesting.

## GLIOCEPHALOTRICHUM J. J. Ellis and Hesseltine

Type Species: *Gliocephalotrichum bulbilium* J. J. Ellis and Hesseltine.

**Generic Description:** Conidiophores colored, simple, septate, bearing a polyverticillate penicillus with primary, secondary and frequently tertiary branches (metulae) in successive whorls, ultimate branches phialides; conidiophores with sterile arms subtend the penicillus; arms arise just below the primary metulae of the penicillus; conidia oblong elliptical, hyaline, and coalesce into mucus-like droplets. (Description from Ellis and Hesseltine, 1962.)

**Diagnostic Features:** *Gliocephalotrichum* is distinguished from similar genera by the whorl of determinate, sterile, hair-like arms which grow out from the conidiophores just below the compact penicillus (Fig. 95).

**Notes:** The genus *Gliocephalotrichum* was erected by Ellis and Hesseltine on a single species *G. bulbilium*, isolated from soil under moss in Louisiana. These workers also list a record of this fungus from soil of the West Virginia University campus. The only other record of this unusual Hyphomycete is that of Christensen *et al.* (1962).

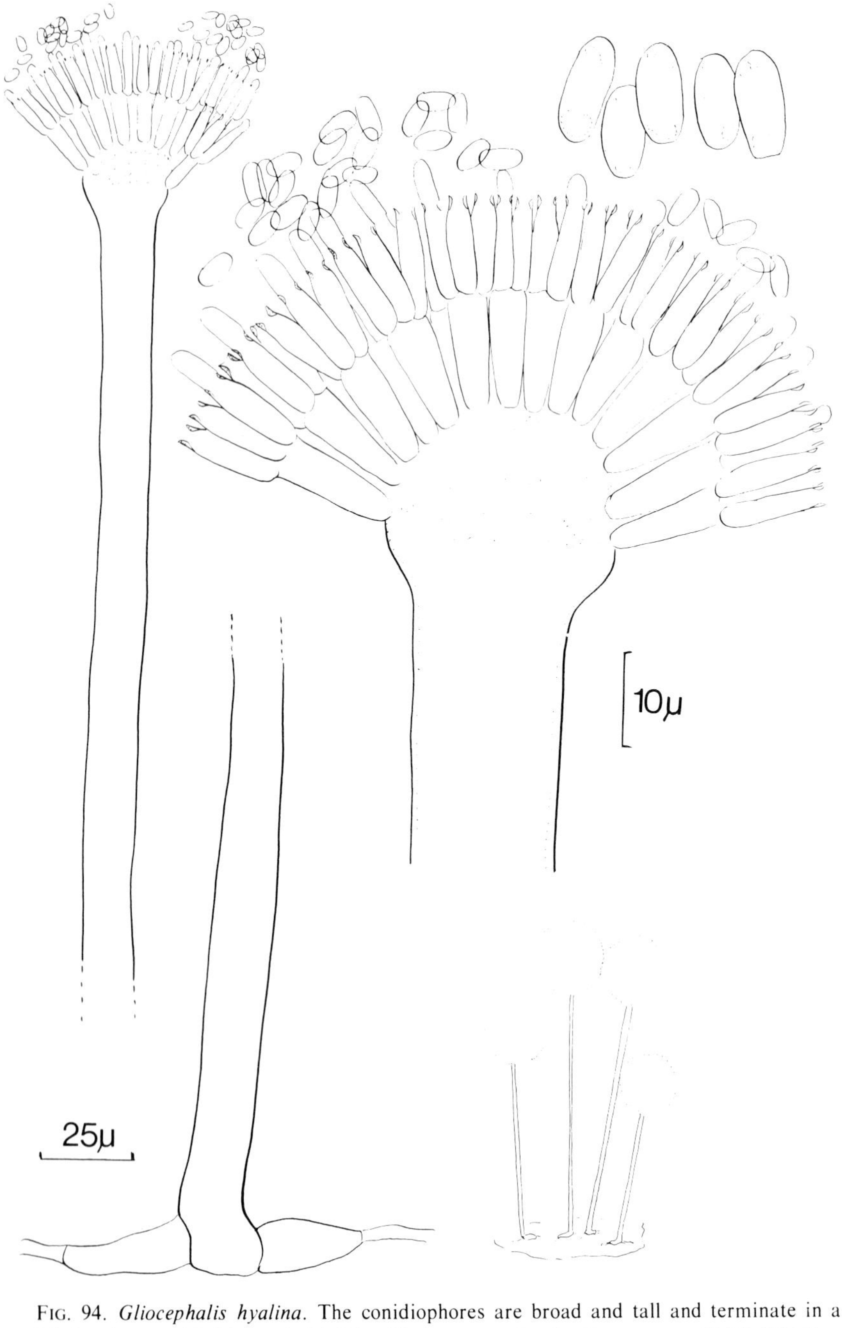


FIG. 94. *Gliocephalis hyalina*. The conidiophores are broad and tall and terminate in a vesicle bearing prophialides and phialides. The spores gather in slimy masses (OAC 10546).

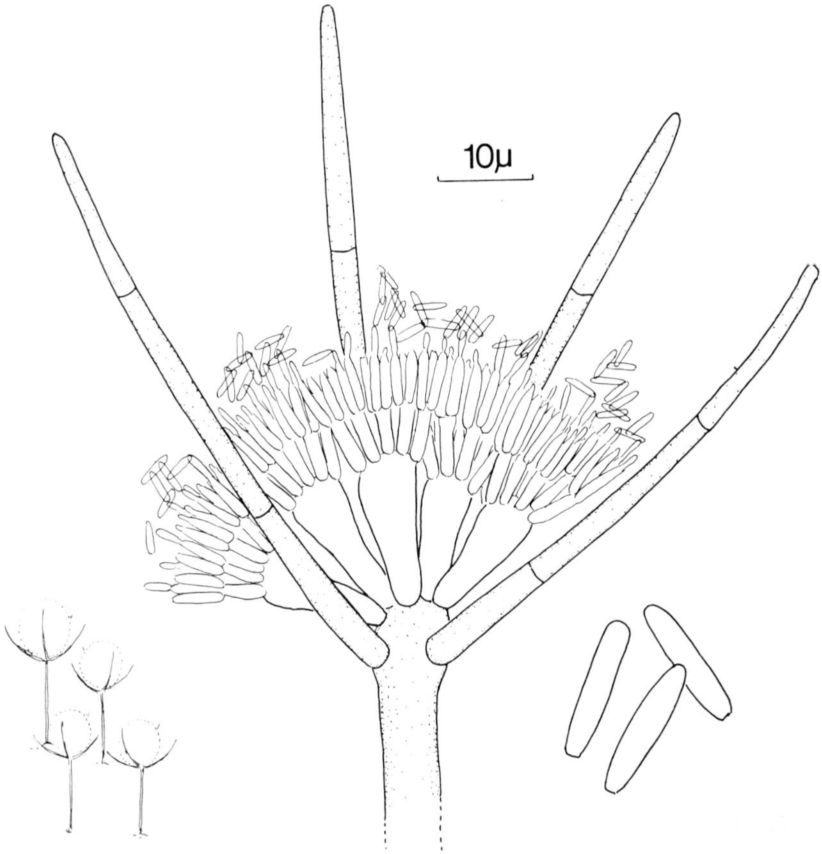


FIG. 95. *Gliocephalotrichum bulbiferum.* The gloeoid spore masses are contained by four, stout, seta-like appendages. The sporogenous cells are arranged in a symmetric poly-verticillate arrangement with the primary series of metulae significantly larger (OAC 10464).

## GLIOCLADIUM Corda

Type Species: *Gliocladium penicillioides* Corda.

**Generic Description:** Conidiophores erect or suberect, arising from the substratum or from the aerial hyphae, septate, hyaline, branching at the apex in a penicillate fashion, ultimate sporogenous cells phialides; phialides sometimes divergent, more often slightly or closely adpressed, bottle-shaped, frequently more convex on one side; phialospores hyaline or pigmented, nonseptate, ovoid or sometimes flattened on one side, sliming down to form gloeoid heads or held together in loose columns.

**Diagnostic Features:** The penicillate heads producing slimy spores in balls or loose columns are diagnostic for *Gliocladium* (Fig. 96).

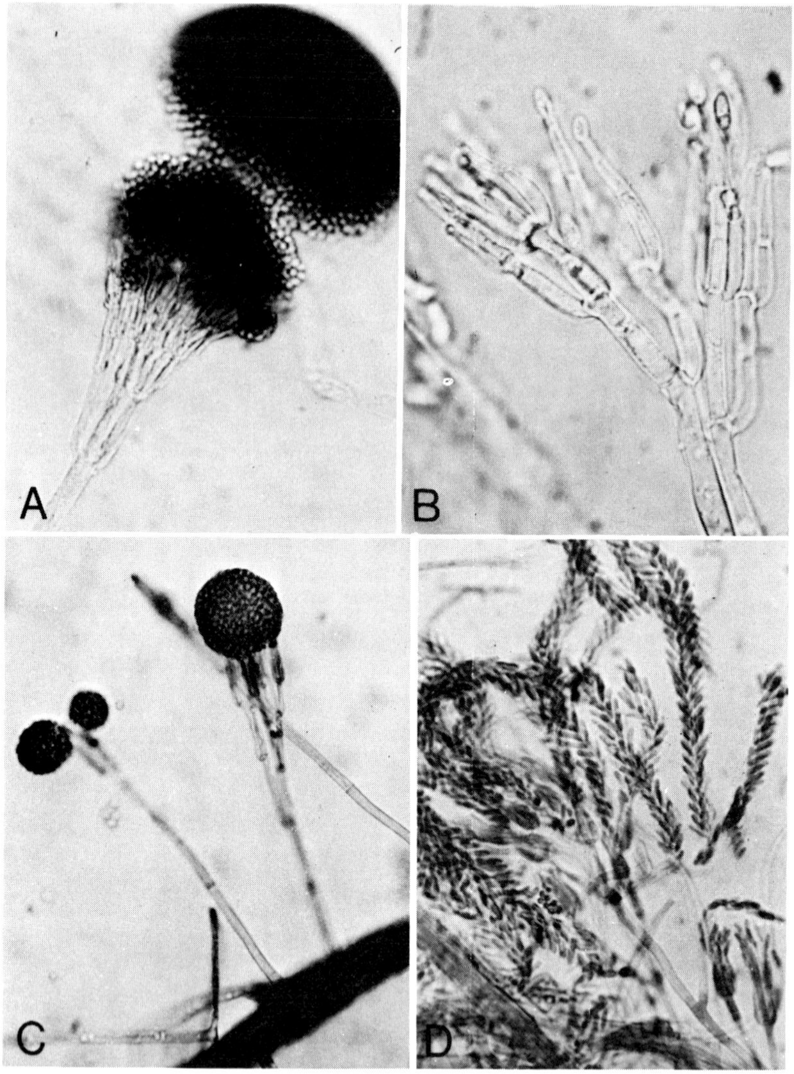


FIG. 96. *A*, *Gliocladium deliquescens*. Conidia aggregate into a dark-green globule on a stout conidiophore. *B*, *Gliocladium roseum*; penicillate habit of the sporogenous cells with closely adpressed phialides. *C*, *Gliocladium* sp. Conidiophore arising from an aerial hyphal rope. *D*, *Gliocladium* sp. (*Clonostachys*-like) with the conidia produced in columns.

**Notes:** *Gliocladium* species are very common in soil, particularly members of the *G. roseum* series. *G. roseum* has been connected to its perfect state *Nectria gliocladioides* by Smalley and Hansen (1957). As pointed out by Booth (1959), in the *Nectria ochroleuca* group of species the conidial states belong to the genus *Gliocladium* and can be distinguished from each other only with difficulty.

An extensive treatment of the genus *Gliocladium* has been published by Morquer *et al.* (1963). Unfortunately, this treatment is not a monograph and many of the described species are not considered.

Corda, in his Prachtflora, described the genus *Clonostachys* based on *C. araucaria.* This is a penicillate organism which he figured as producing columns of elliptical conidia with the long axis of the conidium diagonal to the axis of the column. *Clonostachys*, although still apparently recognized in some of the recent literature (Tubaki, 1963), is sufficiently close to the columnar types found in *Gliocladium* that the distinction between the two genera would be difficult to make. *Clonostachys* might best be considered for the moment under *Gliocladium. Clonostachys araucaria* has been reported frequently from soil in Italy (Peyronel and Dal Vesco, 1955; Mosca, 1964).

The significance of *G. roseum* as a soil fungus was discussed by Pugh and Dickinson (1965) and the role of this same species as a destructive parasite was studied by Barnett and Lilly (1962).

## GLIOMASTIX Guégen

Type Species: *Gliomastix murorum* (Corda) Hughes.

**Generic Description:** Conidiophores short or lacking; sporogenous cells phialides, arising directly from the aerial hyphae or hyphal ropes; phialides sometimes arise in groups of up to three from a short conidiophore; phialides hyaline, smooth or rough; if smooth initially then becoming darker and roughened in some species, especially in the upper portions; phialides straight or frequently bent or sinuate, especially near the apex, apical collarette found in some species; phialospores nonseptate, pigmented, smooth or with patches of dark granular aggregate attached to the spore wall, globose, ovoid, or short-cylindric, sometimes fusiform, produced in chains or in gloeoid heads.

**Diagnostic Features:** The chains or balls of dark amerospores arising from almost hyaline sporogenous cells give the common *Gliomastix* species a distinctive appearance (Fig. 97).

**Notes:** The taxonomy and morphology of *Gliomastix* was given in some detail by Mason (1941) who pointed out the similarities between *G. convoluta* (=*G. murorum*) and its variety *G. convoluta* var. *felina*. In the latter, the spores tend to slime down to form gloeoid heads with the spores tending to be ovoid. In *G. convoluta* the spores tend to hold together in short, tortuous chains and are globose in shape. In both varieties aggregates of dark material cling to the exospore to give it a roughened appearance; this latter character is not always in evidence.

Hughes (1958) pointed out that *G. convoluta* had previously been described by Corda as *Torula murorum* and made the appropriate transfers.

*Gliomastix murorum* (Fig. 98) is commonly isolated from soil by dilution plate techniques. In North American soils, *G. murorum* var.

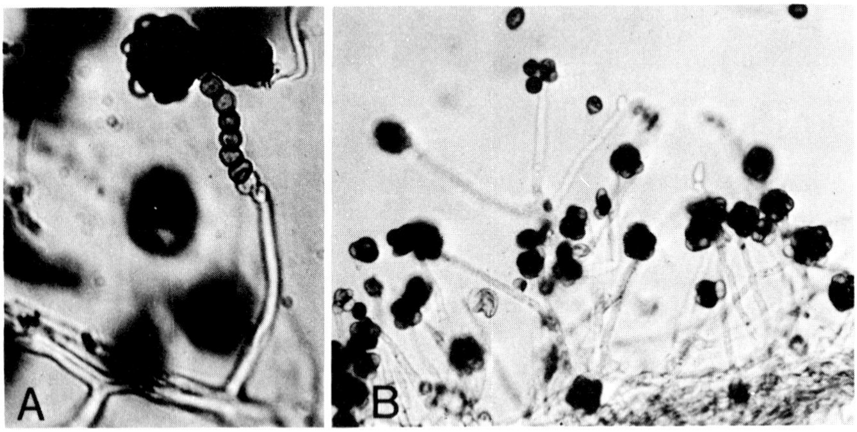


FIG. 97. *A*, *Gliomastix murorum*. Showing tortuous chains of phialospores (OAC 10209). *B*, *G. murorum* var *felina* (OAC 10329). Phialospores slime down to form dark, gloeoid balls.

*felina* is most common, with *G. murorum* being less frequent. Brown and Kendrick (1958) described *G. guttuliformis* from soil in England and *G. luzulae* has been recorded from soil in England as *Fusidium viride* Grove (Gilman, 1957).

A revision of *Gliomastix* has been published by Dickinson (1968), who recognized ten species and two varieties. Details included culture and host morphology and a key to species identification. The relationships of *Gliomastix* to other genera were also considered. Onions and Barron (1967) noted that *Gliomastix* approaches the monophialide species of *Paecilomyces* to the point where it becomes arbitrary whether a species is included in one genus or the other.

## GOIDANICHIELLA Arnaud

Type Species: *Goidanichiella scopula* (Preuss) Arnaud.

**Generic Description:** Conidiophores stout, erect, pigmented, simple, frequently swollen at the base, swollen at the apex to produce a well marked vesicle; vesicle bears prophialides and phialides in the manner of *Aspergillus*; phialides hyaline, become pigmented in age, flask-shaped; phialospores hyaline, nonseptate, globose or ovoid or asymmetric, not catenulate, produced in mucus and gather as glistening balls at the apex of the conidiophore.

**Diagnostic Features:** The glistening, white spore heads produced from a darkly pigmented, *Aspergillus*-like conidiophore are distinctive for *Goidanichiella* (Fig. 99).

**Notes:** The genus *Goidanichia* was erected by Arnaud (1953), with a somewhat abbreviated description based on *G. scopula* (Preuss) Arnaud.

Unfortunately the genus was not validated with a Latin diagnosis. Discovering that the name *Goidanichia* was a later homonym, Arnaud appended *Goidanichiella* as an autographic footnote on certain of the separates. While the name *Goidanichiella* is invalid according to the in-

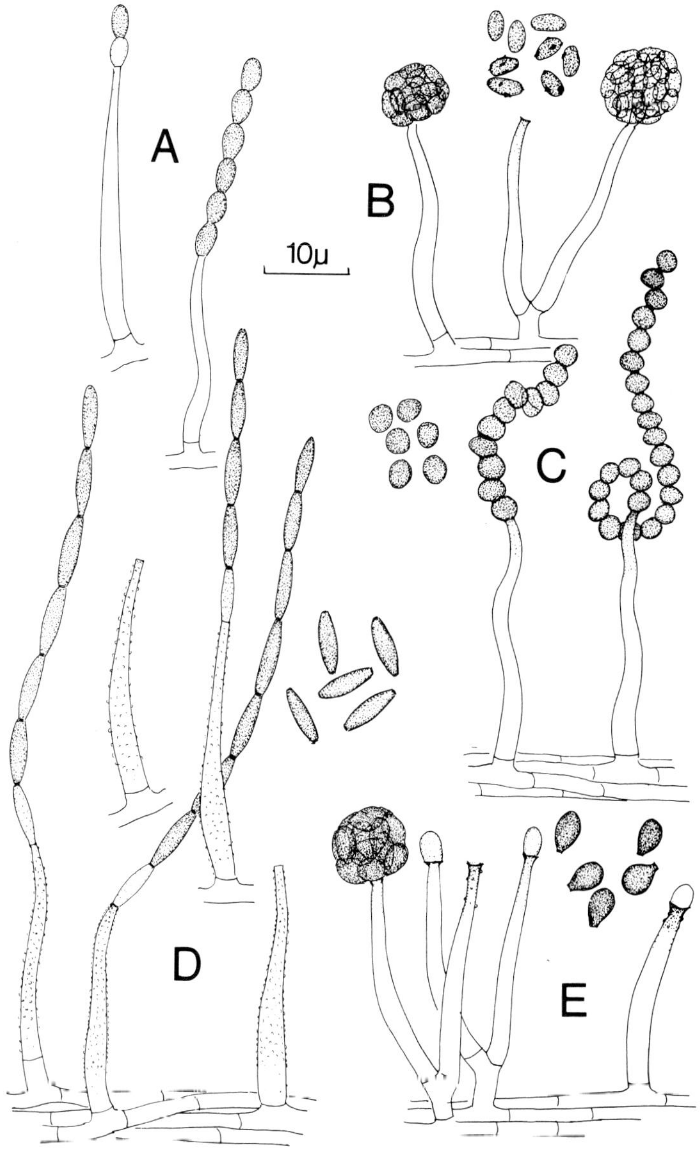


FIG. 98. *Gliomastix. A, G. murorum var. polychroma. B, G. murorum var. felina. C, G. murorum. D, G. luzulae. E, G. cerealis.*

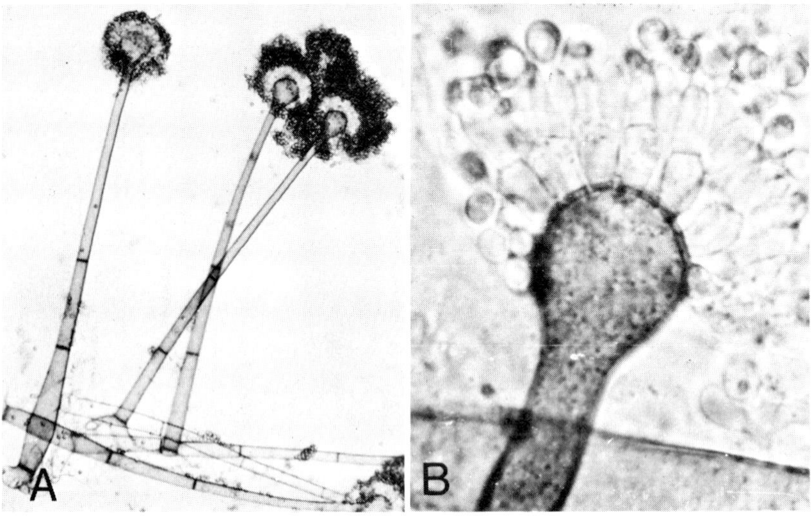


FIG. 99. *Goidanichiella* sp. *A*, the spores are borne in slimy heads on top of stout, dark conidiophores. *B*, the vesicle bears a series of prophialides on which the phialides are borne (OAC 10008).

ternational rules, there is no other suitable genus available at the moment to contain this group of fungi. I have isolated a *Goidanichiella* species only once from soil in Ontario. Embree (1963) described a *Gliocephalis* from potting soil from San Francisco. According to Embree, the conidiophores of his *Gliocephalis* were darkly pigmented. It is possible that the species discovered by Embree would be better considered under *Goidanichiella*. Both of these genera are in need of revision and re-evaluation.

## GONATOBOTRYS Corda

Type Species: *Gonatobotrys simplex* Corda.

**Generic Description:** Conidiophores sometimes repent, more often suberect or erect, simple or sparingly branched, hyaline, septate, swollen at the apex and bearing an apical cluster of conidia which arise simultaneously on short denticles over the surface of a terminal ampulla; conidiophores continuing growth from the apex by successive proliferations which produce an elongate conidiophore with intermittent node-like swellings which bear conidia; conidia large, hyaline, continuous. (Description based on *G. simplex*.)

**Diagnostic Features:** The clusters of large, hyaline, amerospores arising from "nodes" along the length of the conidiophores serve to distinguish *Gonatobotrys* (Figs. 7*B* and 100).

**Notes:** A detailed study of *Gonatobotrys simplex* was given by Drechsler (1950), who noted that, in the earliest stage in the formation of a spore cluster, the distended tip of the conidiophore bears up to 30 minute protuberances. Drechsler described the conidia as originating, developing and maturing more or less simultaneously. He was unable to show any predaceous potential against nematodes, amoebae or testaceous rhizopods or any parasitism against *Alternaria solani* or *A. tenuis*. *G. simplex* was originally reported by Corda to be parasitic on *Helminthosporium tenuissimum*. Barnett (1963), on the other hand, showed that *G. simplex* was parasitic on *Alternaria* and *Cladosporium*.

## GONATOBOTRYUM Sacc.

Type Species: *Gonatobotryum fuscum* (Sacc.) Sacc.

**Generic Description:** Conidiophores stout, dark brown, septate, more or less erect; each conidiophore swollen slightly at the apex to form a terminal ampulla; conidia borne simultaneously over the surface of each ampulla; conidiophore continues sterile growth from the apex of the ampulla and produces a second ampulla at a higher level; a succession of proliferations produce an elongate conidiophore with intermittent ampullae covered with conidia, to give a "noded" appearance to the mature fructification; conidia nonseptate, brown, in simple or branching acropetal chains.

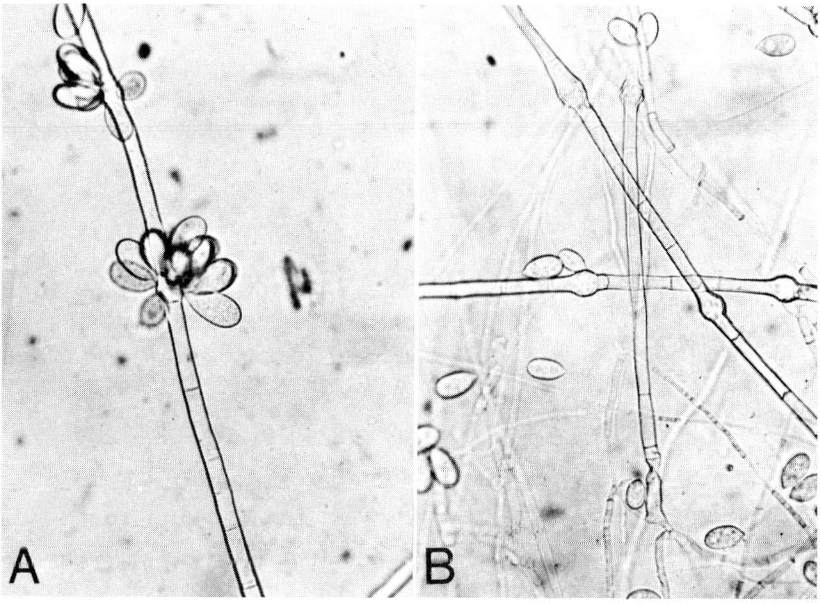


FIG. 100. *Gonatobotrys simplex*. *A*, conidiophore with clusters of spores at the "nodes." *B*, spores removed to reveal swollen ampullae. The conidiophores are simple or sparingly branched.

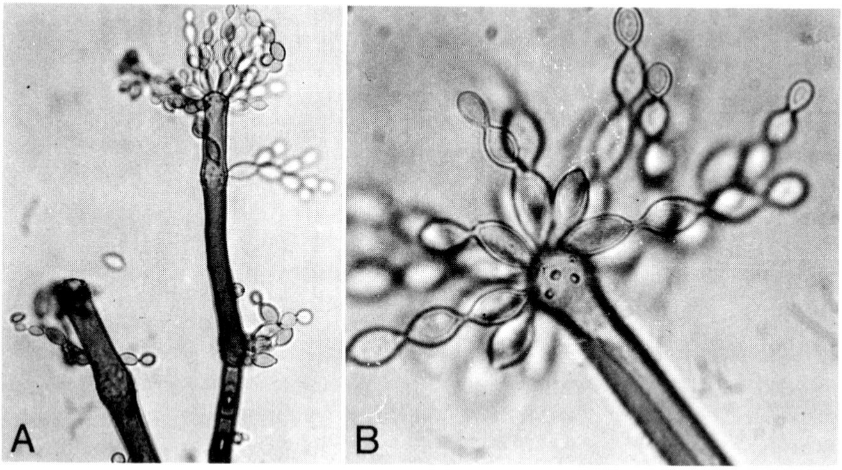


FIG. 101. *Gonatobotryum apiculatum*. *A*, terminal and intercalary ampullae bear blastospores in branching chains. *B*, apex of conidiophore showing that the conidia in the primary series are somewhat larger. Chains secede to leave pronounced denticles on the ampulla (OAC 10622).

**Diagnostic Features:** This genus is characterized by chains of pigmented amerospores which arise from terminal and intercalary ampullae on stout, dark conidiophores (Figs. 7*A* and 101).

**Notes:** As pointed out by Hughes (1953), the genus *Gonatobotryum* belongs in his Section IB, referred to herein as the Botryoblastosporae. When the conidia are in chains, the primary conidia in each chain are produced simultaneously over the surface of the terminal ampulla and 2nd, 3rd, and 4th order conidia, and so on, are produced more or less synchronously.

The genus *Gonatobotryum* is rarely reported from soil. It has been recorded from this source by Williams and Schmitthenner (1956) and by Kaufman and Williams (1964). I recently received an excellent culture of *G. apiculatum* from Mr. G. C. Bhatt, University of Waterloo, Ontario. This species is a most striking one and I have used it to illustrate the genus (Fig. 101) and to exemplify the series Botryoblastosporae (Fig. 7*A*).

## GONYTRICHUM Nees ex Wallr.

Type Species: *Gonytrichum caesium* Nees.

**Generic Description:** Main stalk of the conidiophore unbranched, straight or geniculate, bearing below a number of septa collar-like hyphae from which primary lateral branches arise either along the length of the main stalk or restricted to the distal end, lateral branches may be absent altogether; secondary, tertiary or even quaternary lateral branches, when present, develop in the same way from collar-like hyphae; conidiophores thus appear arborescent or more or less simple, with or without lateral branches above; ends of main stalk and laterals are usually sterile; phialides produced from the collar-like hyphae, more or less flask-shaped with a well marked collarette; phialospores subglobose to oval, slimy, continuous, hyaline to pale brown. (Description based on Hughes, 1951c.)

**Diagnostic Features:** *Gonytrichum* is readily recognized by the false verticils of phialides arising from collar hyphae (Figs. 102 and 103).

**Notes:** A detailed historical account of *Gonytrichum* was given by Hughes (1951c), who was the first to interpret correctly the precise method of phialide development from "collar hyphae." Swart (1959) observed that the production of spores in *Gonytrichum* is very similar to that in *Chloridium*. In both genera more than one spore may be produced simultaneously from the phialide apex. Under drying conditions, the spores in both genera may adhere in columns. Barron and Bhatt (1967) noted that in culture *G. chlamydosporium* failed to develop the typical *Gonytrichum* state after several transfers and appeared as a *Chloridium*.

## GRAPHIUM Corda

Type Species: *Graphium penicillioides* Corda.

**Generic Description:** Fructification a synnema; synnemata stout, darkly pigmented, more or less erect, produced singly or in clusters, with ele-

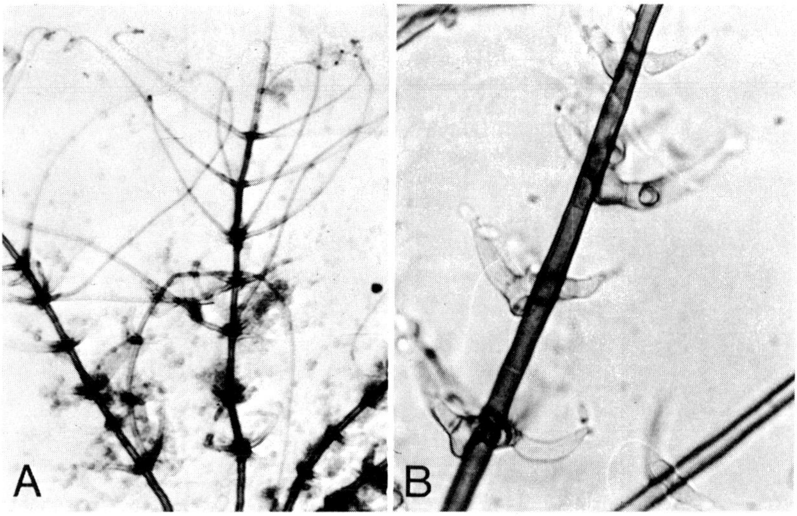


FIG. 102. *A*, *Gonytrichum macrocladum*. The main axis of the conidiophore has verticils of phialides at the lower "nodes" and sterile branches at the upper "nodes" (OAC 10311). *B*, *Gonytrichum chlamydosporium*; false verticils of phialides arise from collar hyphae which grow out from just below the septa (OAC 10032).

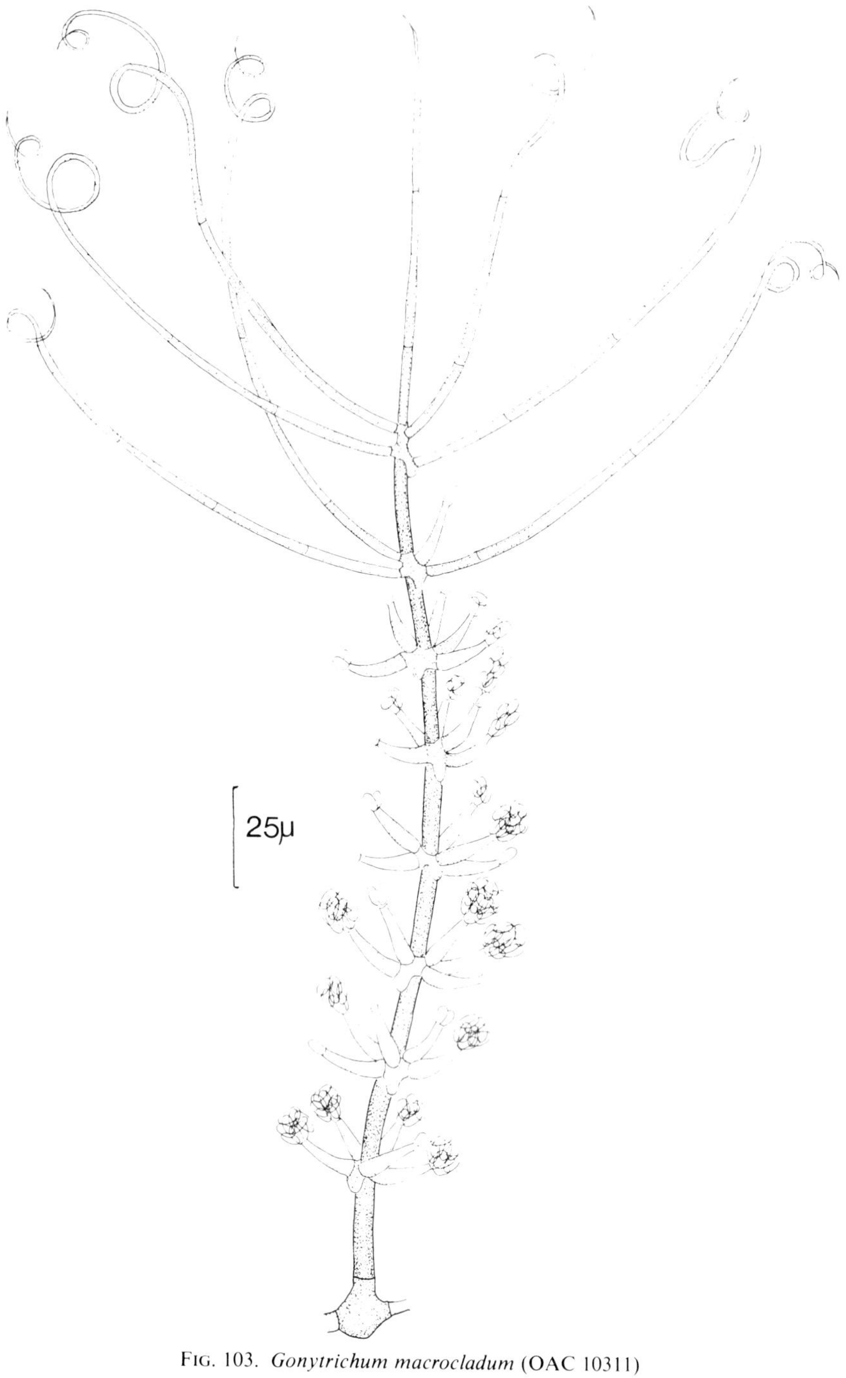


FIG. 103. *Gonytrichum macrocladum* (OAC 10311)

ments of each synnema diverging at the apex to produce a sporiferous head; conidia nonseptate, hyaline, smooth, subglobose to ovoid or short-cylindric, frequently truncate at the attachment point, gathering in large slimy heads.

**Diagnostic Features:** Dark synnemata with glistening masses of amerospores in mucus typifies the popular concept of *Graphium* (Fig. 104).

**Notes:** The generic diagnosis above is intentionally vague. It omits any reference to the sporogenous cells. These have been dealt with rather casually in most of the published descriptions of *Graphium* species where the nature of the sporogenous cell is either omitted or referred to as a phialide. In the "*Graphium*" states of *Petriella*, the sporogenous cells are annellophores. In the "Graphium" states of *Ceratocystis ulmi*, on the other hand, the sporogenous cells appear to be sympodulae. If it is also true, as reported, that some species of *Graphium* have phialides, then it

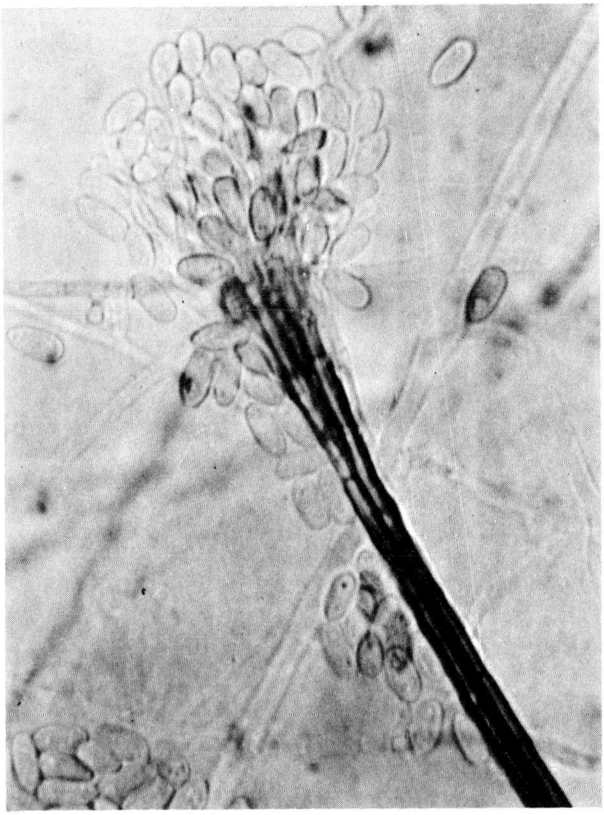

FIG. 104. *Graphium* state of *Petriella guttulata* (OAC 10580). (Reproduced by permission of the National Research Council of Canada, *Can. J. Bot.*, *39:* 837, 1961.)

is clear that at least three basically different forms are contained within the genus. These synnematous forms have mononematous counterparts in the genera *Phialocephala* (phialides), *Verticicladiella* (sympodulae), and *Leptographium* (annellophores). The relationships between the *Graphium* complex and these mononematous forms should be clarified.

*Graphium* species are recorded infrequently from soil (Gilman, 1957). They are commonly found on woody substrata or plant debris (see Morris, 1963). They are frequently found as the conidial states of members of the Microascales. All members of the genus *Petriella* (Barron *et al.*, 1961) have a "*Graphium*" conidial state, as have certain members of the genus *Ceratocystis* (Hunt, 1956). Goos and Timonin (1962) recorded a *Graphium* species from the rhizosphere of bananas in Honduras. In Ontario we have isolated "*Graphium*" species on a number of occasions from soil in cedar swamps. These have all produced annellophores and appear to be related to the "*Graphium*" states of *Petriella*.

## HADROTRICHUM Fuckel

*Hadrotrichum* is commonly found as a foliar parasite of higher plants (Hughes, 1953). The conidiophores are produced in a dense stand on a stroma (sporodochium) and are simple and pigmented. The conidia are produced singly and successively as blown out ends of a succession of new growing points on a sympodially extending conidiophore. The conidia of *H. phragmitis*, the type species, are spherical, nonseptate, pigmented and finely roughened (Fig. 105).

The only record of this genus from soil is by von Szilvinyi (1941), who described *H. lunziense* a new species. Gilman (1957) regarded this species as one of uncertain position, and since no type or other material is available, this would be the best disposition at present.

## HAINESIA Ellis and Sacc.

Type Species: *Hainesia rubi* (West.) Sacc.

**Generic Description:** Fructifications acervulus-like or sporodochium-like, brightly colored, frequently yellowish or pink to red, somewhat gelatinous; conidiophores filiform, simple or branched; conidia hyaline, nonseptate, oblong to suballantoid or sometimes lunate.

**Notes:** The description above of *Hainesia* is vague and could apply equally well to a number of sporodochial genera. It reflects the writer's own uncertainty on the limits of this genus. Saccardo included this genus in the Melanconiales and one would presume, therefore, that its fructification is an acervulus. Grove (1937) also regarded this fungus as melanconiaceous. In Ainsworth (1963) the fungus is cited as a member of the Sphaeropsidales and Barnett (1960) included it in this group, although his

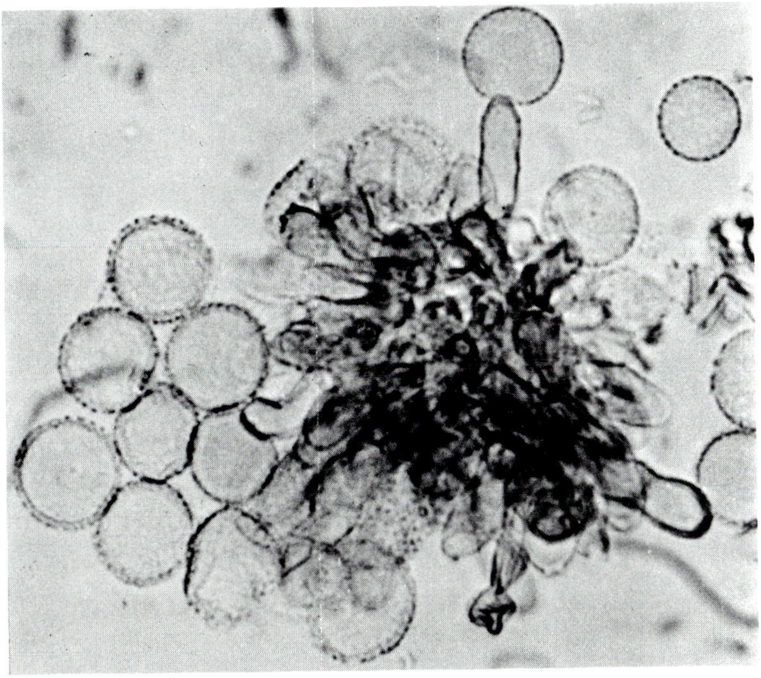

FIG. 105. *Hadrotrichum globiferum* (IMI 96960)

figures of *H. rubi* would pass equally well for an acervulus. Shear and Dodge (1921), in a detailed treatment of *Hainesia lythri*, describe it as sporodochial, but they note that this fungus can also produce a pycnidial state called *Sclerotiopsis concava* in which the conidiophores and conidia are identical with those produced on the sporodochium of *H. lythri*.

*H. lythri* (Fig. 106) was recorded from soil by dal Vesco (1960). This species is the conidial state of *Pezizella lythri* (Shear and Dodge, 1921) a fungus of wide distribution found as a weak parasite of about fifty different host plants in Europe and America.

According to Grove, the generic name *Hainesia* has been much misused. He suggested confining the name to *H. rubi* and similar forms and questioned whether the fungi placed with *H. rubi* in the third and other volumes of the Sylloge really belong there. *H. rubi* is cited by Grove as a hyperparasite on the uredosori of *Phragmidium*.

## HANSFORDIA Hughes

Type Species: *Hansfordia ovalispora* Hughes.

**Generic Description:** Conidiophores hyaline or pigmented, erect or

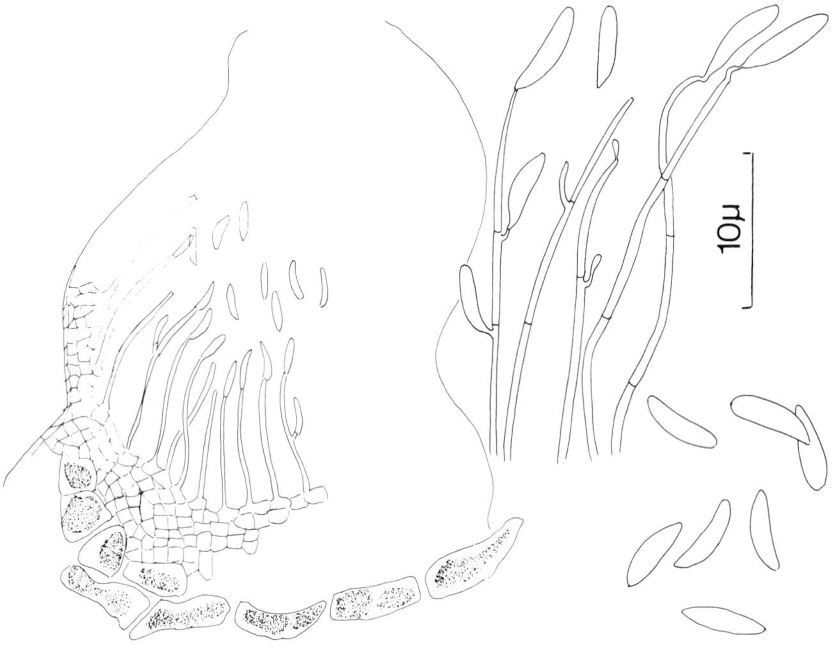


FIG. 106. *Hainesia lythri* (IMI 80261)

repent, straight or bent, lateral branches primarily fertile, sometimes sterile or with a sterile apex, single or in pairs; secondary branches in pairs and unilateral; branches terminate in one to three sporogenous cells; conidia acropleurogenous, arise singly from truncate denticles, continuous, hyaline, dry, spherical or ovate to fusoid, smooth or slightly roughened. (Description based on Hughes, 1951.)

**Diagnostic Features:** The hyaline spores, borne in acropetal succession on conspicuous denticles, are characteristic of *Hansfordia.* The long, tapering, sterile apex found in some species is also diagnostic.

**Notes:** According to Hughes (1951), the main distinguishing character of *Hansfordia* is the denticulate, more or less cylindrical sporogenous cell. The denticles, he noted, are not restricted to the immediate apex, as in *Calcarisporium.* The primary branches are generally solitary on the main axis, but the secondary and tertiary branches are usually in pairs, unilateral and directed upwards. These branches may terminate in two or three sporogenous cells, but of these one or two are lateral and the other the modified continuation of the branch. True verticils of three or more laterals arising at a common level, are absent. This lack of verticillate arrangement of the sporogenous cells distinguishes *Hansfordia* from the closely related *Nodulisporium* and *Calcarisporium. Hansfordia* has not been recorded from soil. It is included here to allow comparisons with related genera from soil.

## HAPLOBASIDION Eriksson

Type Species: *Haplobasidion thalictri* Eriks.

**Generic Description:** Conidiophores arise singly or in groups at the ends of hyphae or as lateral branches, erect, simple, flexuous, subhyaline to brown, forming a vesicle at the apex, lower part of the vesicle dark brown, upper part pale brown; sporogenous cells borne on the upper part of the vesicle, subspherical or ovoid or clavate, subhyaline to brown, smooth or verruculose; conidia formed acropetally in simple or branched chains on the sporogenous cells or directly on the vesicle, often separated from one another by a short isthmus, spherical or subspherical, subhyaline to brown, verruculose. (Description based on Ellis, 1957.)

**Diagnostic Features:** The formation of subspherical conidia in branched acropetal chains from a vesicle or from sporogenous cells on the vesicle is diagnostic for *Haplobasidion* (Fig. 107).

**Notes:** *Haplobasidion* and related genera have been given a detailed treatment by Ellis (1957), who noted that when the sporogenous cells and the conidia become detached from the vesicle, they sometimes leave well defined, annular scars on its surface; the pale, thin-walled upper part of

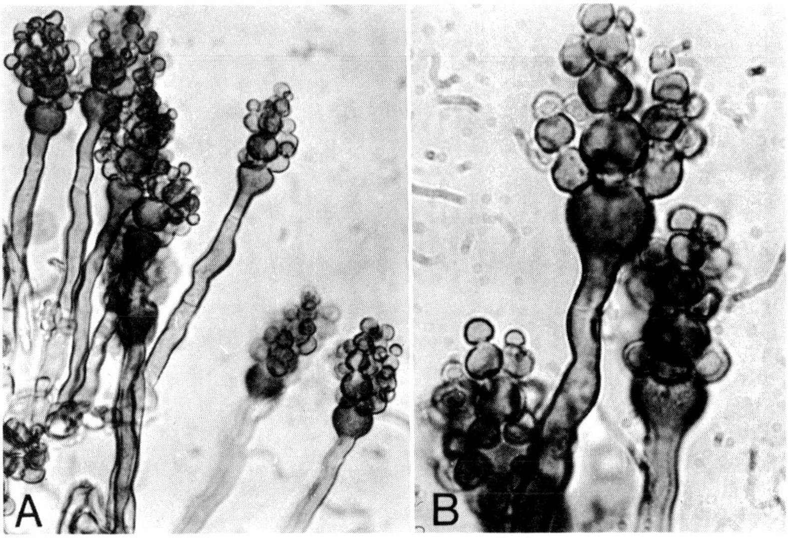


FIG. 107. *Haplobasidion* sp. *A* and *B*, branching chains of dark, spherical conidia are borne on the swollen apex of each conidiophore (IMI 88790).

the vesicle later becomes flattened or invaginated and the conidiophore may proliferate through the apex to form another globose fertile cell at a higher level. *Haplobasidion* differs from *Lacellina* in lacking setae. It shows close relationships to *Periconia* but differs in the acropetal maturation of the conidia.

The only record of this genus from soil is by Williams and Schmitthenner (1956).

### HAPLOGRAPHIUM Berk. and Broome.

Type Species: *Haplographium delicatum* Berk. and Broome.

**Generic Description:** Conidiophores simple, erect, stout, septate, brown; main stalk bears an apical cluster of up to nine oval, pale-brown metulae; primary metulae each bear a cluster of up to eight narrower, hyaline to subhyaline, secondary metulae; a tertiary series of hyaline metulae may also be produced; the ultimate metulae function as sporogenous cells; each sporogenous cell bears an apical conidium produced as a blown out end and three or four subterminal or lateral conidia, chains not formed; spores gather to form a slimy drop. (Description based on Hughes, 1953.)

**Diagnostic Features:** This genus is characterized by slimy masses of blastospores borne singly at the apex of dark conidiophores with penicillate heads (Fig. 108).

**Notes:** A detailed account of the conidiophore structure and spore formation in *H. delicatum* was given by Hughes (1953), who showed by

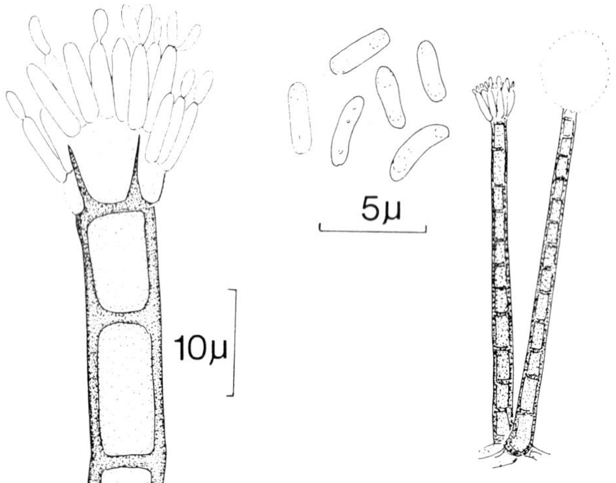


FIG. 108. *Haplographium bicolor*. Erect conidiophores produce several series of metulae bearing solitary blastospores at the apex (IMI 89617).

pure culture techniques that *H. delicatum* is the conidial state of *Hyaloscypha dematiicola.* Hughes considered *H. delicatum* to be homologous with *Cladosporium*, but having a better differentiated conidiophore and condensed and regularly whorled lateral branches. He pointed out that the crowded primary and secondary branches (metulae) have developed into permanent structures and fail to secede at maturity but are equivalent to the ramoconidia of *Cladosporium*. The tertiary metulae may become detached and function as conidia.

*Haplographium* species have been recorded infrequently from soil (Gilman, 1957). It is probable that most of the isolates cited in floristic listings under this genus do not belong here. They are probably *Leptographium*, *Phialocephala* or *Verticicladiella* isolates.

## HARPOGRAPHIUM Sacc.

Type Species: *Harpographium fasciculatum* Sacc.

**Generic Description:** Hyphae aggregated into more or less erect synnemata; synnemata sterile below, fertile in the upper part, simple or branched, cylindric to clavate, darkly pigmented; sporogenous cells polyphialides; spores produced terminally on the sporogenous cell which then elongates sympodially from a point just behind the apex and quickly produces a second apex, and so on; conidia hyaline, nonseptate, fusoid to falcate.

**Diagnostic Features:** *Harpographium* is a synnematal fungus producing fusoid to falcate, hyaline amerospores from polyphialides (Fig. 109).

**Notes:** This genus is rarely recorded from soil. There is a single record at the Commonwealth Mycological Institute of a *Harpographium* species isolated from soil in England by M. N. Okafor. The only published record is that of Meyer (1959), who recovered a member of the genus in the former Belgian Congo.

*Harpographium* has an interesting method of spore formation. After several successive collarettes have been produced on the polyphialide, it may go through a sterile phase of growth then begin to produce a second series of collarettes. This gives the mature synnemata a distinctly zonate appearance.

## HARPOSPORIUM Lohde

Type Species: *Harposporium anguillulae* Lohde.

**Generic Description:** Conidiophores lacking or poorly developed; phialides scattered or clustered on conidiophores or hyphae, hyaline, minute, globose to subglobose or ovoid, produced singly or in groups, frequently in dense clusters; phialides may proliferate through the apex to produce secondary, tertiary and quaternary series; phialospores nonseptate, hya-

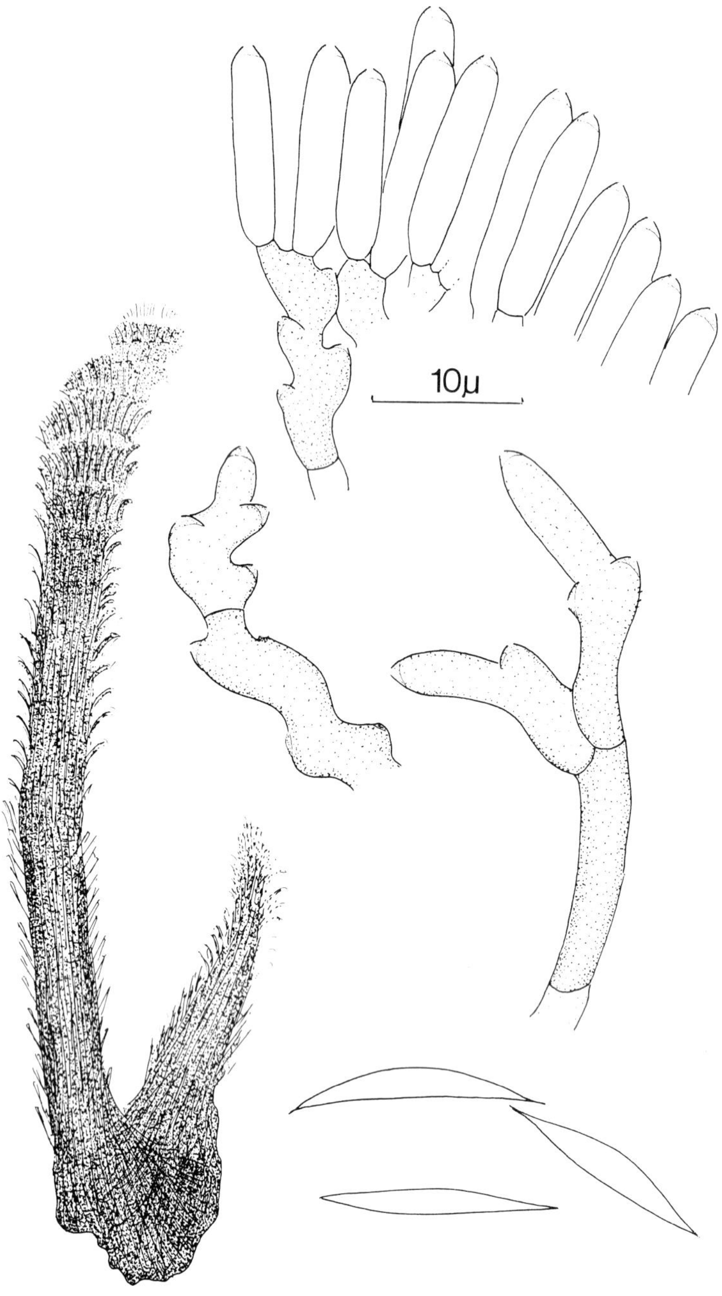


FIG. 109. *Harpographium constrictum*. The falcate phialospores are produced from polyphialides (IMI 74774).

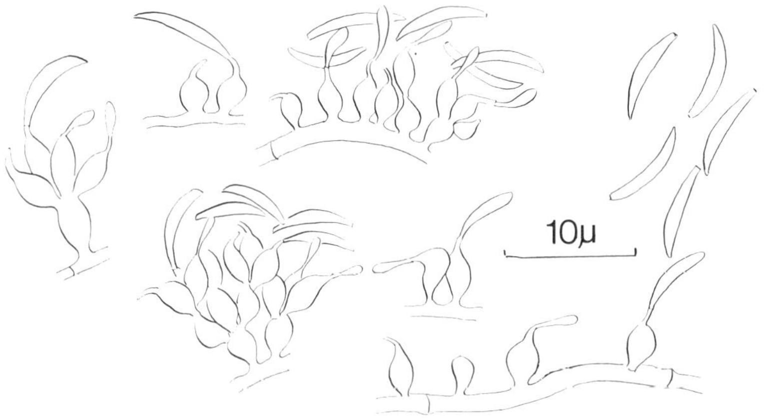


FIG. 110. *Harposporium lilliputianum*. The phialides are swollen and have a short, reflexed, beak-like extension. Frequently the phialides proliferate through the mouth to form secondary or tertiary series giving rise to a sporodochial-like fructification (OAC 10154).

line, usually elongated, curved or hooked, frequently acutely pointed at the distal end, gathering in mucoid masses.

**Diagnostic Features:** This genus is characterized by the tiny, globose to subglobose, sessile phialides producing hyaline, usually curved amerospores (Fig. 110).

**Notes:** *Harposporium* is commonly found as a parasite of nematodes (see Cooke and Godfrey, 1964). It is seldom found in floristic lists of soil fungi but has been recorded from this substratum by Brown (1958). We have isolated a *Harposporium* species which agrees well with *H. lilliputianum* on several occasions from cultivated peat soil (muck soil) in Ontario.

### HELICODENDRON Peyronel

Type Species: *Helicodendron paradoxum* Peyr.

**Generic Description:** Conidiophores simple or much-branched, sometimes lacking, hyaline or fuscous; conidia in chains, coiled in three planes to form a cylindrical, barrel-shaped or ellipsoid spore, occasionally only few seriate, then form an almost disk-shaped body, hyaline or pigmented; sometimes conidia break down into arthrospore-like segments which can function as conidia.

**Diagnostic Features:** Spores coiled in three dimensions and produced in chains are characteristic for *Helicodendron* (Fig. 111).

**Notes:** *Helicodendron* was erected by Peyronel (1918) with *H. paradoxum* as the type. The generic concept was extended by Linder (1929) to

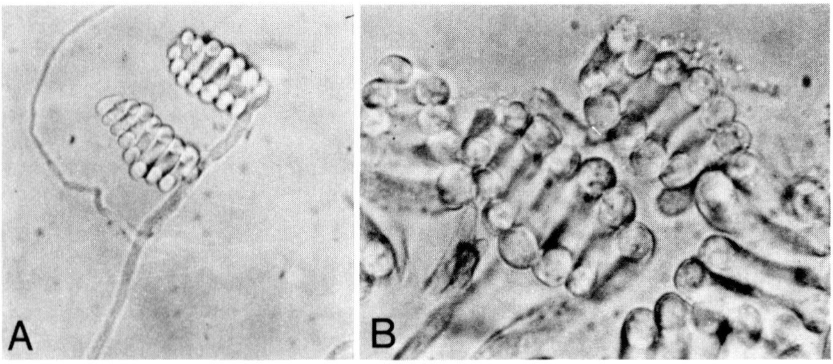

FIG. 111. *Helicodendron tubulosum*. *A*, conidiophores are little differentiated from the vegetative hyphae in culture. *B*, conidia are produced in acropetal succession to form clusters (OAC 9401).

include all species with the spores coiled to form a helix and in which the conidia proliferate to form irregular chains or clusters.

The helicoid conidia are large and distinctive, hyaline to brown in color and borne on slender conidiophores. Inconspicuous phialospore states have been reported for several species of the genus (Glenn-Bott, 1955; Barron, 1961).

This genus is rare in soil. A *Helicodendron* species was recorded by Warcup (1957) and *H. tubulosum* has been reported by Barron (1961). *H. tubulosum* is perhaps the most common and widespread species of *Helicodendron*; it occurs commonly in an aquatic habitat, growing saprophytically on dead leaves on the bottom of ponds and streams. The conidia of *H. tubulosum* are large, measuring up to 30 by 60 $\mu$, and coiled counterclockwise to form a helix of up to 15 gyres (Fig. 6*E*).

## HELICOON Morgan

Type Species: *Helicoon sessile* Morgan.

**Generic Description:** Conidiophores more or less erect, sometimes lacking, simple or branched; conidia hyaline, brightly colored or brown, filaments coiled in three planes to form barrel-shaped, cylindrical or ellipsoid spores, never in chains.

**Diagnostic Features:** In *Helicoon* the spores are solitary and coiled in three planes (Fig. 112).

**Notes:** The genus *Helicoon* is rarely recorded from soil. The only published record is that of Kamyschko (1963), who described *H. spiralis* from soil in the USSR. The limits of *Helicoon* are not clear. I have examined an isolate of *H. multiseptatum*, and in this species the conidia are aleuriospores which are released by rupture of one of the subtending cells.

The type species of the genus, *H. sessile*, is illustrated by Linder (1929) and the conidia could be aleuriospores. Some of the species placed in *Helicoon* by Linder, however, could have sympodulospores. *H. multiseptatum* is similar in many respects to *Helicosporina veronae*, except that in the latter the conidia are coiled in two planes. There is a need for a reevaluation of the helicosporous Fungi Imperfecti on the basis of method of origin rather than on method of coiling.

### HELICOSPORINA Arnaud

Type Species: *Helicosporina globulifera* Arnaud.

**Generic Description:** Mycelium septate, composed of brown or yellow-brown hyphae; conidiophores more or less erect, little distinguished from the vegetative hyphae, simply or sparingly branched, septate, terminal portion coils in two dimensions to form an aleuriospore; conidiophore continues growth sympodially from a point below the spore attachment, successive coiling and proliferation of the conidiophore produces an acropetal series of coiled spores; conidia pale brown, tightly coiled, septate, developing apically and producing successive conidia acrogenously by sympodial proliferation of the conidiophore. (Description based on *H. veronae*.)

**Diagnostic Features:** *Helicosporina* is a Hyphomycete in which the

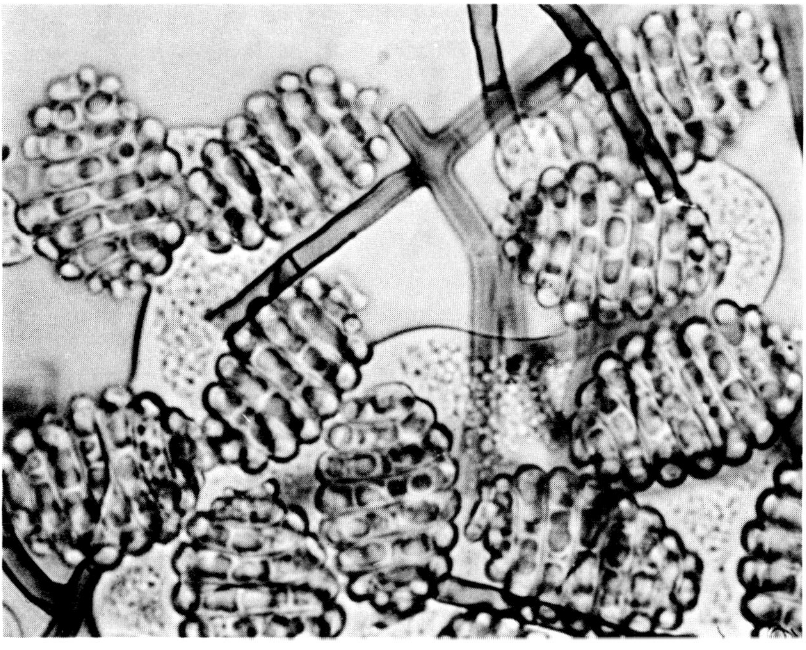

FIG. 112. *Helicoon* sp. (IMI 17261)

conidia develop as aleuriospores by the simple coiling of the apex of the conidiophore or its branches to form a two-dimensional coil (Fig. 113).

**Notes:** *Helicosporina* was erected by Arnaud (1953) based on *H. globulifera*. Neither genus nor species was validated by Arnaud with a Latin diagnosis. Rambelli (1960) recovered a fungus from soils beneath *Eucalyptus* in Italy, which he described as a new species, *H. veronae*. Rambelli validated Arnaud's genus and his own species with Latin descriptions. His generic description of *Helicosporina* was based largely on *H. veronae*. In this species the conidia are terminal aleuriospores which are released by rupture of one of the conidiophore cells near the point of attachment. In *H. globulifera* the description of Arnaud is inadequate. From his diagram of the fungus however, it is possible that the spores are borne on denticles. If such is the case, *H. veronae* and *H. globulifera* may not be congeneric!

## HELICOSPORIUM Nees

Type Species: *Helicosporium vegetum* Nees.

**Generic Description:** Conidiophores conspicuous, dilute brown to fuscous, simple to much branched; conidia hyaline, light-colored or fuscous, filaments slender, hygroscopic, coiled in two dimensions to form a disk-shaped body.

**Diagnostic Features:** The slender, hygroscopic nature of the helicoid spores produced on conspicuous conidiophores is distinctive for *Helicosporium* (Fig. 114).

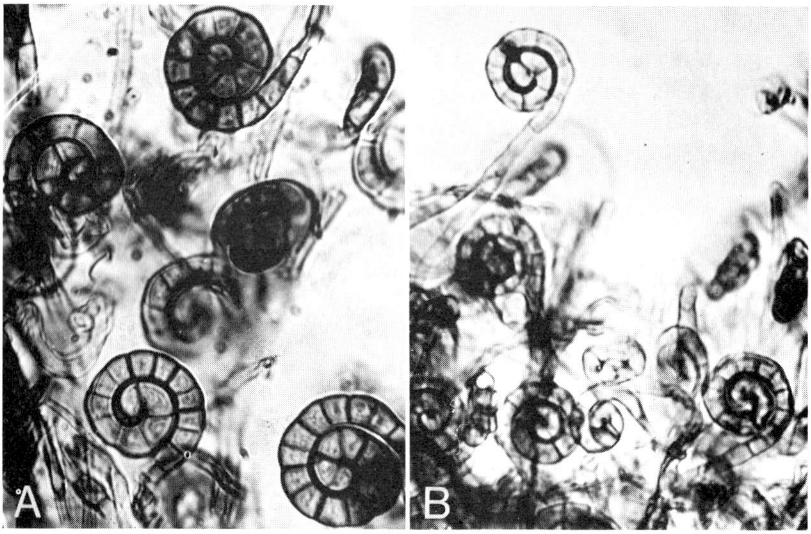


FIG. 113. *Helicosporina veronae*. *A* and *B*, the conidia are formed by coiling of the apex of a sporogenous branch, *i.e.*, aleuriospores (OAC 10328).

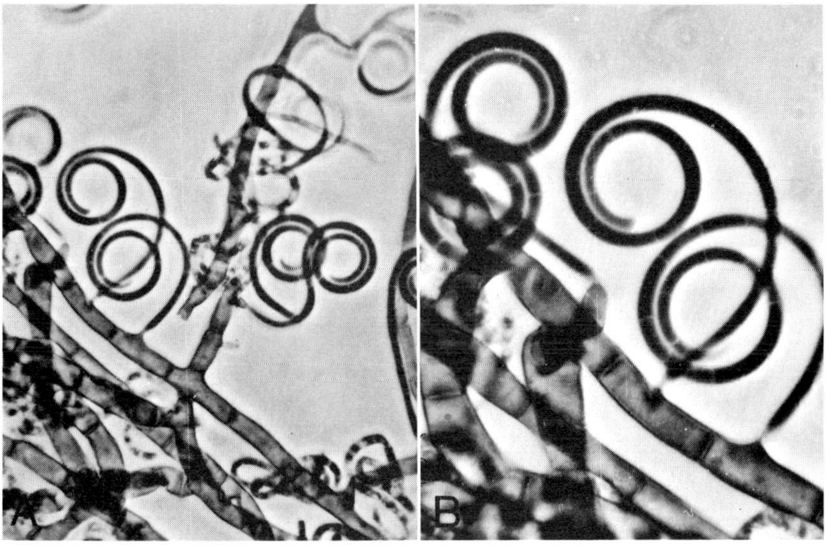


FIG. 114. *Helicosporium lumricoides*. *A* and *B*, the conidia in this species have slender, loosely coiled filaments and are borne on short pegs from the conidiophore or sporogenous cells (IMI 49372).

**Notes:** This genus is separated from *Helicomyces* on the basis of elongate and conspicuous conidiophores and from *Helicoon* on the slender, hygroscopic nature of the conidial filaments. *H. vegetum* is the conidial state of *Ophionectria cerea*. *Helicosporium* is infrequently recorded from soil but has been found by Warcup (1957) in wheatfield soil in England.

### HELMINTHOSPORIUM Link ex Fries

Type Species: *Helminthosporium velutinum* Link ex Fries.

**Generic Description:** Colonies effuse, dark, hairy; mycelium immersed in the substratum; stromata usually present; conidiophores often fasciculate, erect, brown to dark brown; conidia develop laterally often in verticils, through pores beneath the septa, while the tip of the conidiophore is actively growing and growth of the conidiophore ceases with the formation of terminal conidia; conidia formed singly (also in short chains in one species), subhyaline to brown, usually obclavate, pseudoseptate, frequently with a dark brown to black protruding scar at the base. (Description from Ellis, 1961.)

**Diagnostic Features:** In *Helminthosporium* the large, septate porospores are produced apically or laterally in verticils from determinate conidiophores (Fig. 115).

**Notes:** The type species of *Helminthosporium*, *H. velutinum*, was examined by Hughes (1953), who noted that the conidia develop termi-

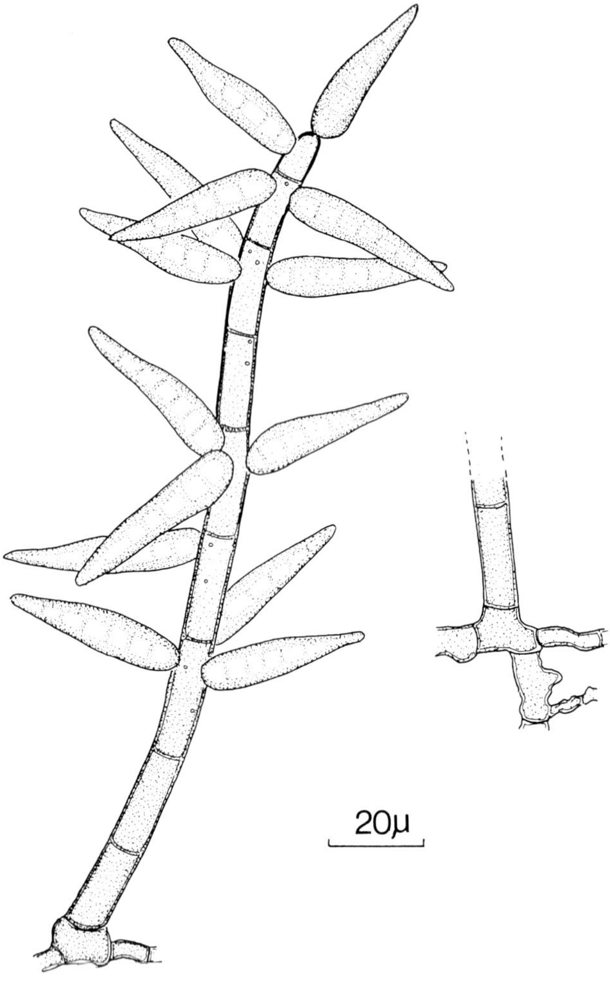


FIG. 115. *Helminthosporium solani*. The conidia are acropleurogenous. The lateral conidia are produced in whorls just below the upper septa of the conidiophore.

nally or laterally in more or less regular verticils below the terminal conidium and below the uppermost septa of the conidiophore. Ellis (1961) pointed out that the lateral conidia are produced while the conidiophore is still growing and that growth of the conidiophore ceases with formation of the terminal conidium. Ellis drew attention to the similarities of *Helminthosporium* and *Exosporium*. In *Exosporium*, however, no conidia are formed laterally beneath the septa. One or more conidia develop through a pore or pores in the thickened apex of the conidiophore. The conidiophore then grows out laterally below the scar, splitting the side wall, pushing the scar to one side, then growing for some distance before producing other conidia through a pore or pores at the newly constituted apex. Ellis pointed out the similarities between spore formation in *Exosporium*, *Bipolaris* and *Drechslera*. Ellis considered it convenient to retain the genus *Exosporium* for species which have this kind of conidiophore development, often well developed stroma and conidia which are very dark, obclavate, pseudoseptate with a thick, dark brown to black, often protruding scar at the base. He noted that many such species are found on old branches especially in tropical regions.

Ito (1930) proposed the name *Drechslera* for those species of *Helminthosporium* with cylindrical conidia that germinate from any cell. *Drechslera* was accepted and extended by Shoemaker (1959), who redefined the genus and at the same time segregated the fusoid-spored forms, which germinate only from the end cells, into a new genus, *Bipolaris*.

Luttrell (1963) gave a detailed comparative account of spore formation in selected species of *Helminthosporium* and related genera. He proposed a number of new terms for taxonomic characters derived from conidium origin, septation, maturation, germination and from conidiophore proliferation studies. Luttrell suggested that these terms may prove to have a broader application in a reorganization of the Deuteromycetes. One of Luttrell's terms, "murogenous," is the descriptive term on which the genus *Murogenella* is based.

Luttrell (1964) disagreed with Ellis (1961) and Shoemaker. While he recognized *Helminthosporium* as a heterogeneous and unwieldy assemblage in need of revision, he considered it desirable at present to maintain *Helminthosporium* in a comprehensive sense. Luttrell would like to defer from changing familiar combinations of many species of economic importance until the taxonomic and nomenclatural problems involved can be more satisfactorily defined and explored. On this basis, Luttrell retained *Helminthosporium* with three subgenera, *viz. Cylindro-Helminthosporium*, *Eu-Helminthosporium* and *Helminthosporium*. Luttrell also suggested that *Cuspidosporium* and *Podosporiella* may be prior names for *Bipolaris*.

The situation is further confounded by the fact that Hughes and Shoemaker prefer to use *Helmisporium* (Link's original spelling) in favor of the well established orthographic variant, *Helminthosporium*. Both Ellis and Luttrell favored conserving the name *Helminthosporium*, if it became necessary to even consider *Helmisporium*. Luttrell suggested further that if *Helminthosporium* must be divided, then common usage should be preserved by conserving *Helminthosporium* with *H. maydis* as the type for species in the *Cylindro-Helminthosporium* and *Eu-Helminthosporium* groups.

For those of us who are innocent bystanders on the taxonomic sidelines, the picture is very confused and the hope of a satisfactory solution seems distant.

## HETEROCEPHALUM Thaxter

Type Species: *Heterocephalum aurantiacum* Thaxter.

**Generic Description:** Conidiophores more or less erect, simple, hyaline, swelling at the apex to form a vesicle, corticating sterile hyphae growing up round the conidiophore as an envelope to give the appearance of a synnema; sporogenous branches radiating out from the vesicle and branching dichotomously or subdichotomously several times, ultimately producing clusters of sporogenous cells at the branch apices; conidia minute, nonseptate, hyaline, oval to oblong, sometimes irregular; sterile sheath extending into the apex and radiating out to give rise to numerous, straight, rigid, bristle-like hyphae; lateral, sterile branches grow tangentially and intertwine to form a basket-like container around the spore mass.

**Diagnostic Features:** This genus is readily distinguished by its bright, orange-yellow spore heads, with radiating setae, borne on synnema-like stalks (Fig. 116).

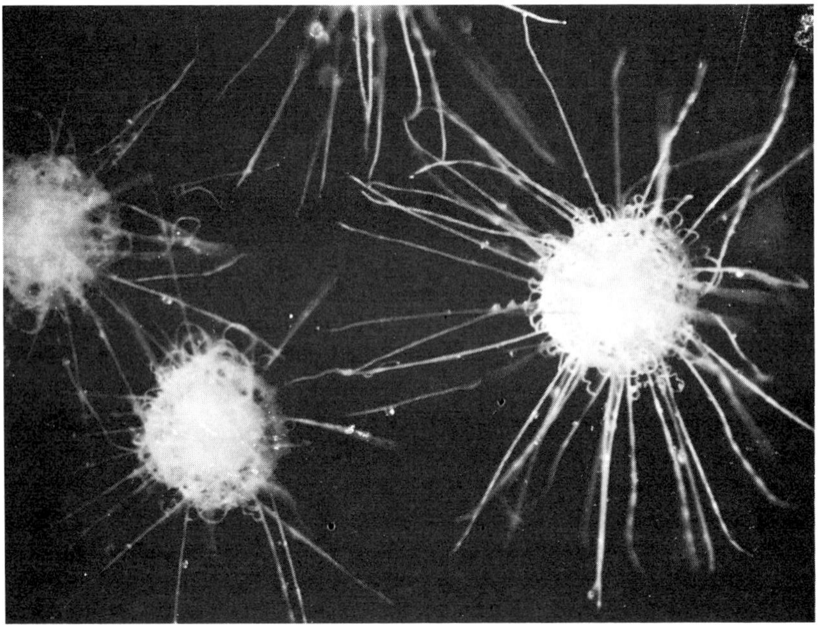

FIG. 116. *Heterocephalum aurantiacum*. The mature fructification simulates the ascocarp of the Gymnoascaceae (IMI 62339).

**Notes:** The genus *Heterocephalum* is one of the most striking of the Hyphomycetes. The mature spore head has the appearance of a cleistothecium of the Gymnoascaceae. The original descriptions of Thaxter (1903) were based on two isolates, one from toad dung collected in Jamaica and the other from goat dung collected in the Phillipine islands.

The genus is monotypic and has been recorded from soil on few occasions. Farrow (1954) recorded it from the Barro Colorado Islands in her studies on tropical soil fungi and Raper and Fennell (1952) recorded it from soil in Liberia. More recently it has been reported from Okinawa by Indoh and Oyatsu (1965), and by Meyer (1959) from soil in the former Belgian Congo.

In most treatments *Heterocephalum* is considered under the Stilbaceae. It was pointed out by Raper and Fennell, however, that this fungus shows affinities to the genus *Aspergillus*. I am not convinced, however, that the sporogenous cells in *H. aurantiacum* are phialides. It has been suggested by both Thaxter and by Raper and Fennell that the complex development of the sterile hyphae is possibly related to spore dispersal.

## HIRSUTELLA Patouillard

Type Species: *Hirsutella entomophila* Patouill.

**Generic Description:** Synnemata cylindric to filiform, terete, usually somewhat attenuated upward, simple or branched, consisting of more or less parallel, septate hyphae; phialides scattered to crowded over most of the synnema, mostly arising as lateral cells or buds or terminating short lateral branches along the outer hyphae of the synnema, occasionally developing on hyphae from the mycelial covering of the host, hyaline, inflated below, abruptly or gradually narrowing into long slender sterigmata; conidia oblong, subcylindric, fusoid to cymbiform, one-celled, hyaline, covered by a persistent mucus, single, or two to several occurring in droplets. (Description from Morris, 1963.)

**Diagnostic Features:** *Hirsutella* is characterized by sporogenous cells which are inflated at the base and taper to a long, narrow apex bearing one or several mucus-coated phialospores (Fig. 118).

**Notes:** *Hirsutella* has been reported only once from soil by Sewell (1959). Members of the genus are commonly found as parasites of insects and have been reported as the conidial states of *Cordyceps*, *Ophiocordyceps* and *Torrubiella* (Morris, 1963).

On its natural substratum, *Hirsutella* is a synnematous fungus and synnemata may also be produced in culture (Loughheed, 1963). Morris reports that the sporogenous cells may also arise directly from the hyphae on insect hosts and this may well be true in some cultures of this genus.

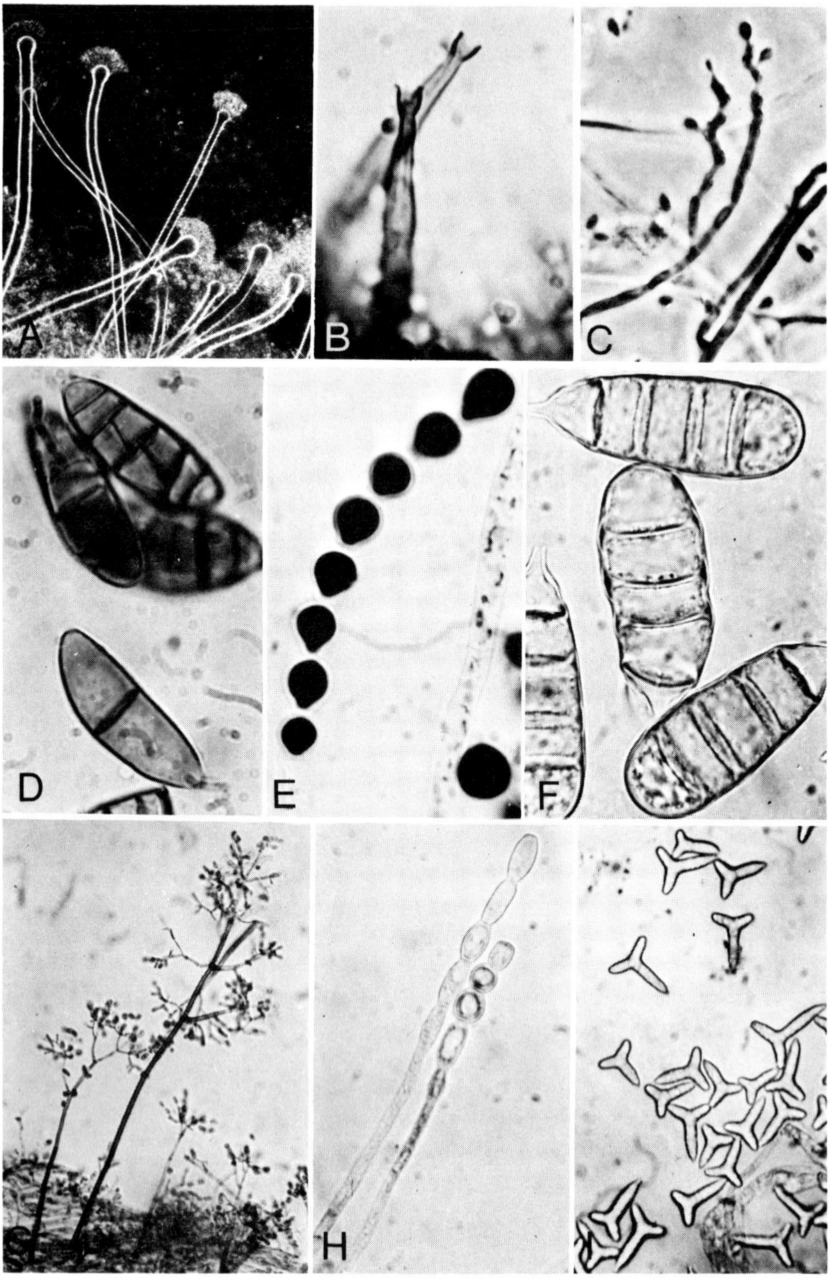


FIG. 117. Miscellaneous. *A*, *Gliocephalis hyalina*. *B*, Conidial state of *Chaetosphaeria mycriocarpa* (see *Chloridium*). *C*, *Geniculisporium* state of *Hypoxylon serpens*. *D*, *Dactylosporium macropus*. *E*, Conidial state of *Monascus purpureus*. These are arthrospores which resemble the annellospores of *Scopulariopsis*. *F*, Conidia of *Cephaliophora tropica*. *G*, *Oidiodendron truncatum*. *H*, *Geotrichum* sp. *I*, Conidia of *Scolecobasidium terreum*.

## HISTOPLASMA Darling

Type Species: *Histoplasma capsulatum* Darling.

**Generic Description:** Conidiophores poorly developed, sometimes lacking, short and simple or sparingly branched, bearing aleuriospores terminally or laterally on the branches; aleuriospores large, sometimes smooth-walled, more frequently tuberculate.

**Diagnostic Features:** The common *H. capsulatum* is distingusihed by the large (10 to 25 $\mu$), guttulate aleuriospores bearing finger-like protuberances (1 to 7 $\mu$ long) over the surface (Fig. 119).

**Notes:** The ornamentation of *Histoplasma* spores is quite striking in appearance. According to Howell (1939), softenings or pits appear in the walls through which papillate protoplasmic outgrowths protrude to form an indefinite number of finger-like processes.

According to Emmons *et al.* (1963), the natural habitat of *H. capsulatum* is soil or composted plant material enriched with fecal or other nitrogenous substances. It was first isolated from soil by Emmons (1949), but has since been recorded frequently from this source from suspected exposure sites of patients suffering from histoplasmosis, and is particularly common in or near poorly constructed chicken houses, barnyard soil, and so on. The disease is a serious intracellular mycosis which involves lymphatic tissues, lung, spleen, liver, adrenals, kidneys, skin, central nervous

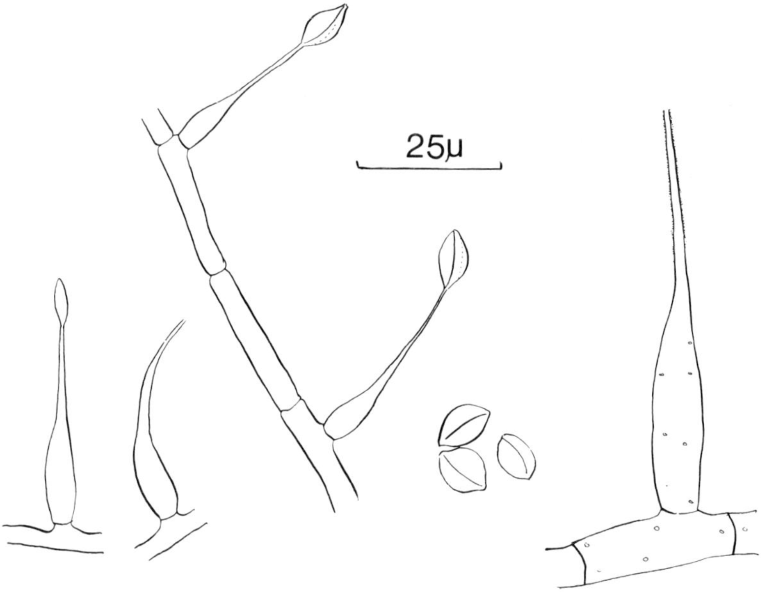


FIG. 118. *Hirsutella* sp. In this soil isolate synnemata were not produced. The phialides have a long tapering neck and produce phialospores with a mucous sheath (IMI 55839).

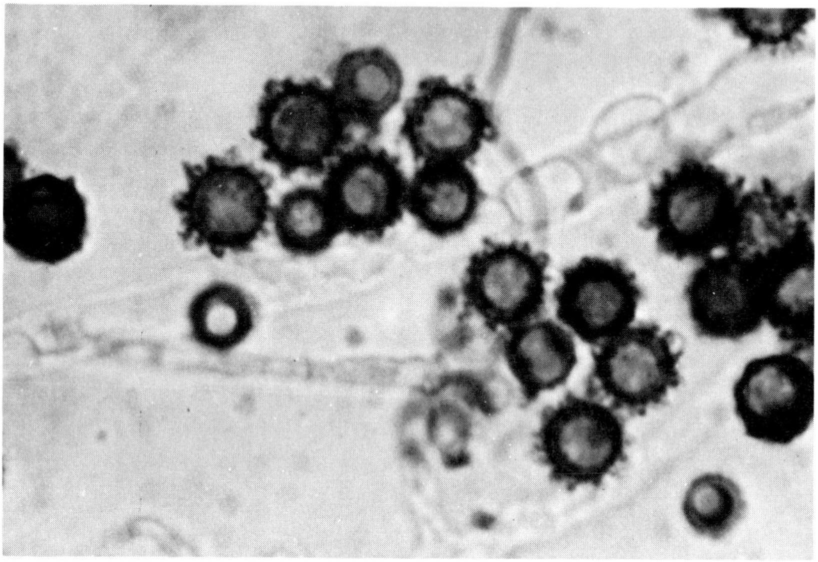

FIG. 119. *Histoplasma capsulatum*. The spherical aleuriospores are characterized by finger-like protuberances protruding over the surface of the spore. (From a slide supplied by Dr. R. Goos.)

system and other organs of the body (Emmons *et al.*). Man is usually infected through inhalation of spores.

Growth and survival of *H. capsulatum* in soil was studied by Goos (1965), who found that this fungus was sensitive to antagonistic biological effects, but that conidia could survive as long as eight weeks in unsterilized soil. Ajello and Cheng (1967) have reported the perfect state of this fungus as belonging to the genus *Gymnoascus*.

## HORMIACTIS Preuss

Type Species: *Hormiactis fusca* and *Hormiactis alba* described.

**Generic Description:** Conidiophores more or less erect, hyaline, septate, consisting of a simple, main axis branching irregularly above, tree-like; conidia blastospores, produced in acropetal succession to form branching chains, hyaline, septate, smooth, connected by a narrow isthmus.

**Diagnostic Features:** This genus is characterized by branching chains of hyaline, septate blastospores with narrow connectives (Fig. 120).

**Notes:** *Hormiactis* is very similar in appearance to *Hyalodendron*. It differs in lacking the typical dark connectives. It is reported from soil infrequently and has only been recorded from England by Stenton (1953) and Chesters (1960). In our laboratory we have isolated a member of this genus from soil only once.

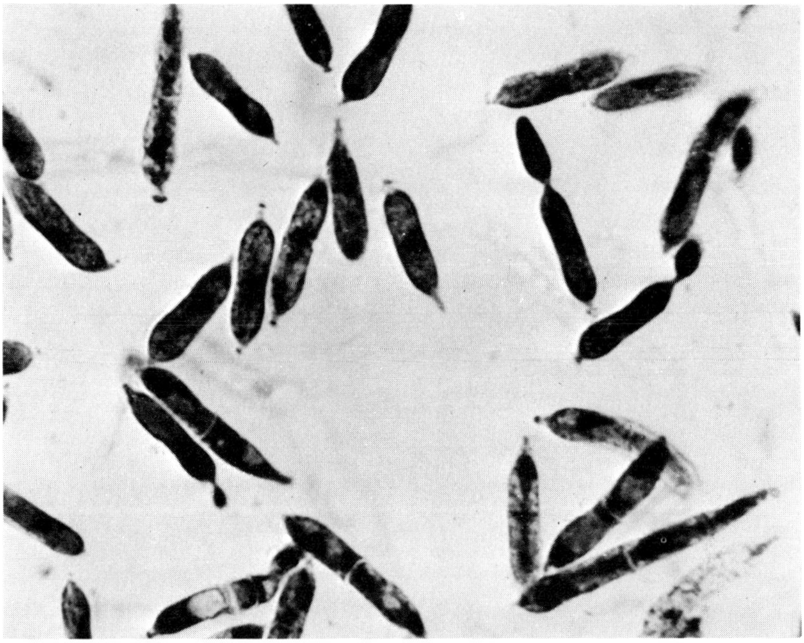

FIG. 120. *Hormiactis* sp. (OAC 10256)

## HUMICOLA Traaen

Type Species: *Humicola fuscoatra* Traaen.

**Generic Description:** Vegetative hyphae hyaline to pigmented; conidiophores undistinguished; conidia aleuriospores, borne directly on the vegetative hyphae or on short lateral conidiophores; conidiophores continuous or sparingly septate, cylindric or slightly inflated, swelling apically to form globose or subglobose aleuriospores; aleuriospores nonseptate, dark brown, globose to ovoid, borne singly or sometimes in short chains; phialospore state associated with some species.

**Diagnostic Features:** This genus is characterized by the large, dark, more or less spherical aleuriospores borne singly on short stalks (Fig. 121).

**Notes:** *Humicola* was established by Traaen (1914) on two species isolated from Norwegian soils. *H. fuscoatra*, the type species, has aleuriospores measuring 6 to 9 by 12 μ, and in *H. grisea* the aleuriospores measure 12 to 17 μ in diameter. *H. brevis* was described from soil by Gilman (1957), and Omvik (1955) described *H. nigrescens* from this same habitat. The taxonomy of the genus *Humicola* is confused and a contribution elucidating the relationships and differentiating the described species on the basis of a comparative study would be welcome. Bunce (1961) described *H. stellatus* as a thermophilic fungus from hay, but more recently Apinis (1963) transferred this fungus to *Thermomyces*.

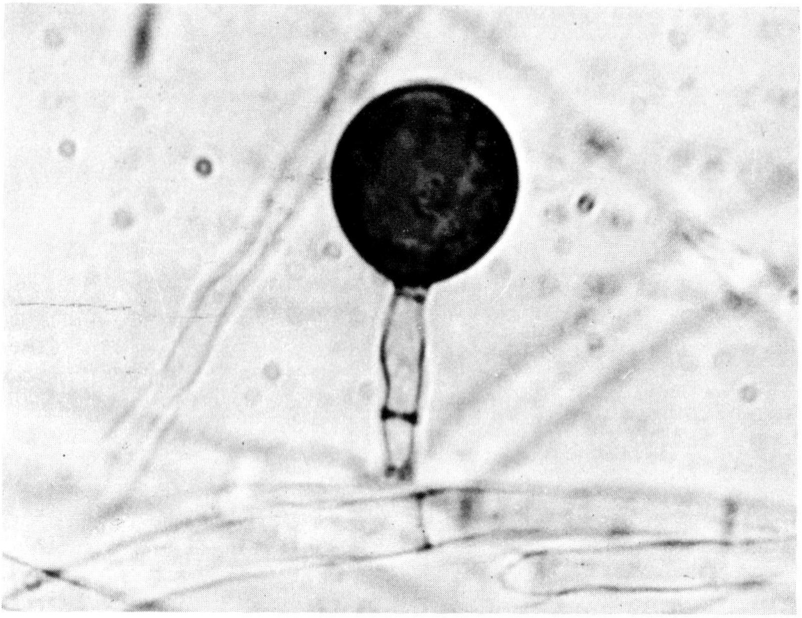

FIG. 121. *Humicola grisea*. The dark, spherical aleuriospores are borne singly on short branches.

There have been a number of nomenclatural problems surrounding *Humicola* and *Monotospora* Corda. These are discussed in detail by White and Downing (1953), and their acceptance of *Humicola* to include these and similar forms seems a satisfactory one.

*Humicola* species are apparently very common in soils, especially those high in organic matter, and they are known to be strongly cellulolytic.

## HYALODENDRON Diddens

Type Species: *Hyalodendron album* and *Hyalodendron lignicola* described.

**Generic Description:** Conidiophores erect or nearly so, hyaline, septate, branching irregularly at the apex to give tree-like heads; conidia blastospores, produced from the ultimate branches in acropetal succession by budding to give branching chains of spores; conidial chains very fragile, breaking up readily into short units of one to several cells, hyaline.

**Diagnostic Features:** This genus is readily recognized as being essentially a hyaline *Cladosporium* (Figs. 122 and 123).

**Notes:** *Hyalodendron* has been reported as the conidial state of certain *Ceratocystis* species by Goidanich (1935), but has rarely been recorded from soil. We have isolated *Hyalodendron* species on only one occasion

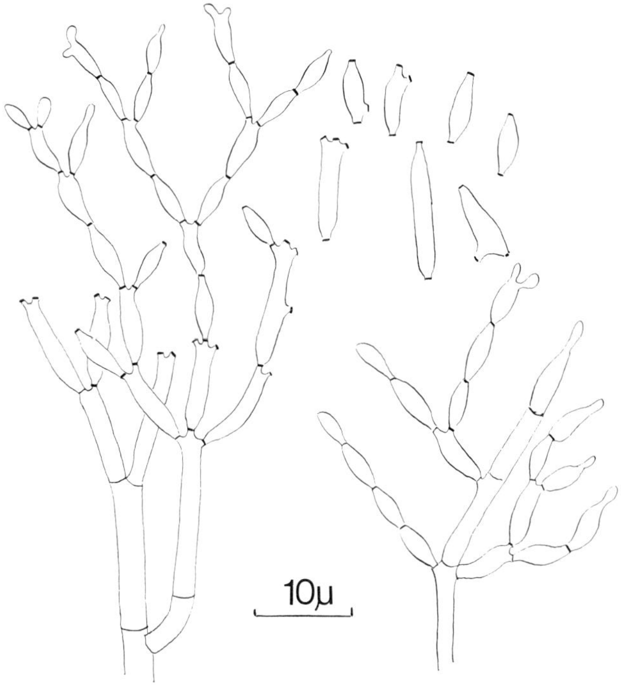


FIG. 122. *Hyalodendron* sp. This isolate from soil is completely hyaline producing white colonies; the similarities to *Cladosporium* are striking. As in Cladosporium, the end walls are highly refractive (OAC 10294).

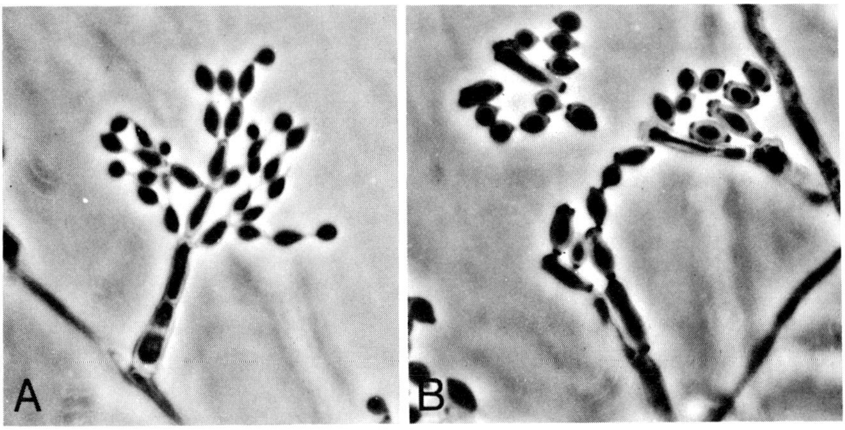


FIG. 123. *Hyalodendron* sp. *A* and *B*, the apparently thick walls are an artifact. The cells are plasmolyzed. (From a slide supplied by Mr. G. C. Bhatt).

in Ontario. This species produced dense, slow-growing colonies in culture which were pure white in color but changed to dirty white or very pale brown in age. The conidia were produced in very fragile chains so characteristic of *Cladosporium* and fell apart at a touch. The spores were hyaline but with dark, highly refractive end walls. An unidentified *Hyalodendron* species was recorded by Meyer (1959) from the former Belgian Congo.

I have not examined any type or authenticated material of the two species originally described under *Hyalodendron* and I am therefore uncertain as to the true nature of this genus. The popular concept of this genus suggests a hyaline *Cladosporium*-like fungus, but this needs confirmation by examination of the type material.

## IDRIELLA Nelson and Wilhelm

Type Species: *Idriella lunata* Nelson and Wilhelm.

**Generic Description:** Conidiophores short, simple, nonseptate, broader and somewhat inflated at the base, tapering towards the apex, conidia produced acropetally in dry heads; apex of sporogenous cell denticulate; conidia lunate to falcate, hyaline, with acutely pointed ends; chlamydospores dark brown, one- to several-celled, sessile or stalked, borne laterally on the hyphae. (Description based on *I. lunata*.)

**Diagnostic Features:** Dry heads of falcate spores borne on short sporog-

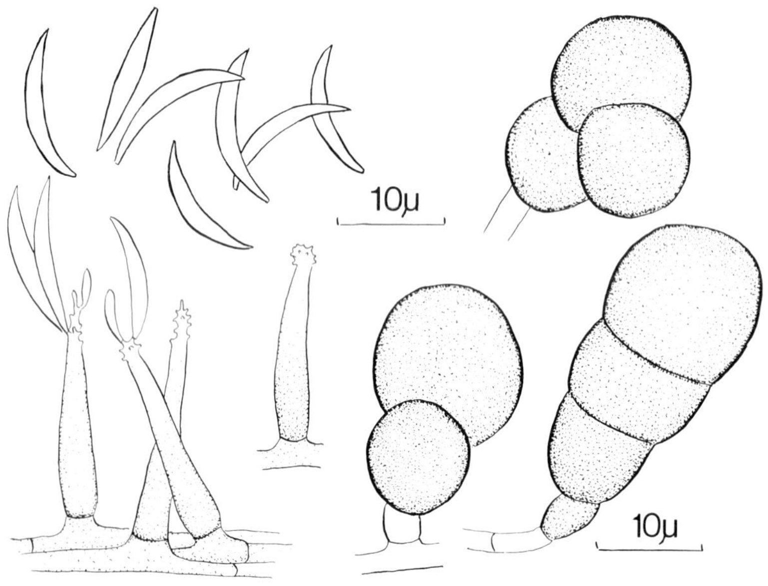


FIG. 124. *Idriella lunata*. Conidiophores may swell up or elongate at the apex with successive spore production. Falcate conidia are borne in acropetal succession (OAC 10094).

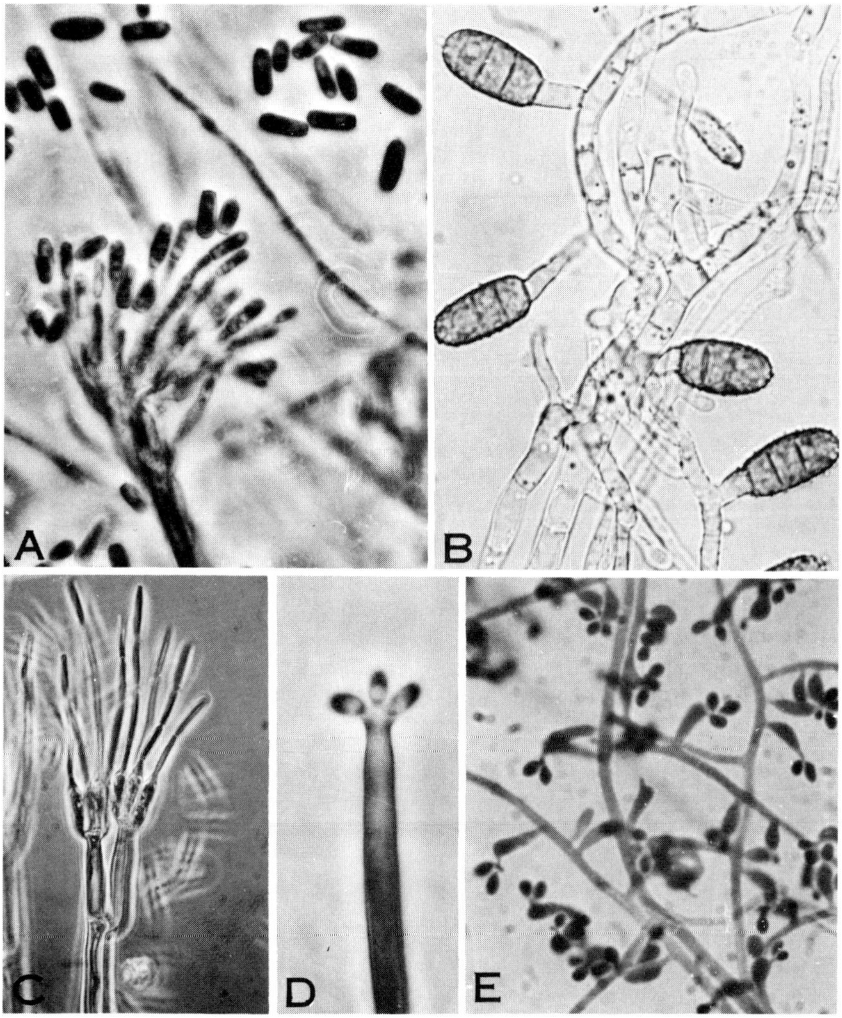


FIG. 125. Miscellaneous. *A*, *Graphium* sp. *B*, *Pithomyces chartarum* (young aleuriospores). *C*, *Thysanophora longispora* (courtesy of W. B. Kendrick). *D*, *Chloridium chlamydosporis*. *E*, *Isaria cretacea*.

enous cells which are denticulate at the apex characterize the genus *Idriella* (Fig. 124).

**Notes:** This genus was recorded originally by Nelson and Wilhelm (1965) from strawberry roots in California. Only the type species has so far been described. We have recovered it on a number of occasions from soils in Ontario. Our records indicate its recovery from greenhouse, agricultural and forest soils on at least 20 separate occasions, so it is probably relatively common.

*Idriella* shows superficial similarities to *Rhinocladiella* and *Rhinocladiella*-like fungi.

## ISARIA

An excellent account of the nomenclature of *Isaria* has been given by Petch (1934). His conclusions are as follows.

> There is another, and more important, aspect of the question, what is *Isaria*. The prevailing idea of the genus since the time of Persoon has been an erect fascicle of mucedinous, amerosporous conidiophores united, in part, into a stalk or clava of varying height. Many species included in *Isaria* do not agree with that conception, but have been placed in the genus merely because they grew in insects. That was per‑eived by Persoon, who adopted the genus *Ceratonema* for *Isaria sobecophila* Ditm. But excluding such obvious misfits, the genus has been a centre for species which agree only in their most general shape. It is a form genus in the widest sense of the term, more general than *Agaricus* when used for all lamelliferous Hymenomycetaceae, and almost as unscientific as *Clavaria* in the sense of Holmskiold.
>
> The unsatisfactory nature of the genus was evident to Fries, who remarked that it embraced species which differed from one another in structure and in the mode of origin of their spores. It would appear that the type of conidiophore is of greater importance in classification than the fact that the conidiophore are united. In identifying an *Isaria*, the type of conidiophore should

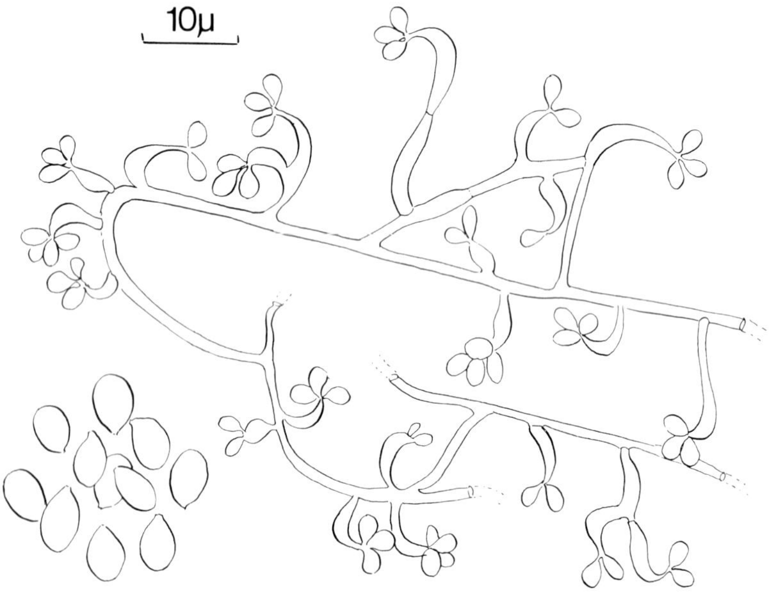


FIG. 126. *Isaria cretacea*. The loose, cortical hyphae of the synnema produce reflexed sporogenous cells which form clusters of conidia at the apex (IMI 112084).

be the primary diagnostic character, and it is almost impossible to interpret the descriptions of very many species of *Isaria*, because the type of conidiophore has not been recorded.

Presuming that the genus *Isaria* is retained, with *Isaria farinosa* as the type species, then *Isaria* is a compound *Spicaria*. Species which have a different type of conidiophore cannot remain in the genus, and a large number of other genera will have to be instituted for them. That, indeed, will happen, whatever species is taken as the type. But the isarioid form is not necessarily an invariable feature of a species. *Isaria farinosa*, for example, frequently occurs as a simple *Spicaria* and conversely, *Beauveria densa* may take the isarioid form. Nor have all isarioid species been consistently placed in *Isaria; Botrytis tilletii*, for example, is an *Isaria* in the wide sense in which the name has hitherto been used.

It would seem simpler to discard the genus *Isaria*, and to distribute the species among the corresponding mucedineous genera. In most species, the type of conidiophore agrees with one of the established genera of Mucedinaceae. The name *Isaria* might be retained as a descriptive term, parallel to sporodochium and synnema, or it might be used to distinguish the isarioid species of a genus, e.g. *Spicaria* (*Isaria*) *farinosa*.

*Isaria* is still in a very confused state. Morris (1963), in his treatment of the synnematous genera of the *Fungi Imperfecti*, cites the lectotype as *Isaria farinosa* Fr. and his generic description is based on this species. *I. farinosa* is a coremial *Paecilomyces* (=*Spicaria*) and this and similar coremial forms have been treated under *Paecilomyces* by Brown and Smith (1957). At the moment this seems the best disposition. Barnett (1960) gave a generalized description of *Isaria* and used *Isaria cretacea* to illustrate the genus. *I. farinosa* and *I. cretacea* are not congeneric. The latter has been recorded from soil on several occasions and is shown in Figures 125*E* and 126.

### ISARIOPSIS Fres.

Type Species: *Isariopsis alborosella* (Desm.) Sacc.

**Generic Description:** Hyphae aggregated into dark synnemata; synnemata more or less erect, sterile below, producing sporogenous cells above at the branch apices; conidia produce in acropetal succession, hyaline to darkly pigmented, cylindric or obclavate, several-septate, truncate at the point of attachment, which corresponds to the pronounced scars left on the sporogenous cells.

**Diagnostic Features:** *Isariopsis* is a coremial Hyphomycete bearing dark phragmospores acrogenously on geniculate sympodulae (Fig. 127).

**Notes:** This genus is rarely reported from soil. The Commonwealth

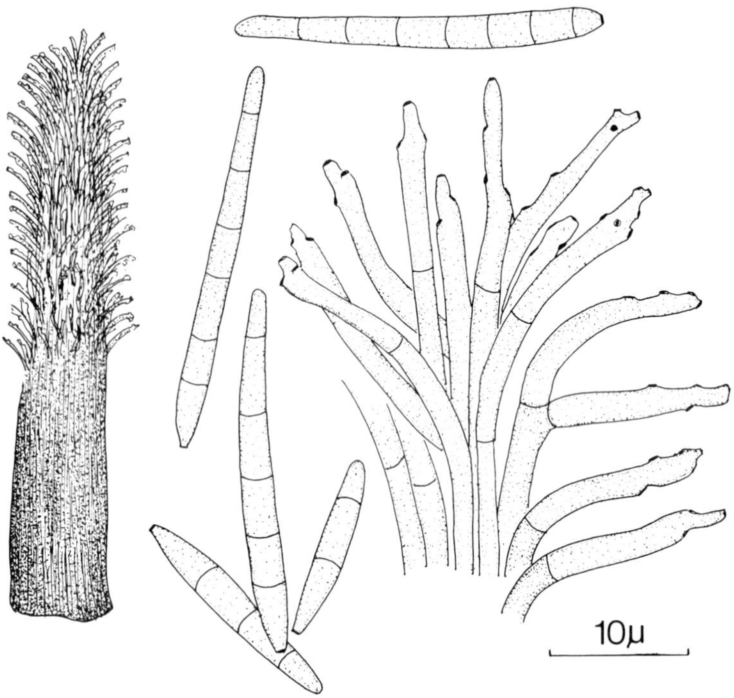


FIG. 127. *Isariopsis helichrysi*. The pigmented phragmospores arise acropetally and leave pronounced scars on the sporogenous cells (IMI 93610). Note the similarities to *Cercospora*.

Mycological Institute, Kew, has a single record of an *Isariopsis* species isolated from soil in a cabbage field in Hong Kong by M. Chu.

## LEPTODISCUS Gerdemann

Type Species: *Leptodiscus terrestris* Gerdemann.

**Generic Description:** Sclerotia small, black, spherical to fusiform; acervuli superficial, yellow or fuscous, shield-shaped, developing radially from a central cell to form a thin stroma one cell-layer thick, often fusing to form irregular plates as large as 200 to 800 $\mu$ in diameter; conidiophores lacking; conidia produced on upper surface of stromatic cells and formed in linear rows in mucus; conidium walls hyaline, cells become pale yellow and finally pale brown in age, once-septate, with a setula at each end. (Description from Gerdemann, 1953.)

**Diagnostic Features:** *Leptodiscus* is typified by two-celled, setulate phialospores borne in mucus on acervuli (Fig. 128).

**Notes:** The genus *Leptodiscus* was erected by Gerdemann (1953) based on *L. terrestris* isolated from the roots of red clover. The generic de-

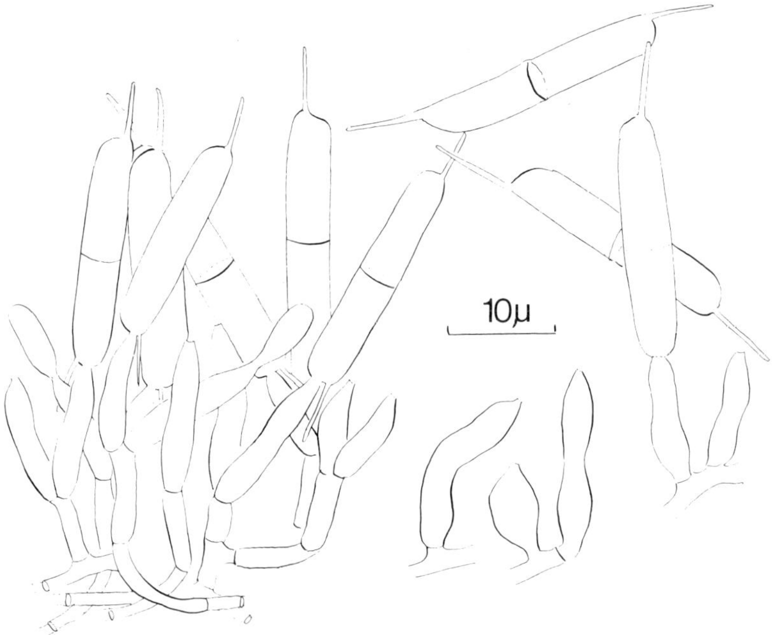


FIG. 128. *Leptodiscus* sp. Two-celled, setulate phialospores arise from short, tightly-packed phialides on a sporodochium (OAC 10112).

scription above is circumscribed around *L. terrestris* and is perhaps a little too precise. *L. terrestris* has also been recovered from tea gardens in Assam by Agnihothrudu (1964). We have isolated a fungus on a number of occasions from organic soils in Ontario which appears to be congeneric with *L. terrestris* (Fig. 128).

McVey and Gerdemann (1960) showed that the function of the setulae in *Leptodiscus* was related to spore dispersal. The setulae are folded over in the developing spore when the mucus material becomes diluted, the setulae spring out violently. Hughes and Kendrick (1963) have observed the same phenomenon in *Menispora*.

### LEPTOGRAPHIUM Lagerberg and Melin

Type Species: *Leptographium lundbergii* Lagerb. and Melin.

**Generic Description:** Conidiophores solitary or in clusters, more or less erect, with simple main axis, stout, septate, dark brown, branched at the apex to produce a penicillus of metulae and sporogenous cells (annellophores); annellophores hyaline, elongate and slightly tapered towards the apex; conidia produced in basipetal succession, hyaline, nonseptate, truncate at the attachment end, rounded at the distal end, gathering in large mucoid heads. (Description based on *L. lundbergii*).

**Diagnostic Features:** *Leptographium* is characterized by having darkly pigmented conidiophores each with an apical penicillus, and by producing annellospores which gather in mucus (Fig. 129).

**Notes:** The genus *Leptographium* was erected by Lagerberg and Melin (1927) on a single species, *L. lundbergii.* There has been much confusion surrounding this fungus and many others which appear superficially the same but are quite unrelated. The precise method of development of the conidia in *Leptographium* was explained and illustrated clearly by Hughes (1953). Details of the method of conidium production and the elucidation of the *Leptographium* complex has been the subject of a number of papers by Kendrick (see *Phialocephala* and *Verticicladiella*).

Because of the uncertainty of the identifications in the earlier literature, it is not possible to say at this time whether or not *Leptographium* is common in soil. We have never recovered it from soils in Ontario. An undisputed *Leptographium* has been figured and described by Meyer (1959). *Leptographium* is frequently found as the conidial state of *Ceratocystis* (Hunt, 1956; Wright and Cain, 1961). Parker (1957) associated *Leptographium* with the Ascomycete genus *Europhium* which is closely related to *Ceratocystis.*

## LIBERTELLA Desm.

Type Species: *Libertella betulina* Desm.

**Generic Description:** Fructification an acervulus; acervulus smooth or uneven, bearing a compact stand of more or less erect conidiophores; conidiophores simple or irregularly branched, hyaline, slender, septate; conidia (phialospores) produced terminally and successively from sporogenous cells and aggregating in slimy masses; broad cirrhi sometimes produced; conidia filiform, nonseptate, more or less curved.

**Diagnostic Features:** Long, thin, curved spores, produced from slender conidiophores on an acervulus are characteristic of *Libertella* (Fig. 130).

**Notes:** *Libertella* has been reported from soil by Johnson and Osborne (1964) and Caldwell (1963). I have only once isolated a species which might be assigned to this genus. In culture, it had a pustular yeast-like growth habit.

*Libertella* is normally a leaf or bark saprophyte (Grove, 1937). The illustration of this genus is based on *L. faginea* Desm. from a collection in Herbarium IMI.

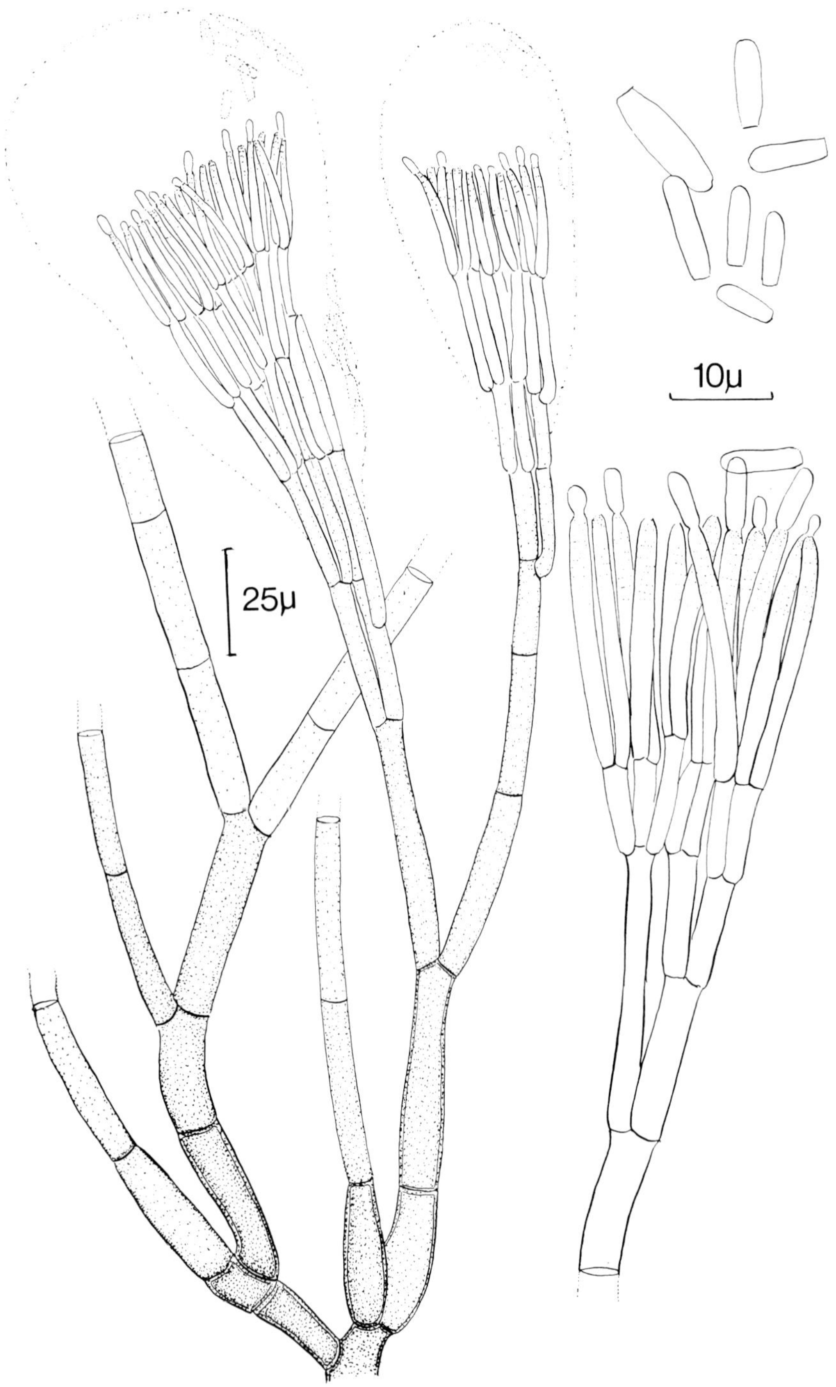


FIG. 129. *Leptographium lundbergii.* The conidiophore terminates in an apical penicillus. The sporogenous cells (annellophores) produce conidia in mucus which envelopes the head in a slimy mass. The conidia have a markedly truncate base characteristic of annellospores (OAC 10343).

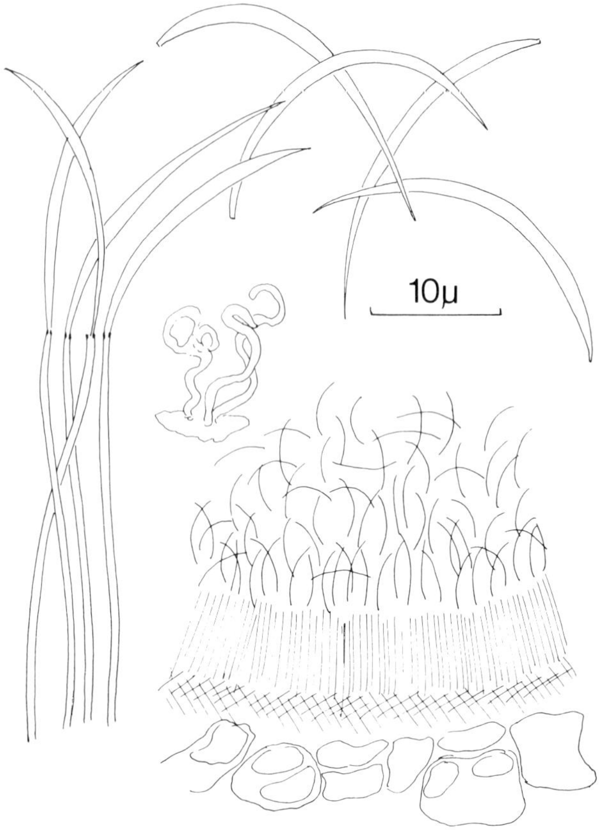


FIG. 130. *Libertella faginea*. The sporogenous cells arise in compact masses from a stroma. Phialospores aggregate to form long, sinuate cirrhi (IMI 60797).

## MAMMARIA Cesati

Type Species: *Mammaria echinobotryoides* Ces.

**Generic Description:** Conidiophores erect or almost repent, not readily distinguished from the vegetative hyphae, straight or slightly curved, usually simple but sometimes bearing one or two short branches, septate, subhyaline to pale brown; conidia aleuriospores, produced directly on the aerial hyphae or "conidiophores," sessile or on short pedicels, persistent, continuous, darkly pigmented, ovoid to flame-shaped, more or less apiculate, flattened at the basal scar, with well marked longitudinal germ slit, borne singly or in groups, sometimes in *Echinobotryum*-like clusters. (Description based on *M. echinobotryoides*.)

**Diagnostic Features:** *Mammaria* is distinguished by its large, dark, apiculate aleuriospores with longitudinal germ slits (Fig. 131).

**Notes:** The generic name *Mammaria* was erected for a single species,

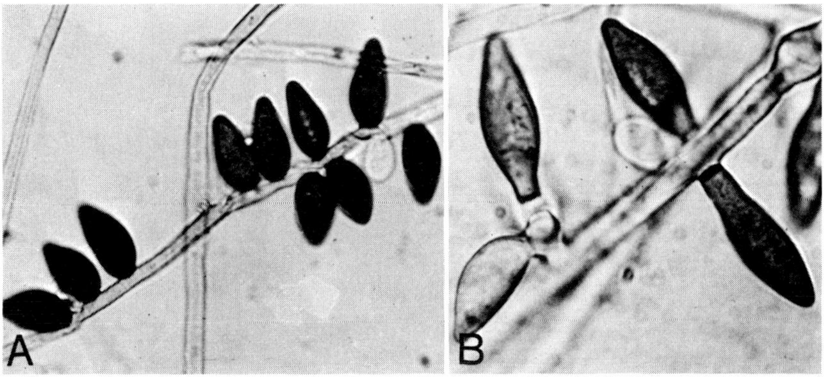


FIG. 131. *Mammaria echinobotryoides*. *A* and *B*, the large, dark aleuriospores are usually sessile and are broader near the base and tapered or somewhat papillate towards the apex (OAC 10220).

*M. echinobotryoides*. The genus was reviewed by Hughes (1957), who gave an account of the morphology and taxonomy of *M. echinobotryoides*. Saccardo (1886) compiled this species under *Trichosporium*, but this latter genus was considered by Hughes (1958) to be a *nomen confusum*.

We have isolated *M. echinobotryoides* from peat soils in Ontario on a number of occasions and it is probably common in organic soils. It may well be that this species has been listed under other genera such as *Trichosporium* or *Echinobotryum*. As the specific epithet suggests, the genus bears a close morphological resemblance to *Echinobotryum*. The drawing and descriptions by Bayliss Elliott (1918) of *Trichosporium murinum* are very close to *Mammaria*.

### **MENISPORA** Persoon

Type Species: *Menispora glauca* Pers.

**Generic Description:** Setae straight or slightly bent to sinuous or variously coiled, thick-walled, sometimes anastomosing, independent or comprising hair-like prolongations of conidiophores; phialides borne terminally and singly on simple conidiophores or terminally and laterally on branched conidiophores (lateral and sessile or lateral and terminal on branches if main axis prolonged into a seta); phialides usually slightly or strongly recurved at the apex, sometimes proliferating to form a polyphialide; collarettes inconspicuous, short, narrow, more or less tubular and not flaring; phialospores hyaline, curved, continuous or three-septate, terminally or subterminally setulate or nonsetulate, more or less cylindrical or tapered towards the apex and the inconspicuous basal scar; phialospores gather in compact, slimy colorless or yellow-green fascicles or confluent masses at the apex of the phialides or parallel with the phia-

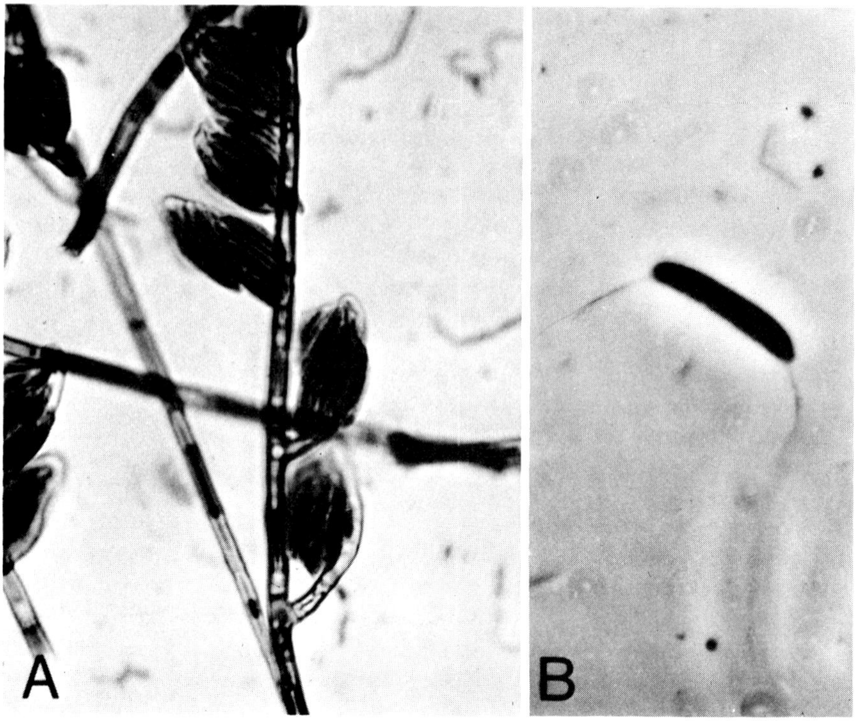


FIG. 132. *Menispora ciliata*. *A*, the phialospores are borne in fascicles at the mouth of reflexed phialides. *B*, the conidia have a single setula at each end (IMI 31466).

lide when this is strongly recurved. (Description based on unpublished data supplied by S. J. Hughes and W. B. Kendrick.)

**Diagnostic Features:** *Menispora* is characterized by slimy fascicles of elongate, curved phialospores, produced from slightly or strongly recurved phialides on well differentiated conidiophores (Fig. 132).

**Notes:** *Menispora* has been critically reviewed by Hughes and Kendrick (1963), who recognized and described in detail six species from North America. A chronological list of all taxa classified in *Menispora* was also given with comments on their possible identity. These workers considered the genus as representing a somewhat heterogeneous group (see also *Menisporella*).

*Menispora* species are rarely recorded from soil. In Ontario we have recovered a *Menispora* only once. Agnihothrudu (1962) recorded a *Menispora* species from tea plantation soils in Assam.

## MENISPORELLA Agnihothrudu

Type Species: *Menisporella assamica* Agnihothrudu.

**Generic Description:** Setae straight or slightly bent, thick-walled, darker and longer than the conidiophore, independent or comprising sterile prolongations of the conidiophores, sometimes lacking; sporogenous cells polyphialides, borne terminally and singly on simple conidiophores or terminally and laterally on branched conidiophores (lateral and sessile or terminal and lateral on branches if the conidiophore is prolonged into a sterile seta); phialides straight, proliferating to produce polyphialides with conspicuous collarettes; collarettes cupulate to funnel-shaped, flaring, somewhat thick-walled and often refringent towards the base, appearing laterally as flat or denticulate scars on older sporogenous

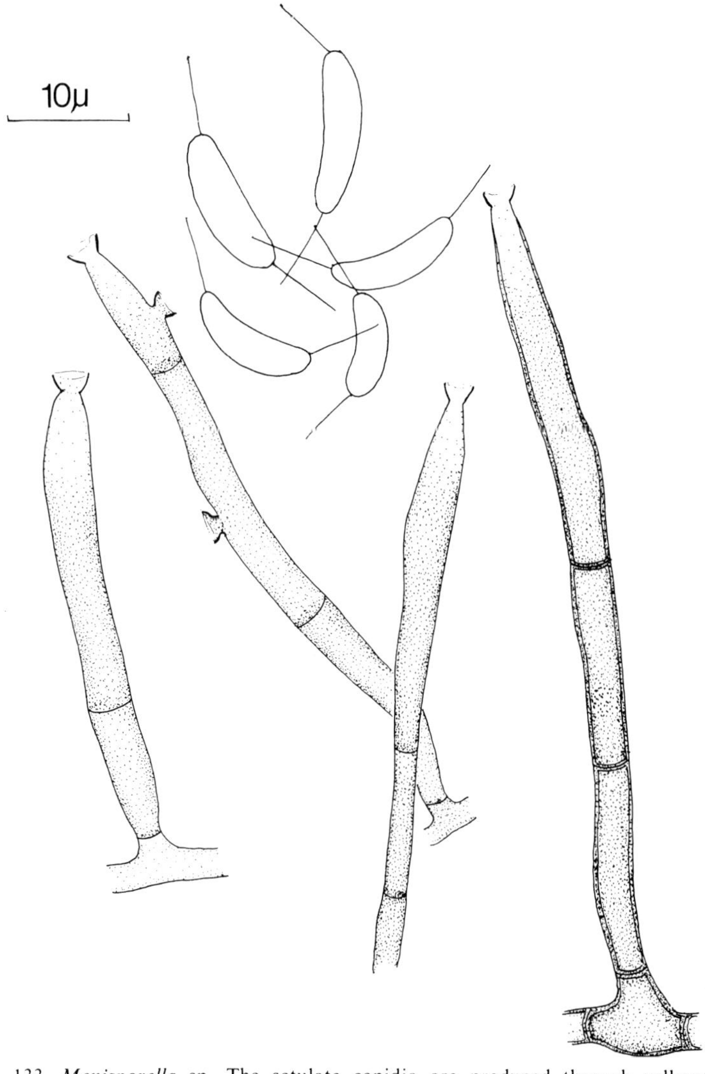


FIG. 133. *Menisporella* sp. The setulate conidia are produced through collarettes. In this species the conidiophores elongate sympodially (IMI 104642).

cells; phialospores hyaline, curved, continuous or one- to three-septate, nonsetulate or terminally or subterminally setulate, more or less falcate and symmetrical or asymmetrical, or botuliform, rounded or pointed at the apex, tapering towards an inconspicuous basal scar, gathering in a compact slimy fascicle at the apex of each polyphialide; fascicles colorless to straw-colored or brown. (Description based on unpublished data supplied by S. J. Hughes and W. B. Kendrick.)

**Diagnostic Features:** *Menisporella* is distinguished from the closely related *Menispora* in having phialides which are never recurved and which possess conspicuous, flaring collarettes (Figs. 133 and 191).

**Notes:** *Menisporella* was erected by Agnihothrudu (1962) based on *M. assamica* isolated from the roots of the tea plant. The genus has not been recorded from soils in any of the published lists. We have isolated from soil in Ontario a Hyphomycete which appears to be congeneric with *Menisporella* and the herbarium of the Commonwealth Mycological Institute records an isolate of *Menisporella* from pasture soil in New Zealand.

The above generic description is based on unpublished data supplied to me by Dr. S. J. Hughes and Dr. W. B. Kendrick who are preparing a manuscript on *Menispora*, *Menisporella* and *Menisporopsis* of New Zealand. Dr. Kendrick has also advised me that *Codinaea* Maire may be an earlier name for species presently disposed in *Menisporella*.

### MENISPOROPSIS Hughes

Type Species: *Menisporopsis theobromae* Hughes.

**Generic Description:** Synnemata composed of two parts, a central emergent seta and an external cortex; seta erect, dark brown, simple, septate, awl-shaped; cortex shorter than the seta, composed of pale brown phialides, compact at the base and loose at the apex; phialides subulate, pale brown; phialospores gathering in mucoid globules, continuous, hyaline, falcate, bearing a single setula at each end. (Description based on Hughes, 1952a.)

**Diagnostic Features:** *Menisporopsis* is distinguished from other Hyphomycetes with setulose conidia in having a central, sterile seta surrounded by a cortex of phialides (Fig. 134).

**Notes:** This genus was originally described from dead leaves of *Theobroma cacao* by Hughes (1952a). It is rarely found in floristic listings, but Meyer (1959) isolated *M. theobromae* frequently from soil in the Congo.

### METARRHIZIUM Sorok.

Type Species: *Metarrhizium anisopliae* (Metsch.) Sorok.

**Generic Description:** Fruit body a sporodochium-like aggregate of

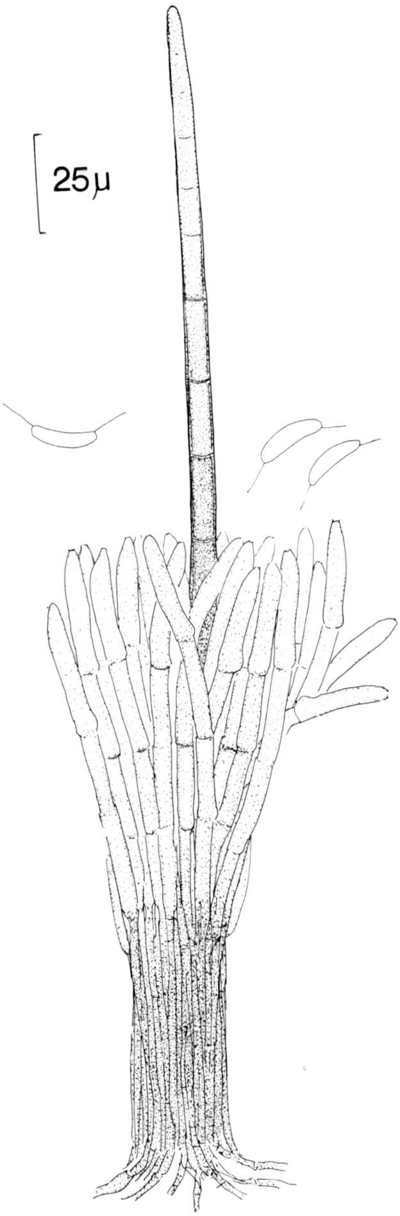


FIG. 134. *Menisporopsis theobromae*. The stout, central seta is sterile. At the base of the seta the conidiophores form a tight cortex which flares at the apex or produce a loose mass of sporogenous cells (IMI 39099).

fairly closely interwoven hyphae bearing a closely-packed mass of conidiophores on a flat or disk-shaped "stroma"; conidiophores simple or more often branch profusely and closely in penicillate fashion, ultimate sporogenous cells phialides; phialides cylindrical, hyaline, taper to a narrow apex; phialospores nonseptate, cylindrical, hyaline or lightly pigmented, produced in chains and adhering in tall, columnar aggregates. (Description based on *M. anisopliae.*)

**Diagnostic Features:** The bright-green columns of spores arising from a pure-white turf give the common *M. anisopliae* a striking appearance in culture (Figs. 135 and 136).

**Notes:** *Metarrhizium* is recorded infrequently from soil, yet our studies on the soil flora of Ontario indicate that *M. anisopliae* is very common in forest soils. We have recovered three culturally distinct types. The most common of these produced bright-green spore columns and agreed with published descriptions of *M. anisopliae.* Less frequently a species was recovered in which the spore masses were olive-brown and agreed with the description of *M. brunneum* Petch. A third type was recovered, on one occasion only, in which the spore columns were purplish and arose from a lemon-yellow turf. Latch (1965) noted that microscopically *M. anisopliae* and *M. brunneum* were very similar and suggested they should be regarded as the same.

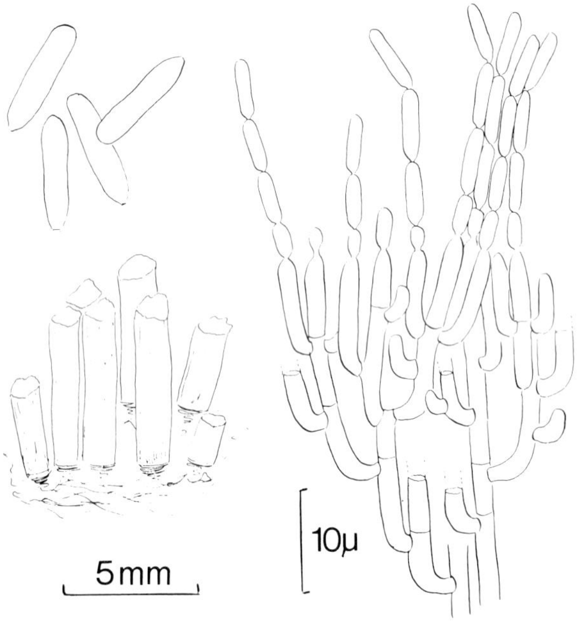


FIG. 135. *Metarrhizium anisopliae.* The sporogenous cells and hyphae form a thin myceloid "stroma" from which tall columns of closely packed conidial chains arise (OAC 10286).

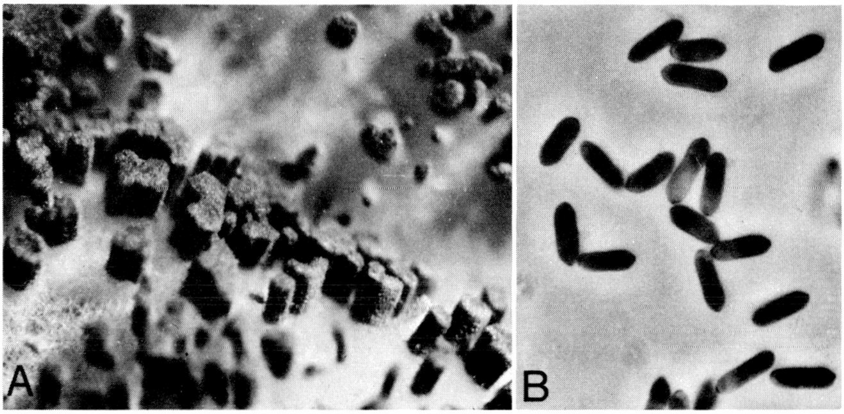


FIG. 136. *Metarrhizium anisopliae*. *A*, sporocarps, showing spores aggregated into columnar masses (reproduced courtesy I. Reid). *B*, conidia (OAC 10286).

Biologically *M. anisopliae* has significance as a parasite causing the "green muscardine" disease of a variety of insects. Details of the taxonomy, morphology and host range of *M. anisopliae* are given by Petch (1930). Apparently there are small-spored strains (5 to 7 by 2.5 to 3 $\mu$) and large-spored strains (10 to 14 by 3 to 4 $\mu$) of this species. We have found only the small-spored strain in Ontario. Attempts have been made by Latch (1965) to use this fungus for the biological control of insect pests in pastures in New Zealand.

## MICROSPORUM Gruby

Type Species: *Microsporum audouinii* Gruby.

**Generic Description:** Colonies cottony or powdery, white to buff or brown with reverse in pink to red or yellow; conidiophores inconspicuous; macrospores (aleuriospores) produced abundantly on more of less short pedicels, terminal, solitary, spindle-shaped, large, thick-walled, smooth or roughened, many-septate; accessory microspore (aleuriospore) state produced in most species; microspores produced singly, directly from the hyphae or on short hyphal branches.

**Diagnostic Features:** The very large (40 to 150 by 8 to 15 $\mu$), thick-walled, septate, macrospores which are usually spindle-shaped characterize *Microsporum* (Fig. 137).

**Notes:** The genus *Microsporum* is one of the keratinophilic fungi classed as dermatophytes (see also *Trichophyton*), although not all members of the genus are pathogenic. The genus is found in few of the floristic listings of soil fungi obtained by standard techniques and we have found a *Microsporum* only once in extensive studies of soils in southern Ontario. There is, however, evidence (Ajello, 1959) to indicate that *Microsporum*

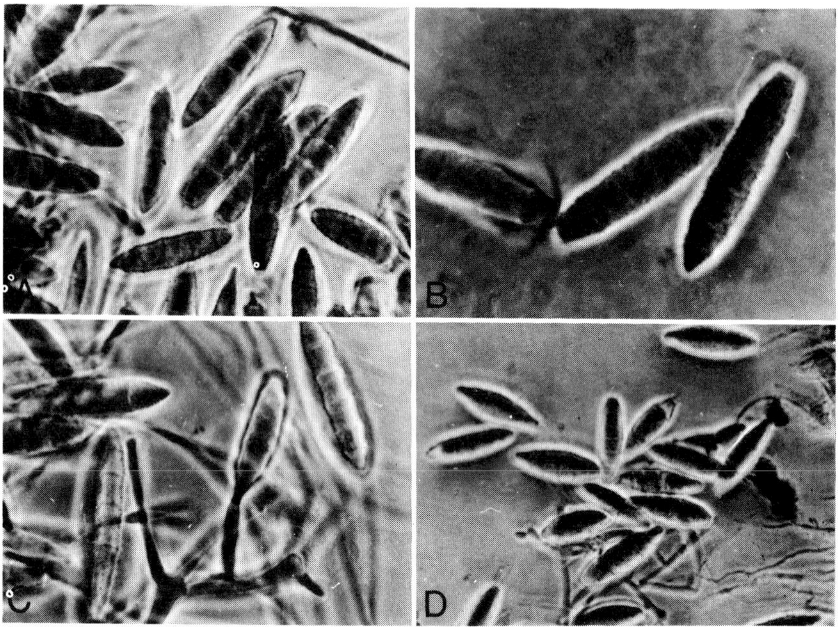


FIG. 137. *A*, *B*, and *C*, *Microsporum fulvum* (IMI 86180). *D*, *M. gypseum* (IMI 80558).

and other dermatophytes are relatively common in soil and can be readily recovered by hair baiting techniques (Vanbreuseghem, 1952).

The genus *Keratinomyces* was erected by Vanbreuseghem (1952a) to accomodate a single species, *K. ajelloi* (Fig. 59*B*), a nonpathogenic keratinophilic fungus. On the basis of the morphology of the macrospores of *Keratinomyces*, there seems no good reason to separate it from *Microsporum*. The perfect states of *Microsporum* are frequently found in *Nannizzia* of the Gymnoascaceae (Stockdale, 1963). Benedek also erected the genus *Veronaia* as the perfect state of *Microsporum audouinii* and *M. canis*. This relationship was questioned by de Vries (1964). Regarding *K. ajelloi*, Dawson and Gentles (1961) reported *Arthroderma uncinatum* as the perfect state. Benedek (1963), on the other hand, found that *Anixiopsis stercoraria* was also the perfect state. In his study, Benedek noted that *A. stercoraria* was not keratinophilic, not keratinolytic, and not pathogenic; on this basis de Vries (1964) questioned Benedek's findings and suggested single ascospore cultures should be made to prove the relationship. Dr. R. Goos has informed me (private communication) that in his laboratory single ascospore cultures of *Anixiopsis* never developed a *Keratinomyces* state.

The nomenclatural controversy surrounding the correct name for the genus, *Microsporum* or *Microsporon*, is discussed by Benedek (1963) who favors *Microsporon*. The genus contains a number of species pathogenic on man and animals and is given extensive treatment in most texts on medical mycology (see Emmons *et al.*, 1963).

## MONILIA Pers.

Type Species: *Monilia fructigena* Pers.

**Generic Description:** Conidiophores little differentiated from the vegetative hyphae, erect or straggling, simple or dichotomously or irregularly branched, hyaline, septate; conidia blastospores, produced in apical succession by apical budding to give branching chains; conidia hyaline to subhyaline, continuous, globose to ovoid, often with isthmus-like connectives between spores. (Description based on *M. fructigena*.)

**Diagnostic Features:** In this genus the branching chains of hyaline blastospores with a bead-like appearance are distinctive (Fig. 138).

**Notes:** Members of the genus *Monilia* are isolated infrequently from soil (Gilman, 1957) and are difficult to identify to species. The genus is heterogeneous and is not a natural grouping. *Monilia sitophila*, the red bread mould, is the conidial state of the Pyrenomycete *Neurospora sitophila*, and commonly grows on sterilized soil in the greenhouse. It has a very rapid growth rate and can be a nuisance as a contaminant in the laboratory. *Monilia* is the conidial state of the Discomycete, *Monilinia fructicola*, the organism causing brown rot of stone fruits.

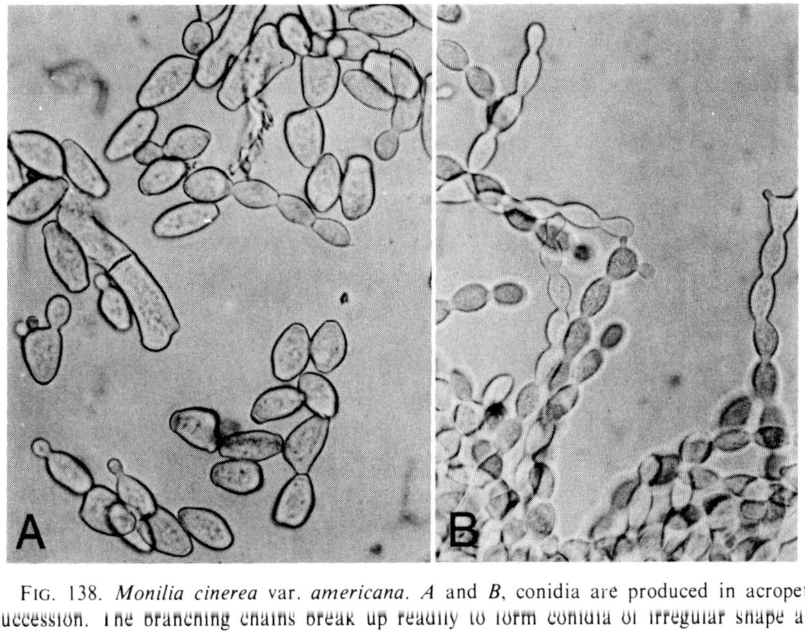


FIG. 138. *Monilia cinerea* var. *americana*. *A* and *B*, conidia are produced in acropetal succession. The branching chains break up readily to form conidia of irregular shape and size (OAC 10146).

## MONOCHAETIA (Sacc.) Allescher.

Type Species: *Monochaetia monochaeta* (Desm.) Allescher.

**Generic Description:** Fruiting bodies black, carbonaceous, usually true acervuli, with or without stromatic area, sometimes pycnidia or pseudo-pycnidia, but usually without a true ostiole and rarely as loose fertile hyphae without a distinct stratum or stroma. Conidia fusiform, straight or curved, four- to six-celled, crowned with a single hyaline setula; exterior cells hyaline or rarely dilutely colored; rarely with contents; intermediate cells equally or variably colored, pale brown to almost black, guttulate; pedicels hyaline, simple, attached to the base of the conidia. (Description from Guba, 1961.)

**Diagnostic Features:** This genus resembles *Pestalotia* in its characteristics, but has a single, hyaline setula at the apex of the conidium.

**Notes:** Members of this genus are commonly found as parasites of higher plants and not frequently found in floristic listings of soil fungi. Guillemat and Montegut (1957) recorded *Monochaetia* from this source. The reader is referred to the notes on *Pestalotia* for additional comments on the genus.

The genus *Monochaetia* is not recognized by Steyaert (1949), who did not consider the single setula as of generic significance. He preferred to dispose of the species under *Monochaetia* in other genera such as *Pestalotiopsis*, *Truncatella* and *Bleptosporium*. Genera related to *Pestalotiopsis* have been studied in some detail by Sutton (1963).

## MONOCILLIUM Saksena

Type Species: *Monocillium indicum* Saksena.

**Generic Description:** Conidiophores lacking; sporogenous cells arising more or less directly from the vegetative hyphae, consisting of a basal, cylindrical stalk terminating in a swollen vesicle with a tapering, somewhat sinuous apical region; conidia hyaline, nonseptate, ellipsoid to obovoid, truncate at the attachment point, produced in fragile chains or sliming down to form balls. (Description based on *M. indicum*.)

**Diagnostic Features:** Slender stalks bearing apical vesicles which produce short chains of hyaline, nonseptate conidia characterize *Monocillium* (Figs. 140 and 191).

**Notes:** The genus *Monocillium* was erected by Saksena (1955) on a single species, *M. indicum*, recovered from a sample of grassland soil near Sagar, India. We have recovered this species on several occasions from sandy agricultural soils in Ontario. The only other record is that of Peyronel and Dal Vesco (1955), who figured a *Phialophora* species as isolate CMT323; this appears to be *M. indicum*.

Barron (1961a) described *M. humicola* from peat soils in Ontario. This species was later shown to be conspecific with the previously described

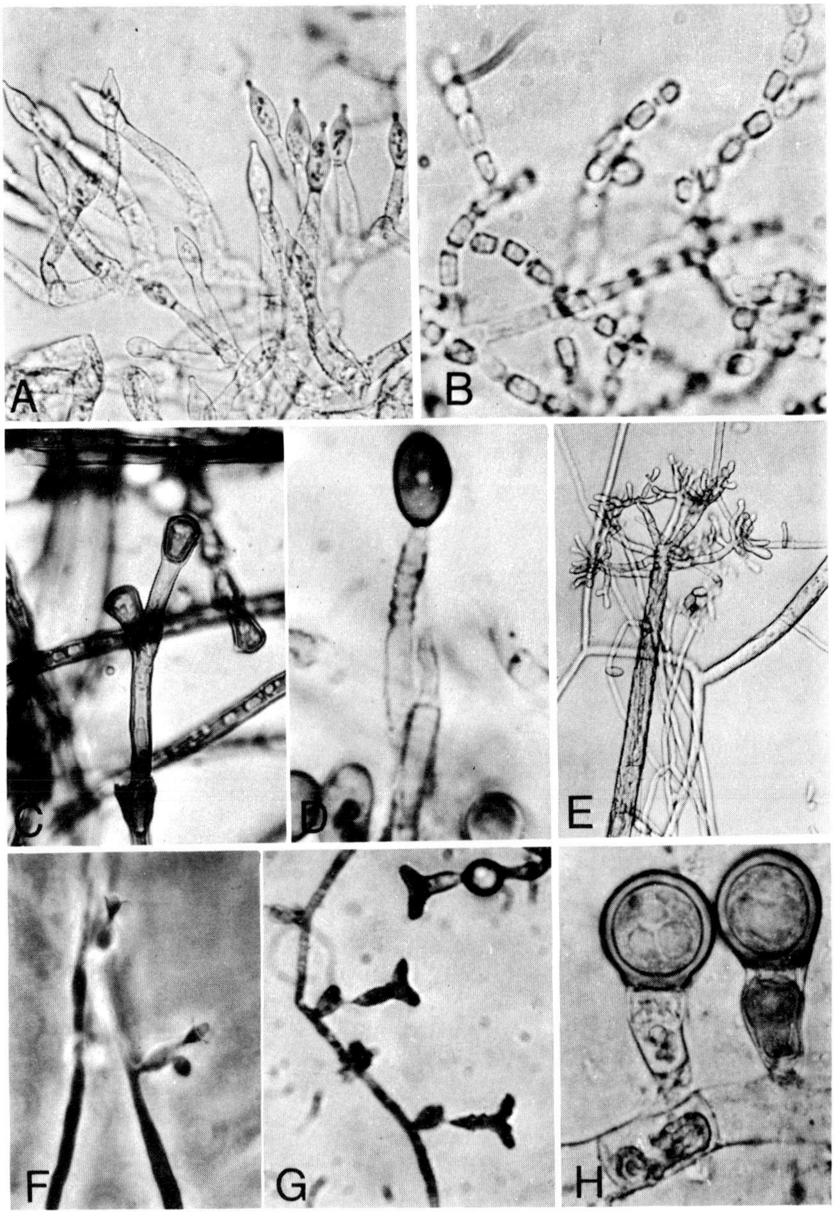


FIG. 139. Miscellaneous. *A*, *Sepedonium ampullosporum*. *B*, *Oidiodendron* state of *Arachniotus striatosporus*. *C*, *Catenularia cuneiformis*. *D*, *Scopulariopsis* sp. showing annellations on the sporogenous cell. *E*, *Amblyosporium botrytis*. *F*, *Phialophora verrucosa*. *G*, *Scolecobasidium terreum*. *H*, *Bactridiopsis* sp.

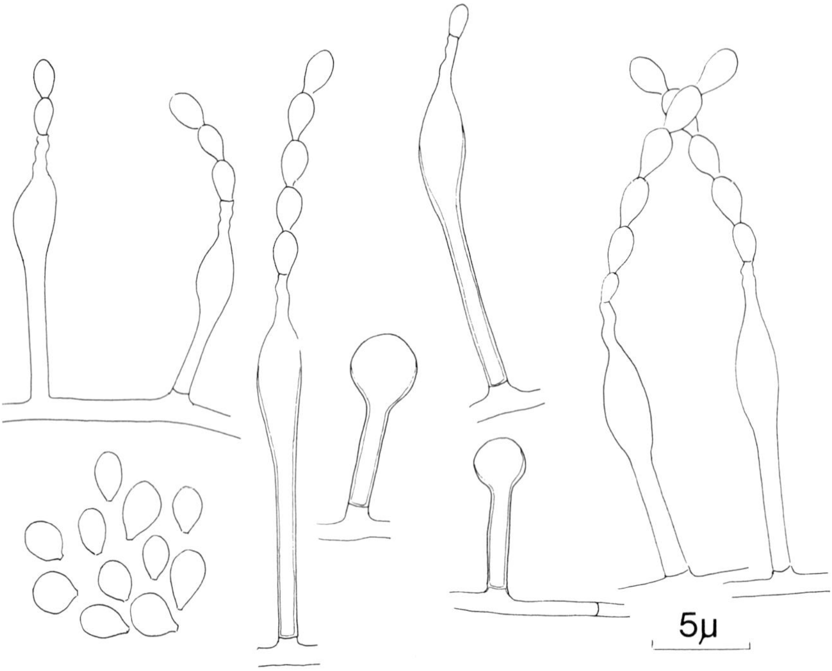


FIG. 140. *Monocillium indicum*. Sporogenous cells consist of a cylindrical stalk bearing a vesicle in the upper part. The conidia are produced in fragile chains and are ovoid and slightly truncate at the attachment point (OAC 10173). (Reproduced by permission of the National Research Council of Canada, *Can. J. Bot.*, *39:* 1573, 1961.)

*Torulomyces lagena* (Barron, 1967). The genus *Torulomyces* was erected by Delitsch (1943), with *T. lagena* as the type species. It was reported by Barron that the sporogenous cells of *Torulomyces* are phialides, but the nature of the sporogenous cell in *Monocillium* was not established.

*Monocillium exsolum* was described as new from Brazilian soil by Batista and Heine (1965). Barron (1967) examined a subculture of the type species from the Centraalbureau voor Schimmelcultures, Baarn and regarded it as a *Metarrhizium*, probably *M. anisopliae.*

## MONODICTYS Hughes

Type Species: *Monodictys putredinis* (Wallr.) Hughes.

**Generic Description:** Conidiophores simple or sparingly branched, short, septate, subhyaline to brown, subcylindric or sometimes markedly inflated, solitary or clustered, sometimes aggregated into pustules; conidia produced singly at the apex of the conidiophores, muriform, sometimes constricted at the septae, smooth or roughened, brown, dark brown or almost black, subglobose to ovoid or subpyriform, sometimes irregular. (Description based on Hughes, 1958.)

**Diagnostic Features:** Darkly pigmented, muriform aleuriospores, borne singly at the apex of simple, undistinguished conidiophores, are characteristic of *Monodictys* (Fig. 141).

**Notes:** *Monodictys* was erected, with *M. putredinis* as the type species, by Hughes (1958), who also transferred nine other species to the genus including *Acrospeira levis* and *A. macrosporoidea*, discussed by Wiltshire (1938). Members of the genus are rarely reported from soil under the name *Monodictys*, but may have been reported previously under either *Acrospeira* or *Stemphylium*. We have isolated a *Monodictys* species on only one occasion from soil in Ontario. The records of the Commonwealth Mycological Institute also list a *Monodictys* species from soil, and *M. austrina* was described by Tubaki and Asano (1965) from Antarctic soil.

## MUROGENELLA Goos and Morris

Type Species: *Murogenella terrophila* Goos and Morris.

**Generic Description:** Hyphae hyaline to subhyaline, septate; conidiophores short, long or wanting; conidia formed singly and terminally, dark, oval to elliptical. (Description from Goos and Morris, 1965.)

**Diagnostic Features:** The genus is characterized by thick-walled,

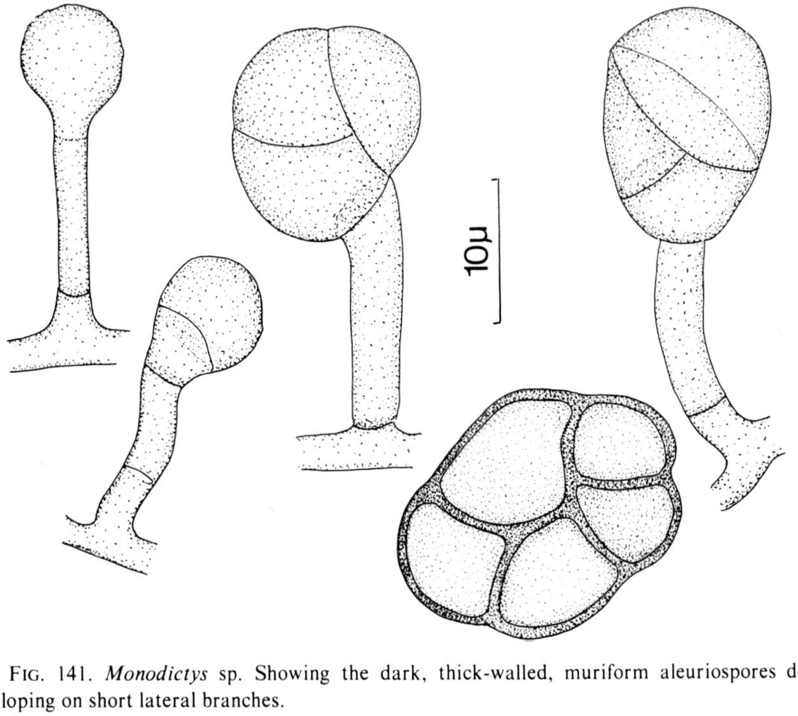


FIG. 141. *Monodictys* sp. Showing the dark, thick-walled, muriform aleuriospores developing on short lateral branches.

darkly pigmented, broadly fusiform, septate aleuriospores which are sessile or borne on short, cylindrical pedicels (Fig. 142).

**Notes:** *Murogenella* was erected by Goos and Morris (1965) on *M. terrophila*, isolated from garden soil in Virginia by mouse inoculation techniques. The authors cite additional records of this fungus from soil, Central Experimental Farm, Ottawa, and from Cherokee oats in Kansas. Vaidehi *et al.* (1967) recorded it from the rhizosphere of paddy in India.

Goos and Morris distinguish their genus from *Coryneum* in that it never produces acervuli and from *Corynespora* in that the spores of the latter are produced in short chains rather than singly, as in *Murogenella*.

## MYCOGONE Link

Type Species: *Mycogone rosea* Link.

**Generic Description:** Conidiophores inconspicuous, simple or branched, borne more or less at right angles to the vegetative hyphae; conidia aleuriospores, initiated as blown out ends of the conidiophores, solitary, two-celled; upper cell larger, warty, slightly pigmented; lower cell smaller, smooth or slightly roughened, hyaline; an accessory *Verticillium* state may be associated with this fungus.

**Diagnostic Features:** *Mycogone* is typified by large, two-celled aleuriospores with larger, pigmented upper cell and smaller, hyaline lower cell (Fig. 143).

**Notes:** *Mycogone* is close morphologically to *Chlamydomyces*. In his

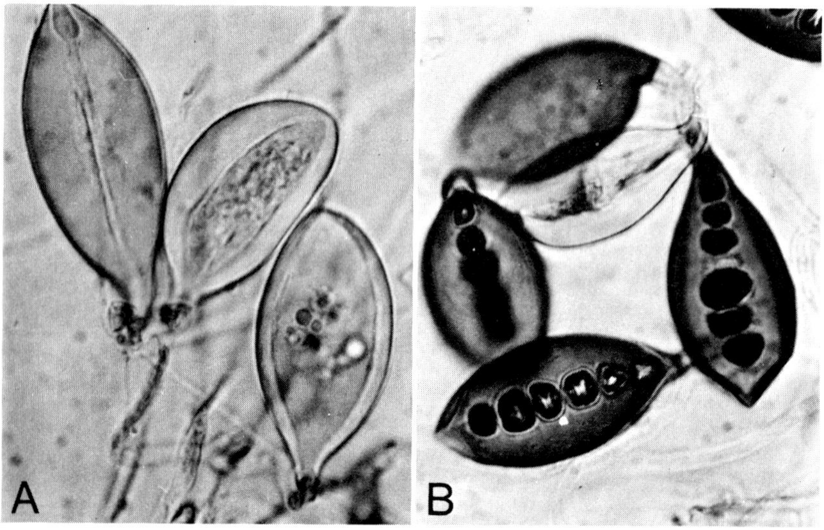


FIG. 142. *Murogenella terrophila*. *A*, young conidia. *B*, older conidia. (ATCC 15140, ex type).

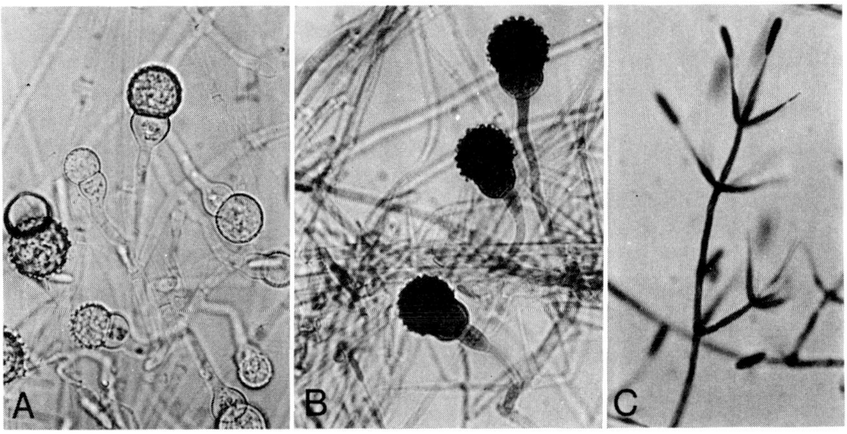


FIG. 143. *Mycogone perniciosa. A*, aleuriospores have a roughened, pigmented, apical cell and smooth, hyaline, basal cell. *B*, aleuriospores stained with aniline blue. *C*, *Verticillium*-like phialospore state (OAC 10589).

consideration of these two genera, Howell (1939) maintained them as distinct. He stated that the aleuriospores of *Chlamydomyces* were essentially one-celled, the two-celled appearance being due to the enlargement of the subtending cell of the sporophore, whereas in *Mycogone* the spores were two-celled almost from their inception. In *Mycogone* the phialospore state is of the *Verticillium* type, whereas in *Chlamydomyces* it is aspergilliform.

*Mycogone* species are rarely reported from soil (Gilman, 1957) and are more commonly found as parasites on the fructifications of higher Basidiomycetes. They are conidial states of the Pyrenomycete genus *Hypomyces*. *M. nigra* found in several lists of soil fungi is now disposed of as *Humicola grisea*. The species chosen for illustration, *M. perniciosa*, is a pathogen of the cultivated mushroom.

## **MYROTHECIUM** Tode

Type Species: *Myrothecium inundatum* Tode.

**Generic Description:** Fructification a sporodochium; sporodochia small, discoid, sessile or with short stalks, often confluent; dark-green to black with spore production; spore mass viscous and green when young, later becoming hard and black; conidiophores more or less erect in a compact layer, septate, hyaline, branched, ultimately surmounted by verticils of slender, hyaline phialides; phialospores hyaline to olive-brown, ovoid, fusoid, or cylindrical, continuous, often guttulate, gathering in slimy masses or loose columns; with fan-shaped, membranous appendage in one species.

**Diagnostic Features:** *Myrothecium* is typified by the viscous dark green masses of phialospores on sporodochia (Fig. 144).

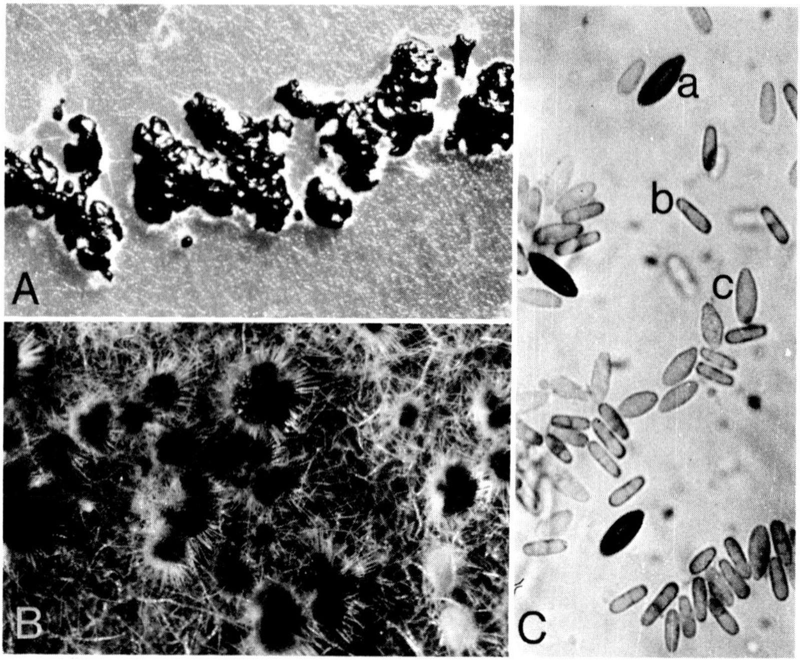


FIG. 144. *A*, *Myrothecium verrucaria*. Sporodochia in petri plate culture producing slimy masses of conidia (OCA 10232). *B*, *M. indicum*. This species is characterized by a fringe of setae round the sporodochium (IMI 103664, ex type). *C*, mixture of spores of three species of *Myrothecium*: *a*, *M. striatosporum* (OAC 10234), conidia are dark and striate; *b*, *M. roridum* (OAC 10194), conidia are cylindric. *c*, *M. verrucaria*, conidia are ellipsoid to fusoid (OAC 10232).

**Notes:** Members of the genus *Myrothecium* are very common in soil. *M. verrucaria*, *M. roridum* and *M. striatisporum* are all relatively frequently recovered, especially from soils high in organic matter. *M. brachysporum* has been described from soil by Nicot and Olivry (1961) and *M. indicum* by Rao (1963). The latter species is quite striking in having a fringe of stiff setae round the sporodochium. The morphology and taxonomy of the genus has been treated by Preston (1943, 1961).

Usually, the spores of *Myrothecium* species tend to slime down to form gloeoid masses, but this is not always true, and I have recovered a number of isolates from soil in which the spores tend to remain in loose chains and aggregate in columnar masses, as in *Metarrhizium*. This being true, there seems little to distinguish *Myrothecium* from *Metarrhizium* on a morphological basis. *Myrothecium* has also been recorded as a plant parasite (Brooks, 1944; Fergus, 1957). Members of the genus are strong cellulase producers. While investigating the degradation of cotton by *M. verrucaria*, Thompson and Simmens (1962) found that the spores of this species had fan-shaped appendages. These were seen best when stained with Loeffler's flagellar stain. The appendages are clear in the photomicrograph accompanying their article, and their presence was confirmed by electron microscopy. Apparently *M. verrucaria* is the only species with appendages. The only association of *Myrothecium* with a perfect state was made by Booth (1959), who reported that *Nectria ralfsii* has a conidial state which could be referred to *Myrothecium*.

## NEMATOGONIUM Desm.

Type Species: *Nematogonium aurantiacum* Desm.

**Generic Description:** Conidiophores indeterminate, more or less erect, hyaline, sparingly septate, unbranched, swelling to produce a terminal ampulla; conidia borne on denticles in simple or branched chains over the surface of the ampulla, hyaline to lightly pigmented, spherical to ovoid, variable in size; main conidiophore axis proliferating through vesicle to produce secondary or tertiary ampullae, giving the mature conidiophore a noded appearance. (Description based on *N. aurantiacum.*)

**Diagnostic Features:** In *Nematogonium* the conidiophores are stout and sparingly branched. The conidia are nonseptate and arise from denticles in simple or branching chains from terminal and intercalary ampullae (Fig. 145).

**Notes:** This genus is rarely recorded from soil. The only isolate identified to species is *N. humicola*, described by Oudemans and Koning (1902). Brown (1958) recorded a *Nematogonium* isolate from sand dunes in Britain.

Hughes (1953) noted that *Gonatorrhodiella parasitica*, the type species of that genus, was not generically distinct from *Nematogonium* and made the new combination *N. parasitica* (Thaxter) Hughes.

## NIGROSPORA Zimm.

Type Species: *Nigrospora panici* Zimm.

**Generic Description:** Hyphae hyaline, becoming more or less pigmented; conidiophores inconspicuous; conidia borne on short pedicels more or less at right angles to the vegetative hyphae; pedicels broader at the base and tapering towards the attachment point of the spore; conidia solitary, nonseptate, black, smooth, flattened in horizontal axis, ovoid to subglobose.

**Diagnostic Features:** The very dark conidia, which are slightly longer in the horizontal axis and borne on very short sporogenous cells, are characteristic of *Nigrospora* (Fig. 146).

**Notes:** Commonly found as a parasite of cereals, *Nigrospora* is infrequently recorded from soils (Gilman, 1957). We have isolated it on only two occasions from agricultural soils in Ontario. In culture the colonies are white at first and the shiny, black spores stand out in sharp contrast

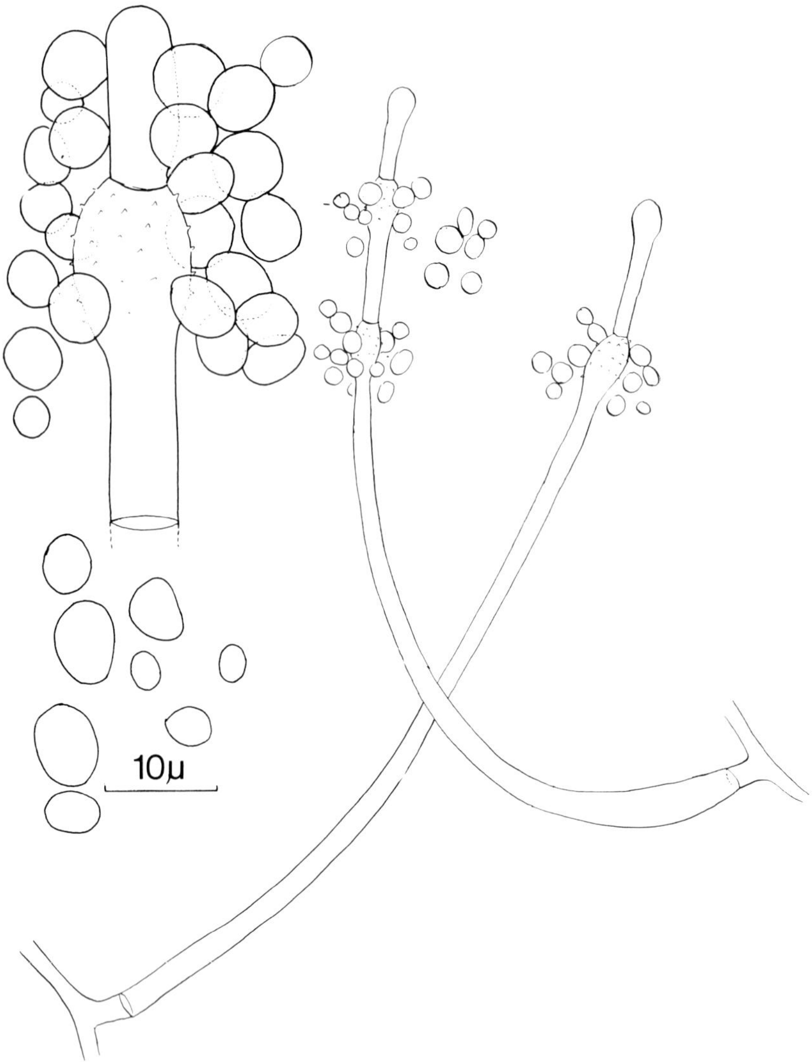


FIG. 145. *Nematogonium aurantiacum.* The conidia are produced in short simple or branching chains from the ampullae. Successive conidia tend to be progressively smaller and a random spore mass will therefore show a great range in size (IMI 18141).

giving the colonies a striking appearance under the binocular dissecting microscope. In older cultures the hyphae darken and, with profuse spore production, the colonies appear black. *Nigrospore oryzae*, one of the commonest species, has been connected by Hudson (1963) to its perfect state, *Khuskia oryzae* of the Sphaeriales.

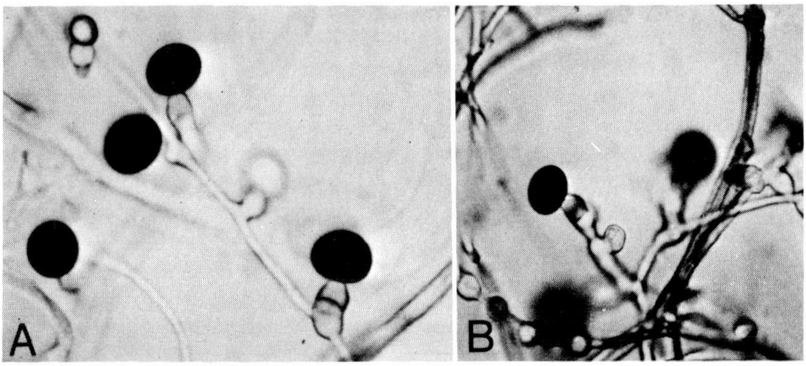


FIG. 146. *Nigrospora oryzae*. *A* and *B*, conidia appear black to the eye. The sporogenous cells are very short and taper towards the spore attachment. The spores have a slightly longer horizontal axis.

## NODULISPORIUM Preuss

Type Species: *Nodulisporium ochraceum* Preuss.

**Generic Description:** Conidiophores erect or suberect, septate, main axis simple or branched, hyaline to subhyaline or darkly pigmented, ultimate branches sporogenous cells; sporogenous cells sympodulae, borne singly or in pairs, frequently in verticillate arrangements; conidia produced singly and in acropetal succession at the apex of the sporogenous cell which increases in length sympodially; conidia secede to leave distinct, highly refractive scars on the sporogenous cells; conidia nonseptate, subhyaline to darkly pigmented, usually with a narrow truncate base corresponding to the point of attachment.

**Diagnostic Features:** *Nodulisporium* is characterized by well developed, branching conidiophores with the sympodulae, arranged in verticillate or subverticillate fashion, bearing prominent scars (Figs. 14*C* and 147).

**Notes:** Members of this genus are reported infrequently. In our studies of Ontario soils, we have isolated at least four species including *N. hinnuleum* described by Smith (1962). Meyer (1959) recorded *N. africanum* and an unidentified *Nodulisporium* species from soil in the Congo. Goos and Timonin (1962) reported a *Nodulisporium* species from the rhizosphere of bananas in Honduras. *N. gregarium* has been recorded from Italy by Badura (1963).

The genus is close morphologically to *Hansfordia*, *Calcarisporium* and *Rhinocladiella* and the separation of an unknown isolate into one or other of these genera may present difficulties. For example *N. griseo-brunni* described from soil by Mehrotra (1965) shows similarities to *Rhinocladiella*.

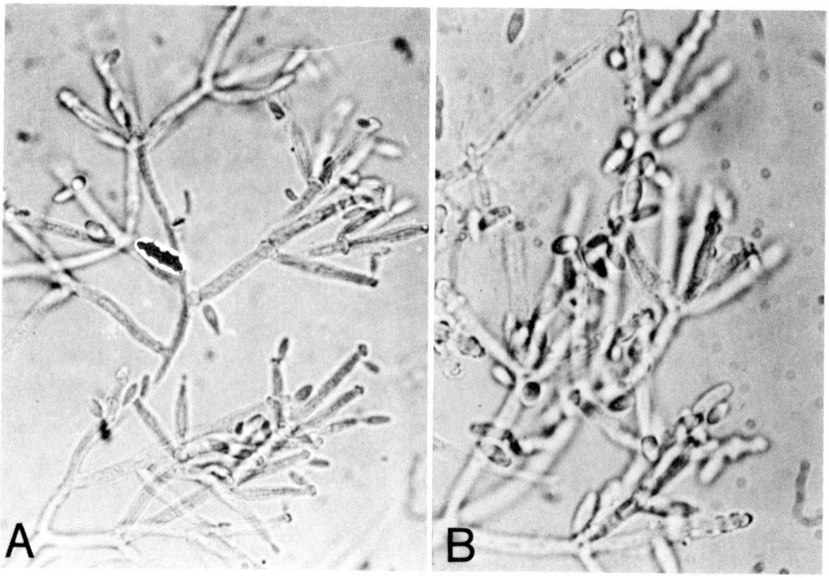


FIG. 147. *Nodulisporium hinnuleum*. *A*, younger conidiophores showing the spores just beginning to develop at the apex of each sporogenous cell. *B*, an older stage where the highly refractive, raised scars can be seen on the sporogenous cell (OAC 10390).

*Acrostaphylus* established by Arnaud (1953), was taken up by Subramanian (1956a) to include the dematiaceous species of *Nodulisporium*. In the light of our present understanding of the taxonomy of Hyphomycetes, however, it would seem more reasonable that the genus should include both hyaline and pigmented species.

*Nodulisporium didymosporum*, described from soil by Nicot (1956), is not a *Nodulisporium* species and is the same fungus identified by de Vries (1962) as *Dactylium fusarioides* (see *Dactylium*) and now disposed as *Fusarium chlamydosporum*.

## OEDOCEPHALUM Preuss

Type Species: *Oedocephalum glomerulosum* (Bull.) Sacc.

**Generic Description:** Conidiophores determinate, more or less erect, usually simple, sometimes sparingly or even verticillately branched, hyaline, septate, main axis or branches enlarged at the apex to produce solitary terminal ampullae; conidia hyaline, nonseptate, subglobose to ellipsoid, developing simultaneously all over the surface of the ampulla on denticles.

**Diagnostic Features:** *Oedocephalum* is readily recognized by its determinate conidiophores bearing nonseptate conidia produced synchronously over the surface of terminal ampullae (Fig. 148).

**Notes:** Members of this genus are reported infrequently from soil (Gilman, 1957). An *Oedocephalum* species was reported by Sewell (1959) from heathland soil and Rall (1965) recorded *O. glomerulosum* from alpine soils in Wyoming.

Members of the genus are reported as the conidial states of certain Basidiomycetes such as *Fomes annosus* (Bakshi 1950, 1952), but are also commonly found as the conidial states of Discomycetes. *Pyronema omphalodes* has an *Oedocephalum* conidial state (Schmidt, 1910). Webster *et al.* (1964), studying Discomycetes of burnt ground, described and figured the *Oedocephalum* states of *Peziza praetervisa* and *P. anthracophila.*

## OIDIODENDRON Robak

Type Species: *Oidiodendron tenuissimum* (Peck) Hughes.

**Generic Description:** Vegetative hyphae hyaline or pigmented, simple or united into funiculose strands; conidiophores produced singly or in clumps arising from substratum, aerial hyphae or hyphal ropes; conidiophores slender, pigmented, smooth or rough, usually simple, sometimes branching into two or several main trunks, 100 to 250 μ tall in most species, rarely up to 500 μ, sometimes short or lacking; conidiophores branching repeatedly at the top to form a head of delicate, hyaline, interlacing sporogenous hyphae; sporogenous hyphae forming numerous short segments by basipetal septation; maturing from the tip back towards the main conidiophore axis into branching chains of unicellular arthrospores; arthro-

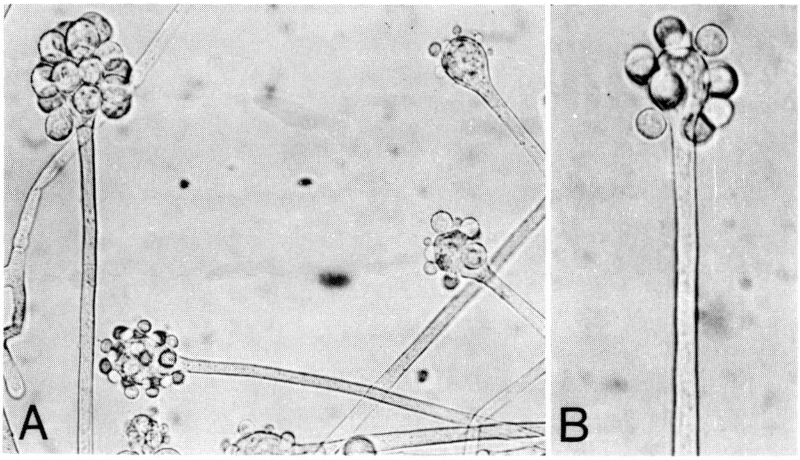


FIG. 148. *Oedocephalum* sp. *A* and *B*, various stages in the development of the conidia from the terminal ampullae. The conidia develop more or less simultaneously but are not exactly synchronized on all of the ampullae (OAC 10352).

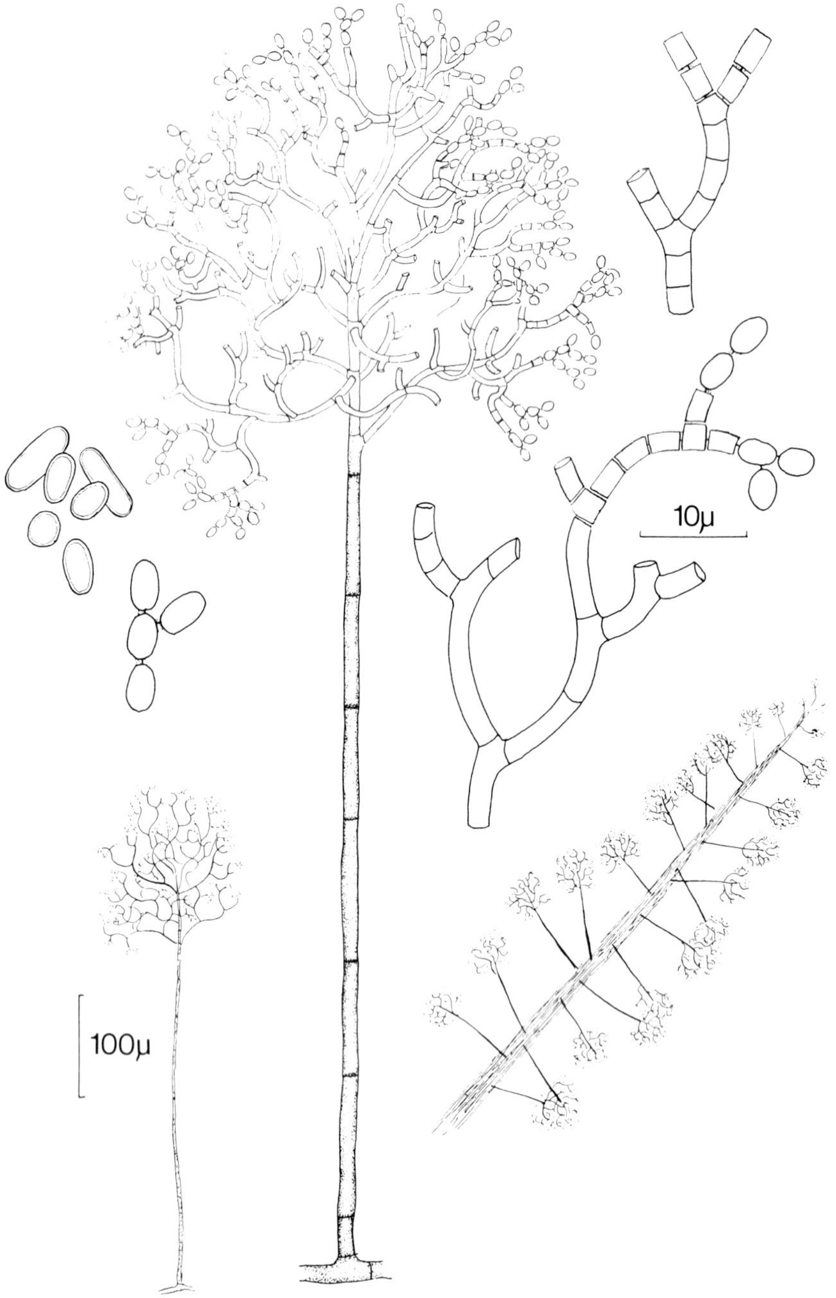


FIG. 149. *Oidiodendron maius*. Showing conidiophores with tree-like appearance. The head of sporogenous hyphae septates and matures in basipetal fashion to give branching chains of arthrospores (Reproduced by permission of the National Research Council of Canada, *Can. J. Bot.*, *40:* 589, 1962).

spores smooth, or more often slightly or strongly roughened, hyaline or pigmented, globose, subglobose, ovoid, or cylindric, falling free at maturity to leave the conidiophores as naked trunks with stumps of branches still apparent.

**Diagnostic Features:** *Oidiodendron* is characterized by the production of pigmented conidiophores which branch, tree-like, at the apex to produce a head of fertile hyphae. Arthrospores develop from the fertile hyphae by basipetal septation and "rounding off" (Figs. 117*G*, 149, and 150).

**Notes:** *Oidiodendron* is apparently widespread in its occurrence in soils and there are frequent records in the literature. A review of the genus by Barron (1962) covers most of the common species.

The perfect state of *Oidiodendron* is found in certain Gymnoascaceae. Orr *et al.* (1963) report that *Toxotrichum spinosum* (= *Myxotrichum spinosum*) has an *Oidiodendron* conidial state. Barron and Booth (1966) recently described *Arachinotus striatosporus* as having an *Oidiodendron* state (Fig. 139*B*).

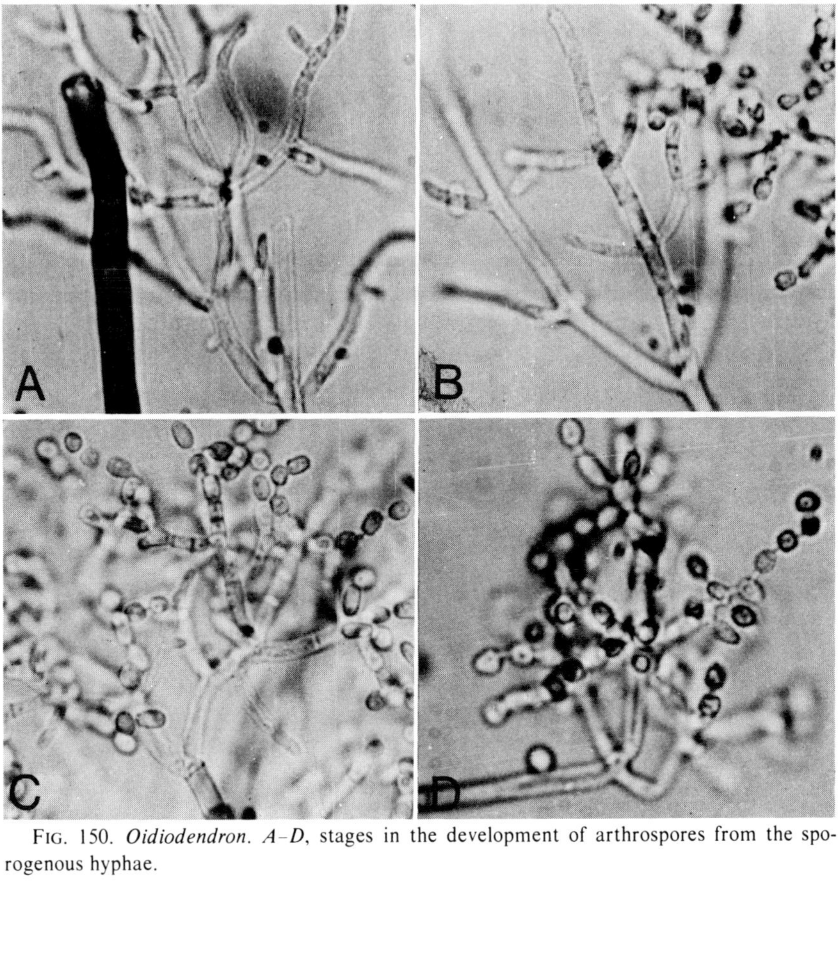


FIG. 150. *Oidiodendron*. *A–D*, stages in the development of arthrospores from the sporogenous hyphae.

## OLPITRICHUM Atkinson

Type Species: *Olpitrichum carpophilum* Atk.

**Generic Description:** Conidiophores usually stout and more or less erect, septate, simple or irregularly branched, ultimate cells of branches or main axis sporogenous; sporogenous cells cylindrical or frequently inflated in the spore bearing region, with prominent denticles; increasing in length by sympodial growth; conidia produced singly and acrogenously, large, hyaline to lightly pigmented, continuous, smooth, ovoid to ellipsoid.

**Diagnostic Features:** Large, hyaline or subhyaline spores borne on pronounced denticles at the apex of stout conidiophores are characteristic of *Olpitrichum* (Figs. 151 and 152).

**Notes:** In his treatment of the genus *Oidium*, Linder (1942) included in his list of synonyms the following genera: *Alysidium* Kunze and Schmidt, *Rhinotrichum* Corda, *Amphiblistrum* Corda, *Physospora* Fr. and *Olpitrichum* Atkinson. According to Hughes (1958) *Oidium* = *Sporotrichum*, and *Alysidium* = *Acladium* (see Hughes, 1953). Hughes apparently could find no type material of *Rhinotrichum simplex*, the type species of *Rhinotrichum*. Until the taxonomy of this group of fungi is clarified, I have reverted to the name *Olpitrichum* for those species related to *Acladium* in which the conidia are restricted to one or few swollen terminal cells on the conidiophore.

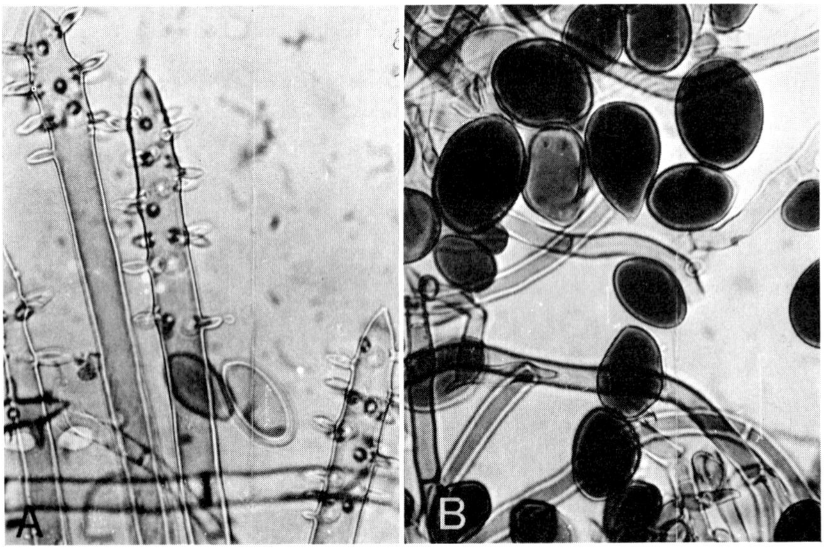


FIG. 151. *Olpitrichum macrosporum*. *A*, sporogenous cells have conspicuous, flame-shaped denticles. *B*, aleuriospores are thick-walled and pigmented (IMI 7068).

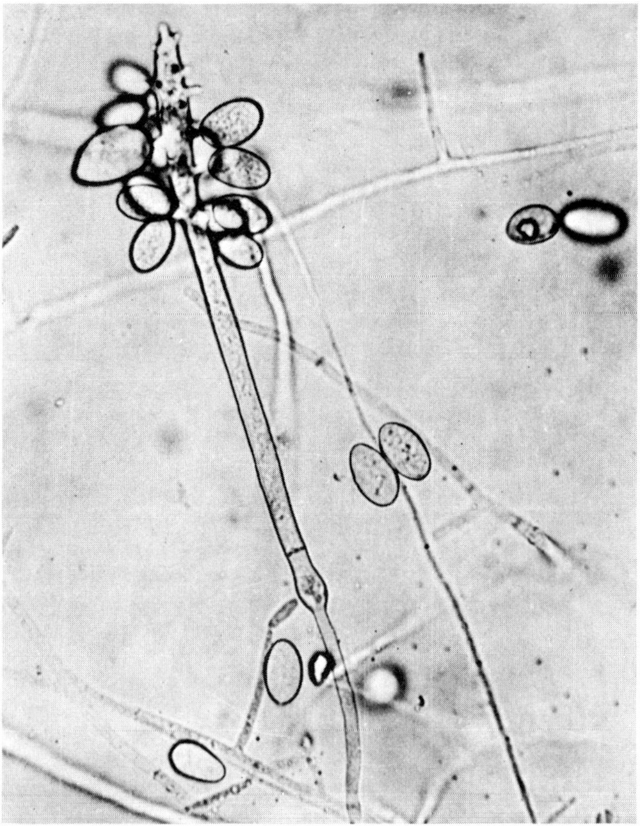

FIG. 152. *Rhinotrichum tenellum.* The apex of the sporogenous cell is slightly swollen and the conidia are produced on conspicuous denticles.

## OSTRACODERMA Fr.

The *Ostracoderma* state of *Peziza ostracoderma* Korf (= *Plicaria fulva* Schneider) is perhaps one of the most commonly encountered but least known of fungi from soil. It is extremely common in greenhouses growing on sterilized soil or vermiculite, in pots and flats. The fungus produces rapidly spreading colonies over the surface of the substratum. At first white, they quickly turn yellowish and finally cinnamon-colored in age. Eventually umber-brown apothecia may be produced in old flats or pots. The taxonomy of this fungus was discussed in detail by Korf (1960), who commented largely on the apothecial state, and by Fergus (1960), who characterized the developmental states of the asexual state with some nice photomicrographs.

The conidiophores of this species are stout, hyaline, erect and septate, with an unbranched main axis. At the apex, up to 12 elongate, divergent,

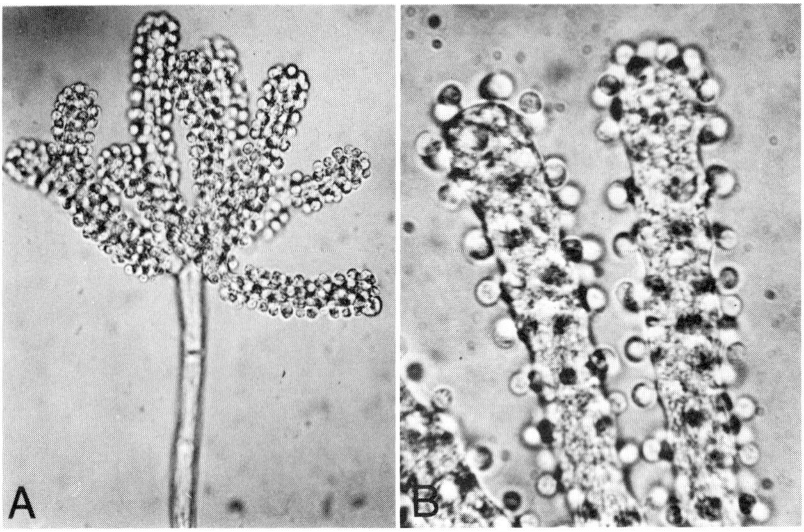


FIG. 153. *Ostracoderma* state of *Peziza ostracoderma*. *A*, apex of conidiophore has divergent elongate ampullae. *B*, ampullae with young conidia developing synchronously (OAC 10269).

sporogenous ampullae form a spore-bearing head. The sporogenous ampullae usually arise in pairs from a short, dichotomous branch of the main axis. Over the surface of each ampulla, the conidia develop simultaneously on slender denticles. They enlarge rapidly and at maturity the entire ampulla is covered with a compact layer of spherical, lightly pigmented conidia (Figs. 8*A* and 153).

As pointed out by Korf, the conidial state of *Peziza ostracoderma* was the fungus described by Wolf (1955) as *Mycotypha dichotoma*. The apothecial state was originally described by Schneider (1954) as *Plicaria fulva*. Korf suggested that this fungus would be better considered under *Peziza*. The specific epithet of the basionym could not be transferred because of an earlier homonym, and, on the basis of its conidial state, Korf named the fungus *P. ostracoderma*. Korf noted that a similar conidial state was described by Wolf as *Rhinotrichum trachycarpa*, the conidial state of *Peziza trachycarpa*.

## PAECILOMYCES Bainier

Type Species: *Paecilomyces varioti* Bain.

**Generic Description:** Conidiophores well developed, simple or branching, septate, hyaline to pigmented, lacking in some species, with the sporogenous cells borne more or less directly on the vegetative hyphae; sporogenous cells phialides, borne singly, in pairs, in verticils, or in penicillate heads, sometimes irregularly arranged; conidia in chains, tending to slip or slime down in some species, hyaline or darkly pig-

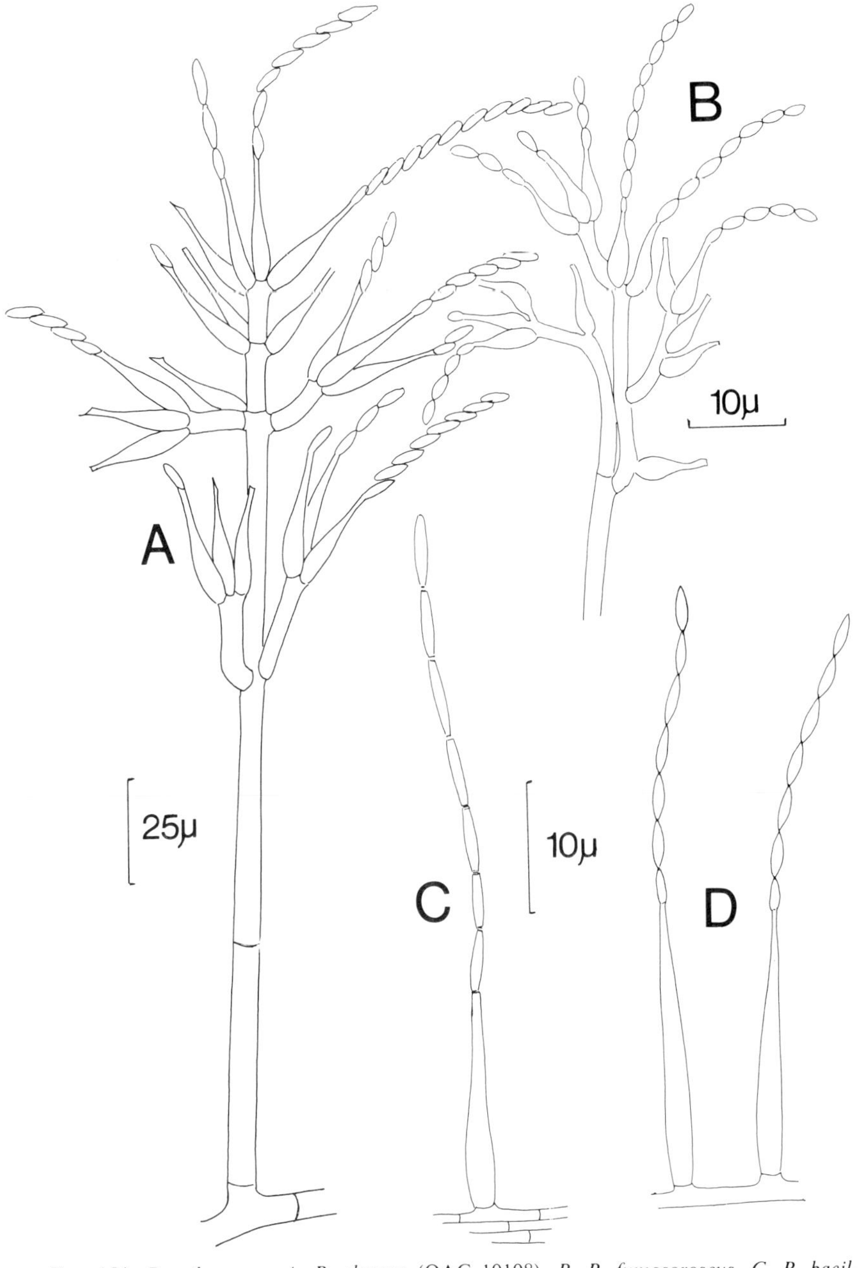


FIG. 154. *Paecilomyces*. *A*, *P. elegans* (OAC 10108). *B*, *P. fumosoroseus*. *C*, *P. bacillisporus* (IMI 113161). *D*, *P. terricola* IMI 96734).

mented, smooth or roughened, sometimes fluted, ovoid to fusoid, sometimes spherical; chlamydospores or aleuriospores produced by some species.

**Diagnostic Features:** *Paecilomyces* is perhaps best distinguished by its sporogenous cell. The phialide is usually swollen towards the base, tapers gradually towards the apex and bears long chains of ovoid to fusoid spores (Fig. 154).

**Notes:** This is a large and heterogenous genus and therefore difficult to characterize in simple terms. Its members show great variability in culture morphology, growth rate, color and temperature response. Many are strongly coremial and have been classified from time to time in *Isaria*. The genus has been monographed by Brown and Smith (1957) who recognized 27 species, but almost as many species again have been described. The genus shows relationships to *Penicillium*, *Verticillium*, and *Gliocladium*. Brown and Smith stated that *Paecilomyces* species are never green; this, however, is not true, and we have several isolates of *Paecilomyces* from soil in Ontario which are green in culture.

*Paecilomyces* includes a number of species in which the main conidiophore axis, characteristic of most members of the genus, is lacking. In these forms, the sporogenous cells are borne singly either directly from the vegetative hyphae or less often in groups of two or three on very short conidiophores. For convenience, these forms were grouped together by Onions and Barron (1967) as the monphialide species of *Paecilomyces*. Onions and Barron recognized 10 such species and noted that they approach the genera *Gliomastix* and *Cephalosporium* very closely.

A number of species produce so-called "macrospores" (Brown and Smith, 1957). This term is somewhat misleading in the sense used for *Paecilomyces*. In *Fusarium* and *Cylindrocarpon* the large septate phialospores are referred to as macrospores, the small rarely septate phialospores as microspores. In *Microsporum* and other dermatophytes, the large septate aleuriospores are referred to as macrospores; the smaller nonseptate aleuriospores are microspores. It might be better to restrict the use of macrospore and microspore to situations in which tow distinct morphological forms of a similar spore type, *e.g.* phialospores *or* aleuriospores, are produced by the same fungus. In *Paecilomyces*, the "macrospores" are chlamydospores or terminal aleuriospores.

The conidial states of *Byssochlamys* belong to *Paecilomyces*. Certain *Cephalotheca* species (*e.g. C. sulphurea*) have asexual states belonging to *Paecilomyces*. The thermophilic ascomycete *Thermoascus* (= *Dactylomyces*) has a *Paecilomyces*-like conidial state.

## PAPULASPORA Preuss

Type Species: *Papulaspora sepedonioides* Preuss.

**Generic Description:** Conidiophores and conidia lacking, reproductive

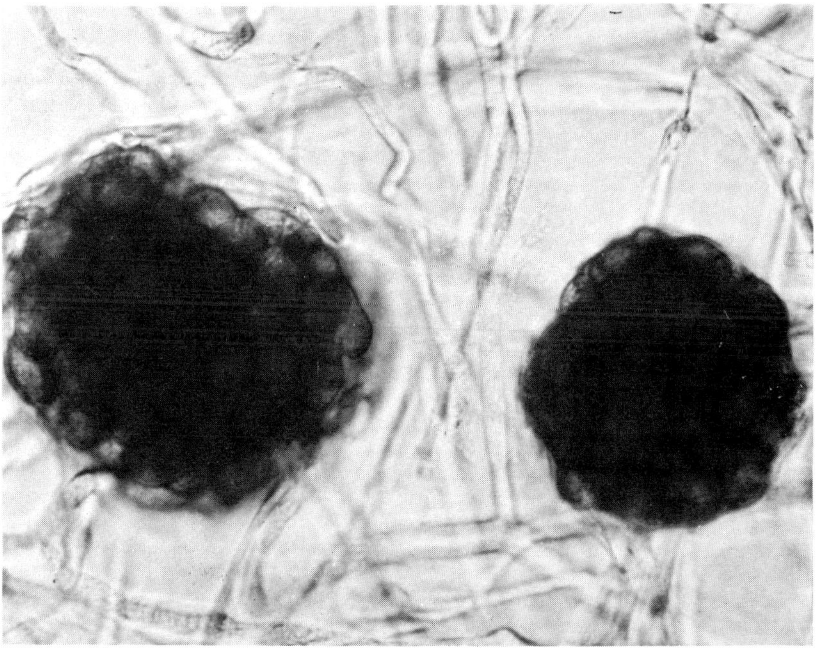

FIG. 155. *Papulaspora* sp. The "bulbils" consist of irregular masses of pigmented cells (OAC 10172).

units consisting of irregular clusters of cells (bulbils); bulbils without organization, frequently pigmented, pale brown or orange, often appearing like microsclerotia; vegetative hyphae hyaline.

**Diagnostic Features:** In this genus no true spores are formed. The hyaline vegetative hyphae bearing "bulbils" are characteristic of the genus (Fig. 155).

**Notes:** *Papulaspora* species are reported infrequently from soil (Orpurt, 1964; Dal Vesco, 1957; Williams and Schmitthenner, 1956; Mosca, 1964). Members of the genus are the imperfect states of Basidiomycetes or Ascomycetes. *P. byssina* is a mycoparasite of the commercial mushroom producing a disease known as "brown plaster." *Papulaspora* has been the subject of a number of papers by Hotson (1917, 1942) to which the reader is referred for details.

## PENICILLIUM Link

Type Species: *Penicillium expansum* Link.

**Generic Description:** Conidiophores usually conspicuous, more or less erect, sometimes aggregated into synnemata, hyaline, rough or smooth, septate, sometimes branching at or near the apex; branches divergent or adpressed to the main conidiophore axis; series of branches giving char-

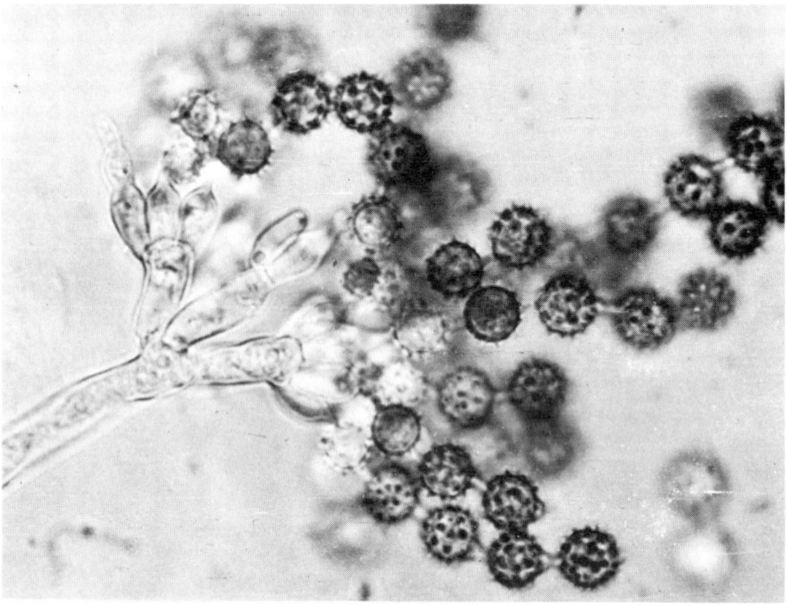

FIG. 156. *Penicillium megasporum.* In this species the conidia are darkly pigmented and echinulate (OAC 10384).

acteristic brush-like penicillus; sporogenous cells phialides, borne in groups directly at the apex of the conidiophore (monoverticillata) or on primary or secondary branches (biverticillata); phialides typically flask-shaped, hyaline; conidia borne in long chains, globose to ovoid, sometimes bacillar, hyaline to darkly pigmented, smooth or roughened; sclerotia produced in some species.

**Diagnostic Features:** The long chains of phialospores borne on .flask-shaped phialides in a brush-like head (penicillus) is typical of *Penicillium* (Fig. 156). Most species have a well developed conidiophore and some form synnemata.

**Notes:** The genus *Penicillium* needs little introduction and the reader is referred to the Raper and Thom (1949) monograph on the genus for details on the biology of this fungus. It is perhaps the most ubiquitous of all fungi and is found from the equator to the polar regions, although perhaps it favors the temperate and colder regions.

The genus merges with *Aspergillus* and *Paecilomyces* and there is still some confusion and controversy surrounding the fringes where they intergrade. The genus *Eladia* is closely related to *Penicillium*.

## **PERICONIA** Tode ex Schweinitz

Type Species: *Periconia lichenoides* Tode.

**Generic Description:** Conidia typically catenulate, sometimes borne singly, nearly always rough-walled, lacking any obvious hilum, produced from both mycelial hyphae and from macronematous conidiophores; conidial chains often branched, usually arise from recognizable sporogenous cells, develop in acropetal succession, but mature from the apex backwards.

**Diagnostic Features:** Spherical, nonseptate, darkly pigmented spores produced in branching chains which develop acropetally but mature basipetally make most species of *Periconia* distinctive (Fig. 157).

**Notes:** The genus *Periconia* has been treated in some detail by Mason and Ellis (1953) and by Subramanian (1955). Most species of the genus are covered in these publications but there are a number of additional species which have been described in more recent literature (see the Index of Fungi of the Commonwealth Mycological Institute).

Records of *Periconia* from soil are not uncommon, especially in recent listings. *P. paludosa* has been recorded by Johnson and Osborne (1964), *P. igniaria* by Brown (1958) and unidentified species by a number of workers. *P. macrospinosa*, originally described from diseased sorghum roots, has been recorded from soil by Warcup (1957) and by Peyronel and Dal Vesco (1955). We have isolated this species on two occasions from Ontario soils and an unidentified *Periconia* species on several occasions.

One of the difficulties in studying soil isolates of *Periconia* is that frequently in culture the macronematous conidiophores are suppressed and the conidia arise more or less directly from the vegetative hyphae.

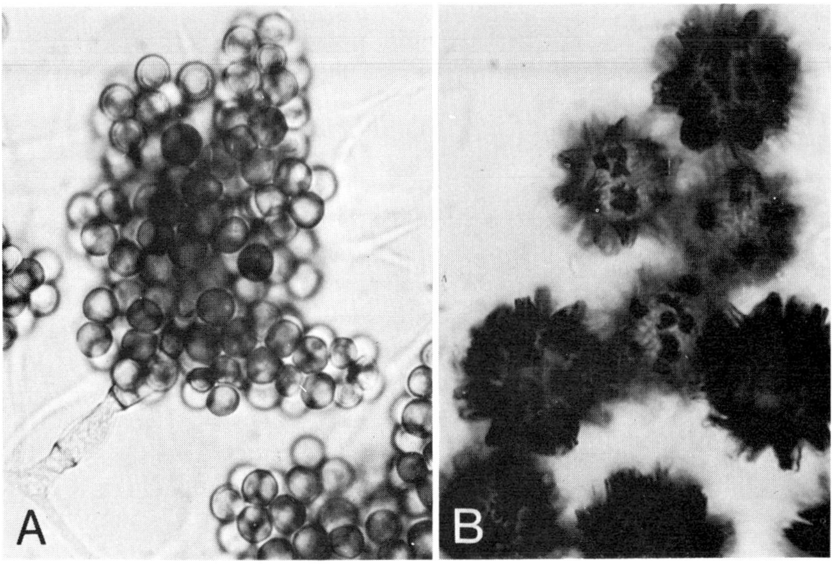


FIG. 157. *A*, *Periconia* sp. The spores are produced in branching chains (OAC 10263). *B*, *Periconia macrospinosa*. This species is characterized by the unusual laminated "spines" on the spore (OAC 10100).

## PERICONIELLA Sacc.

The genus *Periconiella* was erected by Saccardo to accomodate a single species, *P. velutina* (Wint.) Sacc., transferred from *Periconia*. Mason and Ellis (1953) do not pass comment on *P. velutina* and I have not seen this species.

*Periconiella echinochloae* was described by Batista from *Echinochloa polystachya* in Brazil. This same species was recovered from sand in the Sahara by J. Nicot and a subculture was deposited in the collection of the Commonwealth Mycological Institute (IMI 67731).

The conidiophores of *P. echinochloae* are erect, stout, dark brown and septate. The main axis is simple, but branches above to produce an apical penicillus of metulae and sporogenous cells (Fig. 158*B*). The sporogenous cells produce conidia (blastospores) in acropetal succession and these are formed in branching chains. The conidia are long-ellipsoid, thick-walled and roughened. In culture the main conidiophore axis may be suppressed and the conidia develop from micronematous conidiophores little differentiated from the vegetative hyphae (Fig. 158*A* and *C*).

As described above, *P. echinochloae* falls within the generic description of *Periconiella* given by Saccardo (1886) as follows.

> Hyphae fertiles assurgentes, simplices, atrae, septatae sursum in ramulos ascendentes desinentes capitulumque formantes. Conidia ovato-oblonga, continua, fuliginea.

## PESTALOTIA de Not

Type Species: *Pestalotia pezizoides* de Not.

**Generic Description:** Fruiting bodies black, carbonaceous, varying

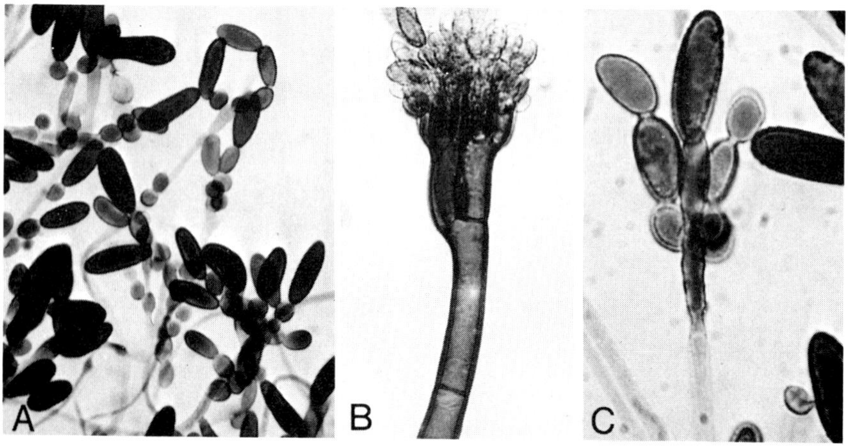


FIG. 158. *Periconiella echinochloae*. *A* and *C*, in culture, conidiophores are suppressed and the conidia develop in acropetal chains from the vegetative hyphae or slightly modified hyphal cells. *B*, this conidiophore from plant material has an apical penicillus of metulae and sporogenous cells (IMI 91387).

from simple acervuli without stromatic area to stromatic apothecoid structures, pycnidia and pseudopycnidia, rarely with a true ostiole and rarely as loose fertile, hyphae without a distinct stratum or stroma. Conidia fusiform, straight or curved, four- to six-celled or loculate and crowned with two or more, rarely one and more, simple or branched setulae, their extremities sometimes spathulate or knobbed, sometimes arising from the slope or base of the apical cells; exterior cells hyaline or rarely dilutely colored, rarely with contents; intermediate cells equally or variably colored, pale brown to almost black, guttulate; pedicels hyaline, simple, rarely branched, attached to the base of the conidia. (Description from Guba, 1961.)

**Diagnostic Features:** *Pestalotia* is characterized by the production of an acervulus bearing dark phragmospores, with hyaline end cells each bearing two or more setulae (Figs. 159 and 160).

**Notes:** *Pestalotia* and related genera are recorded infrequently from soil (Gilman, 1957), although members of this genus are commonly found as parasites of higher plants. A comprehensive monograph on *Pestalotia*, and the closely related *Monochaetia* has been written by Guba (1961). This treatment also includes a number of similar forms under "Miscellanea."

There is some controversy regarding the systematics of *Pestalotia*. Steyaert (1949) reserved the name *Pestalotia* for the single species *P. pezizoides* (the type species). He included the remainder of the species involved into two genera *Truncatella* and *Pestalotiopsis*. *Truncatella* contained the species with four-celled spores with a hyaline epispore which envelopes the two colored cells. *Pestalotiopsis* included species with five-celled conidia. Each of the genera was further subdivided on the basis of number of setulae into Monosetulatae, Bisetulatae, Trisetulatae and Multisetulatae. For the full details of the controversy surrounding this group of fungi the reader is referred to the works of Guba (1956, 1961) and Steyaert (1949, 1955, 1956).

The writer is not prepared to give an authoritative opinion on this situation, but has followed Guba as a matter of convenience, in that a comprehensive manual for identification has been presented in his monograph. Developmental studies in *Pestalotiopsis* and related genera, which show the precise method of origin of the conidia, have been presented by Sutton (1961, 1963).

### PHAEOTRICHONIS Subram.

Type Species: *Phaeotrichonis crotalariae* (Salam and Rao) Subram.

**Generic Description:** Repent hyphae brown, branched, septate; conidiophores not sharply distinguished from the hyphae, erect, septate, with swollen apex, each bearing apically and singly one conidium; conidia acrogenous, brown, elongate, fusiform, with a long appendage at the

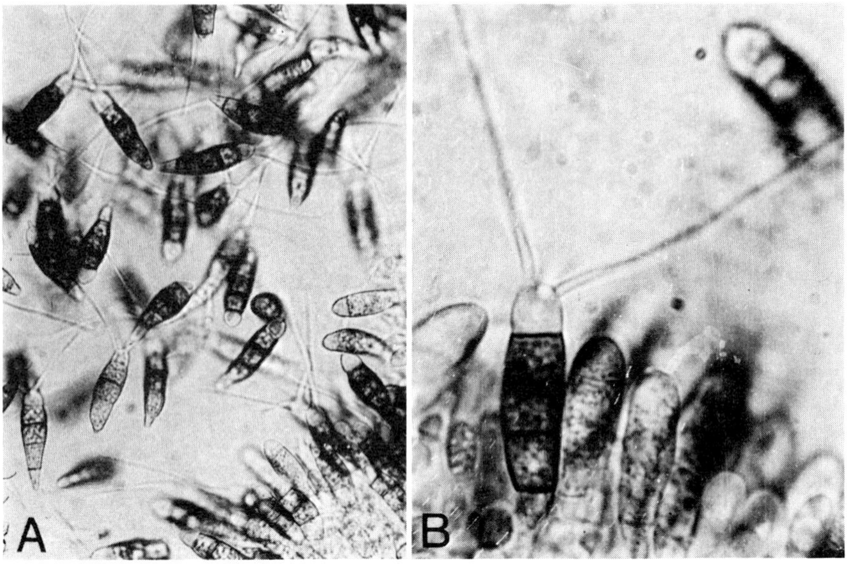


FIG. 159. *A* and *B*, this species has five-celled conidia. It would be classified as *Pestalotia* by Guba or *Pestalotiopsis* by Steyaert.

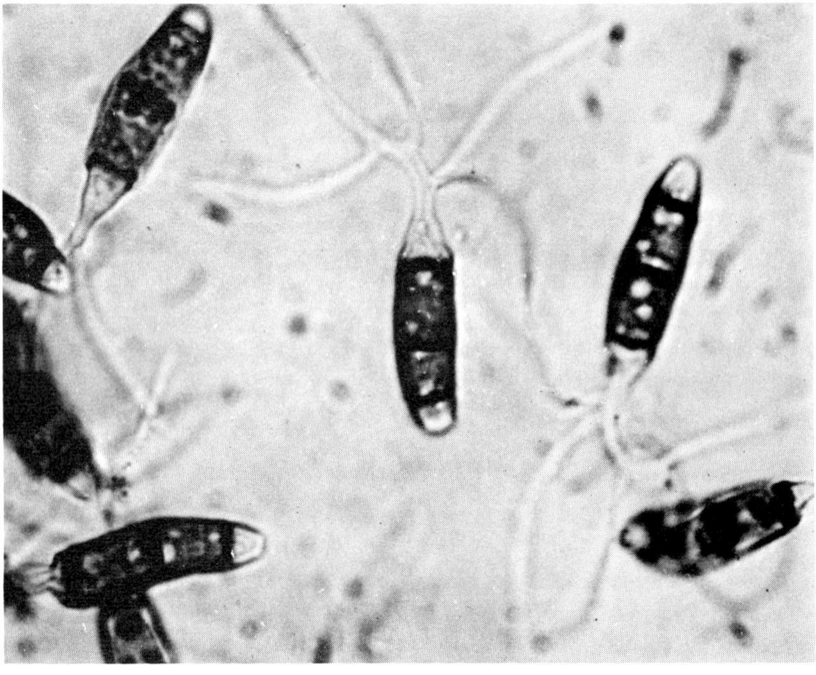

FIG. 160. This species has four-celled conidia. It would be classified as *Pestalotia* by Guba or *Truncatella* by Steyaert.

apex, many times transversely septate. (Description from Subramanian, 1956.)

**Diagnostic Features:** The large, pigmented, septate porospores, with a long, apical appendage, are borne acrogenously on simple, sympodially extending conidiophores (Fig. 161).

**Notes:** This genus is rarely reported from soil. The Herbarium of the Commonwealth Mycological Institute contains a single isolate of a *Phaeotrichonis* species from grassland soil in India. *Phaeotrichonis* pro-

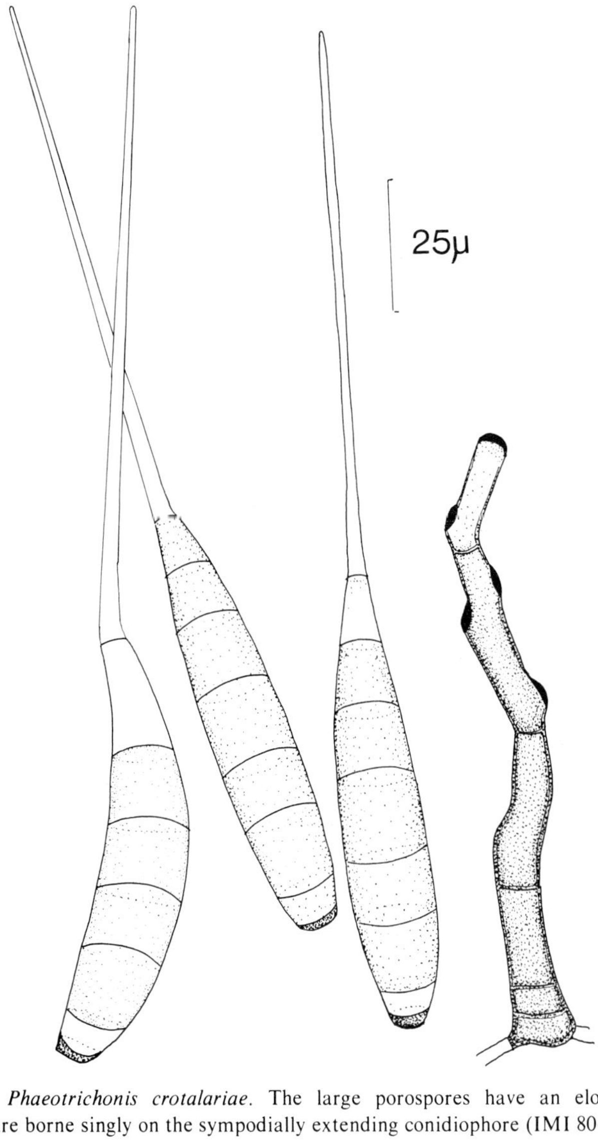


FIG. 161. *Phaeotrichonis crotalariae.* The large porospores have an elongate apical "beak" and are borne singly on the sympodially extending conidiophore (IMI 80606).

duces porospores and the conidia show similarities to *Alternaria*, but, as pointed out by Ellis (in Lucas and Webster, 1964), in *Phaeotrichonis* the spore body and beak are clearly distinct, the conidia are darker, never in chains and never form any longitudinal or oblique septa.

### PHIALOCEPHALA Kendrick

Type Species: *Phialocephala dimorphospora* Kendrick.

**Generic Description:** Conidiophores darkly pigmented, solitary or occasionally clustered, with a long stripe bearing a complex sporogenous head at its apex; head consists of one to several series of metulae, with the ultimate series bearing the numerous phialides; phialides often possessing well marked collarettes and produce numerous small spores; phialospores nonseptate, hyaline to dilute brown, usually aggregate in mucus.

**Diagnostic Features:** *Phialocephala* is characterized by the production of stout conidiophores bearing penicillate heads in which the ultimate cells are phialides. The phialospores are produced in mucous heads which may be white at first but frequently darken in age (Fig. 162).

**Notes:** The genus *Phialocephala* was erected by Kendrick (1961a), who separated this genus from other *Leptographium*-like fungi on the basis of the spore ontogeny. The generic description above is essentially that of Kendrick, modified slightly on the basis of several papers published more recently on the genus by Kendrick (1963, 1964).

Members of this genus are not common in soil. *P. phycomyces* was recorded by Meyer (1959) in the Congo. *P. dimorphospora* was found in

FIG. 162. *Phialocephala phycomyces*. *A*, the conidiophores arise from aerial ropes with the conidia in slimy heads. *B*, the conidiophores are penicillate (IMI 58648).

peat soil in Ireland by Dickinson (private communication), and by Gams (1962) from soil in the Delamere Forest in England. We have not isolated this genus from Ontario soils, but have recovered several isolates of a *Verticicladiella* species which on casual inspection could be mistaken for a *Phialocephala*.

## PHIALOMYCES Misra and Talbot

Type Species: *Phialomyces macrosporus* Misra and Talbot.

**Generic Description:** Conidiophores erect, very tall, hyaline, septate, simple or branched, bearing phialides in monoverticillate clusters at the apex of the conidiophores and branches; phialides flask-shaped; phialospores very large, continuous, darkly pigmented, lemon-shaped, verrucose, produced in short chains. (Description based on *P. macrosporus*.)

**Diagnostic Features:** Very large, dark, warty phialospores borne on flask-shaped phialides in monoverticillate groups characterize *Phialomyces* (Fig. 163).

**Notes:** The genus was erected by Misra and Talbot (1964) to include a single species, *P. macrosporus*. The species is a striking one in appearance. The conidiophores are up to 1 mm in length and the spores are 16 to 20 by 20 to 26 $\mu$. The monoverticillate whorls are reminiscent of *Stachybotrys*. The only two records of this fungus known are from forest soil in Gorakhpur in India and thermophilic soils, Rotorua, New Zealand.

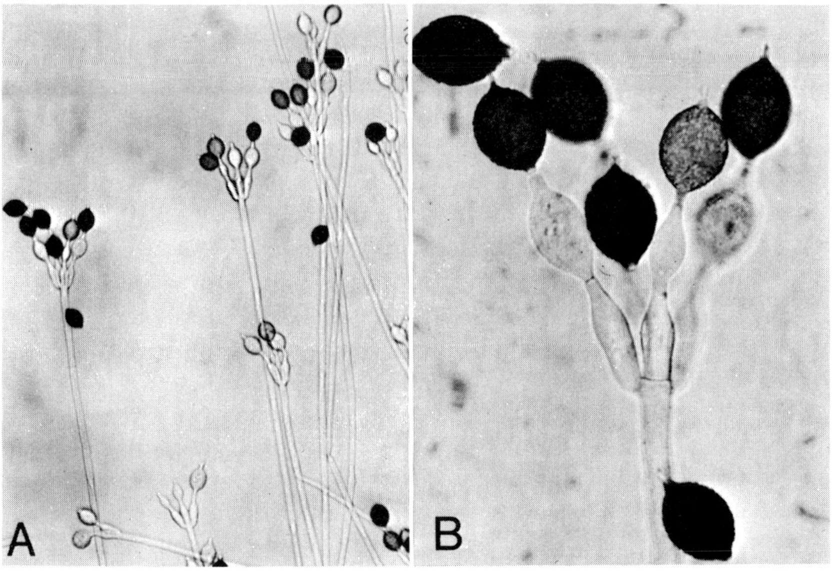


FIG. 163. *Phialomyces macrosporus*. *A*, the tall, hyaline conidiophores have the phialides arranged in an apical cluster (monoverticillate). *B*, the large, rough, lemon-shaped phialospores are produced in short chains (IMI 110130).

## PHIALOPHORA Medlar

Type Species: *Phialophora verrucosa* Medlar.

**Generic Diagnosis:** Conidiophores short or lacking, frequently with the sporogenous cells arising directly from the vegetative hyphae; sporogenous cells (phialides) usually short, cylindrical or inflated, frequently flask-shaped, produced singly or in groups, rarely in penicillate arrangements, usually with a pronounced apical collarette; collarette sometimes lacking; phialospores nonseptate, hyaline or pigmented, spherical to ovoid, produced within collarette and gathering in large gloeoid balls.

**Diagnostic Features:** In most *Phialophora* species conidia are produced singly through a collarette at the apex of short, pigmented phialides and gather in gloeoid balls (Figs. 139 and 164).

**Notes:** Members of this genus are readily identified in most cases. The apical collarette is lacking in some species and the phialospores are produced directly at the apex of somewhat tapering phialides. Such forms have sometimes been included under the genus *Margarinomyces* based on *M. bubakii* Laxa (see Barnett, 1960). A number of workers, however, have pointed out that this is undesirable and that these forms are best considered under a broader concept of the genus *Phialophora.* Wang (1966) has shown that the sporogenous cells in *P. jeanselmei* (= *Torula jeanselmei*) are annellophores rather than phialides, as in *Phialophora.* The common species of the genus are described by Moreau (1963), and Wang (1965) in her treatment of fungi of pulp and paper.

*Phialophora* species are relatively common in soils, especially forest soils. They are frequently isolated from wood, woodpulp and wood products and other cellulosic materials. Some species have been reported as parasitic on humans, causing chromoblastomycosis (see Emmons *et al.*, 1963).

LeGal and Mangenot (1961) have shown that the conidial states of certain *Mollisia* and *Pyrenopeziza* species belong to *Phialophora.* Dr. R. F. Cain, University of Toronto, has advised me in a private communication that there are a large number of Ascomycetes with *Phialophora* or *Phialophora*-like conidial states.

## PHIALOTUBUS Roy and Leelavathy

Type Species: *Phialotubus microsporus* Roy and Leelavathy.

**Generic Description:** Conidiophores simple or with primary, secondary and tertiary branches, bearing phialides at the tips; phialides narrower towards the base, broad and round at the distal end; the distal ends prolonged into long or short, narrow, hyaline, tube-like structures when bearing conidial chains; conidial chains long, with hyaline interconnections; conidia hyaline, becoming yellowish brown at maturity, double-walled and smooth, inner wall dark, outer wall hyaline. (Description from Roy and Leelavathy, 1966.)

**Diagnostic Features:** The authorities of this genus regard the double-walled conidia borne on phialides with tubular extensions as diagnostic.

**Notes:** This genus, described by Roy and Leelavathy (1966), was originally recovered from the rhizosphere of *Dichanthium annulatum*. Roy and Leelavathy noted that *Phialotubus* resembled *Phialomyces*, but I think that there is little possibility of confusing these two genera. The differences between *Phialotubus* and *Paecilomyces* or even *Penicillium* are not so obvious. Roy and Leelavathy point out that the fungus differs from *Paecilomyces* in that the long projection of the sporogenous cell is not bent away from the main axis of the conidiophore. This is a feature of some species of the genus *Paecilomyces*, but it is by no means diagnostic

FIG. 164. *Phialophora verrucosa*. Note the prominent dark collarette characteristic of this species (OAC 10508).

for the genus as a whole. *Paecilomyces* and *Penicillium* are large and variable genera, and the precise differences between these two genera and *Phialotubus* would be worth elaborating.

## PIRICAUDA Bubak

Type Species: *Piricauda paraguayense* (Speg.) Moore.

**Generic Description:** Conidiophores undistinguished, appear as arched branches in a semicircular loop from which further successive similarly curved branches develop in various planes, giving a tangle of curved hyphae; loops are closely septate and pale brown to brown; conidia porospores develop singly from a pore on the arched conidiophore, ovoid, brown, minutely roughened, muriform, apical cell prolonged to form a straight or slightly curved appendage several times as long as the spore. (Description based on Hughes, 1960.)

**Diagnostic Features:** Muriform porospores, each with a long, apical prolongation, are distinctive for *Piricauda* (Fig. 165).

**Notes:** There is some confusion surrounding *Piricauda*. The genus was considered in some detail by Moore (1959). It was pointed out by Hughes (1960), however, that in the original descriptions of *P. paraguayense* and its synonyms, as well as in Moore's redescriptions, the long, attenuated conidium appendage was regarded as the conidiophore. This misinterpretation of the appendage as a persistent conidiophore suggested that the spores were aleuriospores rather than porospores, as shown by Hughes. As a result, a member of quite unrelated aleuriosporous fungi such as

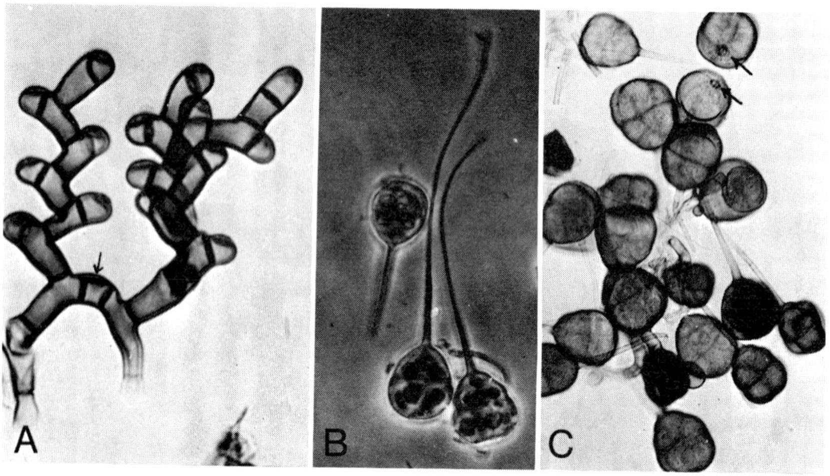


FIG. 165. *Piricauda paraguayense. A, arrow,* the site of the first-formed conidium which arose from a primary sporogenous loop. Sympodially elongating conidiophores have developed from each side of the loop. *B,* conidia are muriform with a conspicuous apical appendage. *C,* the pores (*arrows*) can be seen in the base of some of the spores. (DAOM 66392).

*Pithomyces* were incorporated into *Piricauda* by Moore. The above generic description is therefore more or less restricted to *P. paraguayense* as described by Hughes.

The genus may never have been recorded from soil. The sole isolation is that of Upadhayay, who recorded *P. nodosa* from Brazil. This species as figured by Upadhayay is not a *Piricauda* and may be a *Monodictys*.

## PITHOMYCES Berk. and Broome

Type Species: *Pithomyces flavus* Berk. and Br.

**Generic Description:** Mycelium subhyaline to brown, smooth or rough-walled; conidiophores inconspicuous, consisting of short, subhyaline pedicels arising laterally, more or less at right angles to the hyphae; conidia aleuriospores formed singly as blown out ends at the apex of each conidiophore, subspherical, oval, obovoid, clavate, oblong, pyriform or obpyriform, pale brown to dark brown, occasionally smooth, but more commonly verruculose or echinulate, with up to eight transverse septa, longitudinal septa produced in some species; conidia becoming detached through rupture of the wall of the sporogenous cell.

**Diagnostic Features:** The common species of the genus have dark, septate, rough-walled aleuriospores borne singly on short stalks (Figs. 125*B* and 166).

**Notes:** The genus *Pithomyces* has been treated in detail by Ellis (1960). *P. chartarum* is perhaps the most common species of the genus and is best known as the causal agent of facial eczema in sheep in New Zealand (Dingley, 1962). *P. chartarum* has been recorded from soil by Goos (1963), and in Ontario we have isolated both *P. chartarum* and *P. maydicus*. An interesting account of the discovery of *P. chartarum* in Britain is given by Gregory and Lacey (1964), who discovered it in a spore trap and, following up a concentration gradient of spores, eventually located the source in the debris of *Holcus lanatus*.

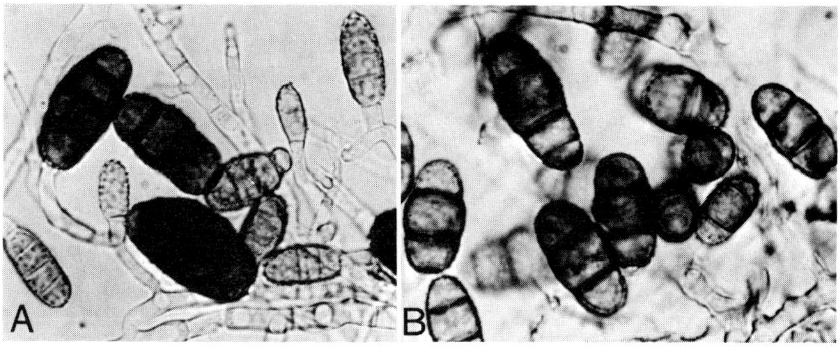


FIG. 166. *A*, *Pithomyces chartarum*. Muriform aleuriospores develop from short lateral branches (OAC 10145). *B*, *P. maydicus*. In this species the conidia are phragmospores (OAC 10257).

*Pithomyces* species may well be more common in soil than the records indicate. It has possibly been mistaken for *Stemphylium* due to the superficial similarities of the spores. The conidia of *Stemphylium*, however, are porospores.

## PLEUROPHRAGMIUM Cost.

Type Species: *Pleurophragmium bicolor* Cost.

**Generic Description:** Conidiophores brown, simple or branched, main axis and branches terminating in sporogenous cells; sporogenous cells geniculate after spore production; spores formed singly at successive apices as sporogenous cell extends sympodially, sometimes appearing as apical clusters; conidia hyaline or brown, one- or many-septate, truncate and frequently darker at the point of attachment.

**Diagnostic Features:** *Pleurophragmium* is characterized by septate, pigmented conidia borne on a geniculate sympodula (Fig. 167).

**Notes:** *Pleurophragmium* is close to *Rhinocladiella* and includes *Rhinocladiella*-like fungi in which the conidia are septate. Reports of this genus from soil are rare. Meyer (1959) figured three species of *Pleurophragmium* from soil in the former Belgian Congo but two of these resembled *Rhinocladiella* in having nonseptate conidia. Some species presently contained within *Rhinocladiella*, however, occasionally produce septate spores, thus the distinction between the two genera becomes a little vague.

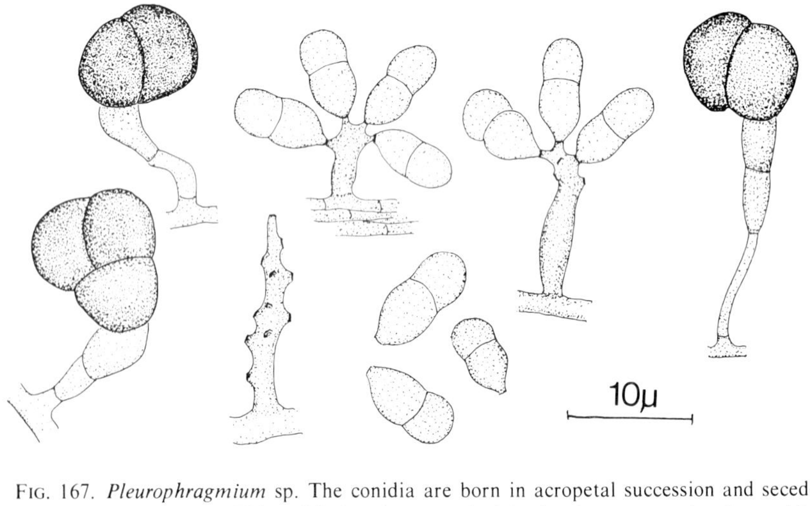


FIG. 167. *Pleurophragmium* sp. The conidia are born in acropetal succession and secede to leave pronounced scars. This soil isolate is not typical in that in most species the conidia are phragmospores (IMI 67833).

## POLYSCYTALUM Riess

Type Species: *Polyscytalum fecundissimum* Riess.

The genus *Polyscytalum* was reviewed by Ciferri and Caretta (1962), who gave the following emended diagnosis.

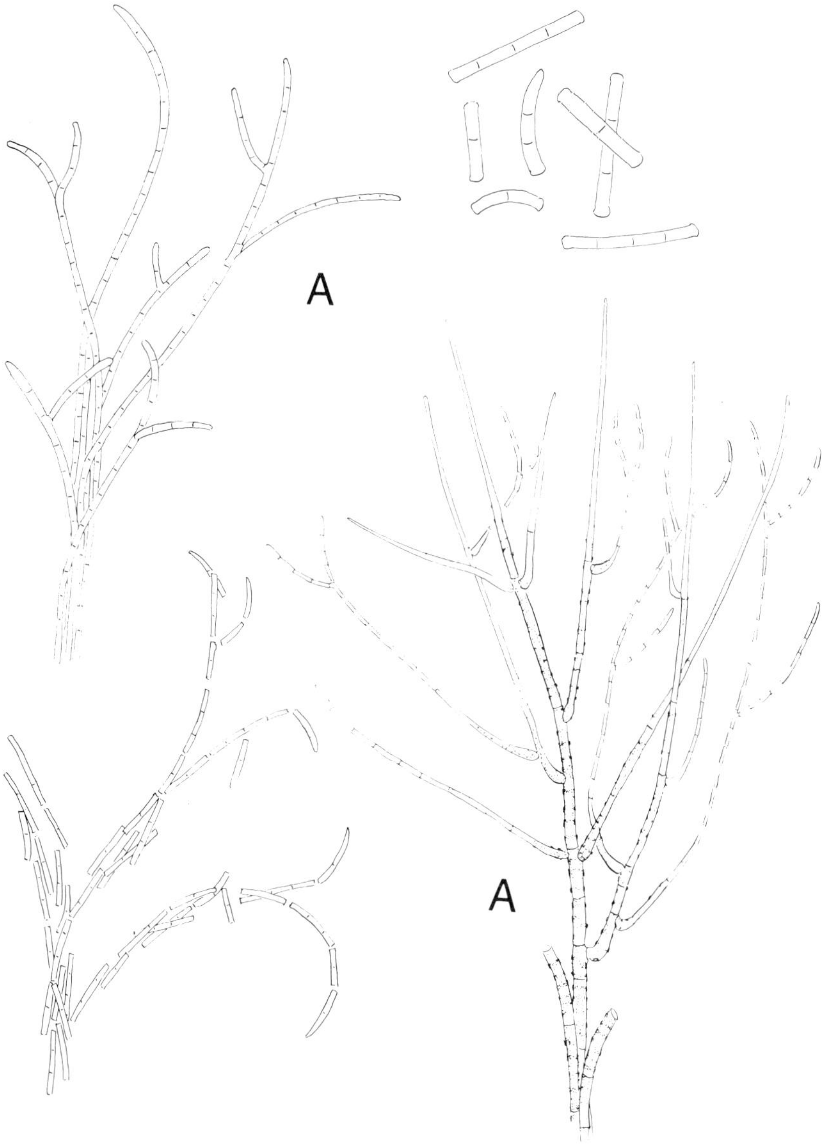


FIG. 168. *A*, *Polyscytalum* sp. The conidia are arthrospores. (See following part *B*.)

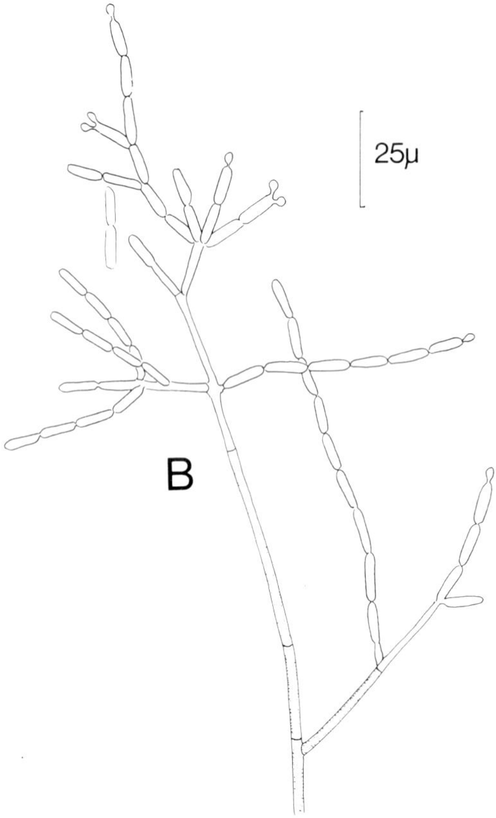


FIG. 168. *B*, *Polyscytalum pustulans*. The dark conidiophores give rise to branching chains of hyaline, cylindrical blastospores (IMI 47720). (See preceding part *A*.)

Fertile hyphae erect, subrigid, branched, subhyaline to pale-brown; spores (arthrospores) elongate-cylindric, septate in the middle with rounded ends, in branching chains, borne in series from the apex of the fertile branches.

Ciferri and Caretta studied the taxonomy of *Polyscytalum* in relation to *Geotrichum* and *Cylindrium*. Their opinion was based on authentic material of a variety of the type species *P. fecundissimum* var *macrosporum* Riess. They regarded *P. griseum*, *P. chymophilum* and *P. saccardianum* as synonyms of *Geotrichum candidum*. They also found a Brazilian isolate of *P. murinum* to be conspecific with *Cylindrium heteronemum*, which they redisposed as *Fusidium heteronemum*.

According to the concept of Ciferri and Caretta the species shown in Figure 168*A* would be a *Polyscytalum*. In this, the more or less erect, brown conidiophores produce branching chains or ropes of hyaline sporogenous hyphae which at maturity break up into units of one to several cells.

There is another concept of *Polyscytalum* shown in Figure 168*B*. In this, the conidiophores are again more or less erect but at the apex produce branching chains of blastospores rather than arthrospores.

## PSEUDOBOTRYTIS Krzemieniewska and Badura

Type Species: *Pseudobotrytis terrestris* (Timonin) Sub.

**Generic Description:** Conidiophores erect, slender, tapering slightly towards the apex, unbranched, septate, dark brown, apex slightly inflated and bearing up to 12 sporogenous cells; main conidiophore axis may extend by proliferation through the apex, to give a second whorl of sporogenous cells; sporogenous cells unbranched, nonseptate, irregularly swollen at the apex, bearing spores on pronounced denticles arising from the apical vesicle; conidia produced singly, continuous or one-septate, with well marked attachment point, ovate to short-cylindric, sometimes indented at the septum, pale to dark brown. (Description based on *P. terrestris* and *P. bisbyi.*)

**Diagnostic Features:** The apical cluster of sporogenous cells on a tall, simple conidiophore with each sporogenous cell bearing an apical cluster of spores on denticles makes *Pseudobotrytis* distinctive (Figs. 15*A* and 169).

**Notes:** *Pseudobotrytis* was erected by Krzemieniewska and Badura

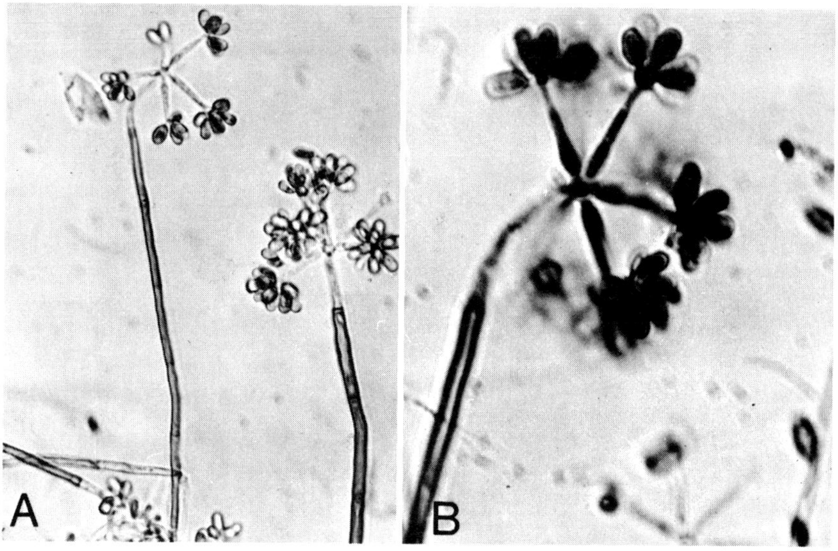


FIG. 169. *Pseudobotrytis bisbyi*. *A* and *B*, the spores arise in apical clusters from the apex of the radiating sporogenous cells (OAC 10355, ex type).

(1954) with *P. fuska* as the type. Unaware of this publication, Morris (1955) described the same fungus in a new genus, *Umbellula*, into which he transferred *Spicularia terrestris* of Timonin (1940). Subramanian (1956b) resolved the matter by making the combination *Pseudobotrytis terrestris* (Timonin) Sub.

Records of *Pseudobotrytis* are rare from soil, but both *P. terrestris* and *P. bisbyi* have been discovered in soil by Timonin (1961). *P. terrestris* was recorded by Stotzky *et al.* (1962) and by Meyer (1959).

## PYCNOSTYSANUS Lindau

Type Species: *Pycnostysanus resinae* Lindau.

**Generic Description:** Conidiophores aggregated into stout coremia; coremia sterile below, branching above to form a spore-bearing head,

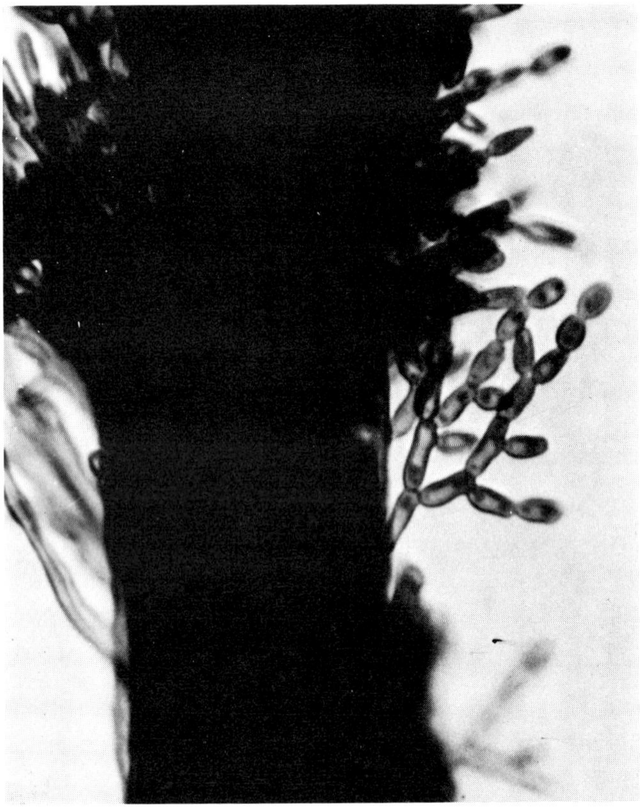

FIG. 170. *Pycnostysanus resinae*. The blastospores are produced in *Cladosporium*-like branching chains from a synnema. (From slide supplied by Dr. E. Morris).

darkly pigmented; conidia blastospores, produced in acropetal succession by budding to form branching chains, continuous, smooth, dark.

**Diagnostic Features:** *Pycnostysanus* is a synnematal fungus producing branching chains of blastospores (Fig. 170).

**Notes:** This genus is rather easily recognized. It appears superficially like a coremial *Cladosporium* and is therefore readily distinguished from the other coremial Hyphomycetes from soil. The genus is rarely reported from soil (Gilman, 1957), and the only record I am aware of in the recent literature is that of Orpurt (1964).

## PYRICULARIA Sacc.

Type Species: *Pyricularia grisea* (Cooke) Sacc.

**Generic Description:** Conidiophores more or less erect, simple or rarely sparingly branched, septate, hyaline to lightly pigmented, ultimate cells sympodulae; conidia borne singly and terminally at the apex of the conidiophore with successive conidia being produced in acropetal succession by sympodial extension of the sporogenous cell; abscission of the conidia leaves pronounced denticles on the spore-bearing apex; conidia ellipsoid or more often pyriform, broader and truncate at the attachment point, tapering towards the distal end, mostly one- or two-septate, hyaline to lightly pigmented.

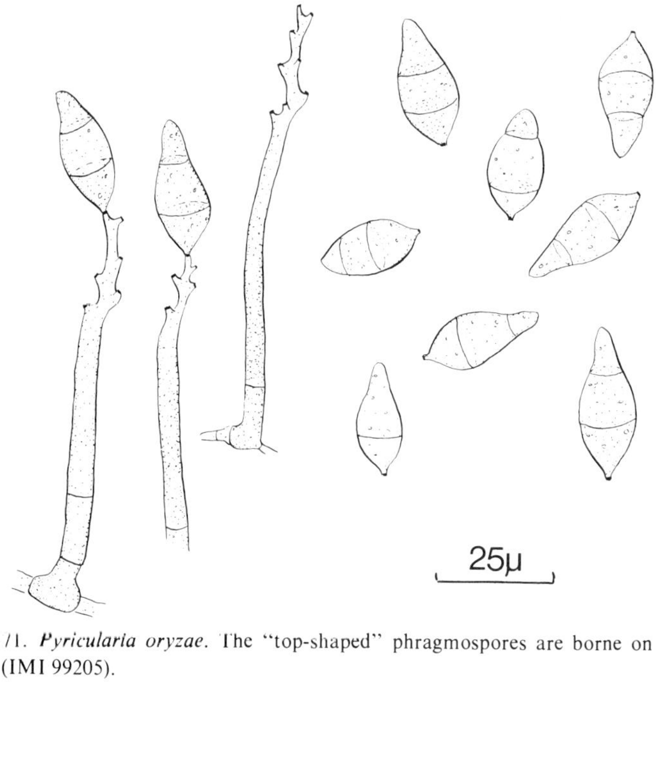


FIG. 171. *Pyricularia oryzae*. The "top-shaped" phragmospores are borne on prominent denticles (IMI 99205).

**Diagnostic Features:** The pyriform, lightly pigmented phragmospores borne in acropetal succession on simple conidiophores with truncate denticles are characteristic of *Pyricularia* (Fig. 171).

**Notes:** *Pyricularia* is commonly found as a parasite of graminaceous hosts. The only record from soil is that of Brown (1958). The genus approaches the *Cercospora/Cercosporella* complex in its morphology, but is readily distinguished on spore shape.

## RACODIUM Pers. ex Wallr.

Hughes (1958), using Persoon 1801 as starting point, cites *R. aluta* Pers. as the lectotype of the genus and regarded this as Mycelia Sterilia. According to de Vries (1952), Fries assumed *Racodium rupestre* Persoon as type of the genus, but he did not regard this species as representing a fungus. De Vries suggested that the genus *Racodium* would therefore become doubtful and the name invalid. He recommended that the name *Racodium cellare*, however, be retained because it is the current name (used since 1801) and there is no doubt as to what fungus Persoon was dealing with. *R. cellare* (Fig. 172) is the only record of this genus from soil (in the records of the Commonwealth Mycological Institute).

Schanderl (1936) transferred *R. cellare* to *Cladosporium*, but de Vries disagreed with this disposition on the following basis.

> *R. cellare* is easily distinguishable from *Cladosporium* colonies. The growth is much slower at 18°C. The colony is hairy and dark green, never with powdery conidial areas. The conidial structures are scattered through the mycelium. The conidial heads are umbellate. Long sterigma-like conidia could be found which were densely covered by single oval or ovate, 1-celled conidia arising from minute hila on the surface of the conidiophore thus forming a structure which resembled the *Acrotheca* form of *Fonsecaea pedrosoi*. This character has never been observed in any *Cladosporium* species investigated by the writer.

*Fonsecaea pedrosoi* is now generally referred to as *Phialophora pedrosoi*, one of the causal agents of chromoblastomycosis of man (Emmons *et al.*, 1963). This is an interesting polymorphic fungus which produces three different conidial states, a *Phialophora* state, a *Cladosporium* state and a *Rhinocladiella*-like state in which the conidia are produced singly and acrogenously on a sympodially extending conidiophore. This latter would be the *Acrotheca*-like state referred to by de Vries for *Racodium cellare*. *Cladosporium* is associated with a Rhinocladiella-like state in *P. pedrosoi*.

I would agree with Schanderl and interpret the "*Acrotheca*-like" conidial state of *R. cellare* as an indication of affinities with *Cladosporium*

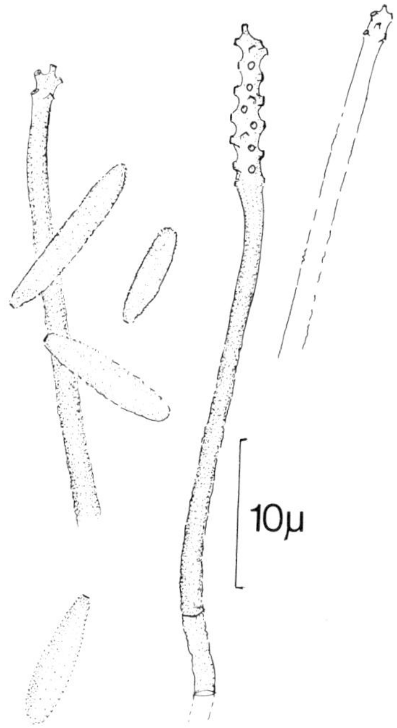


FIG. 172. *Racodium cellare*. The conidiophores are little differentiated from the vegetative hyphae. The conidia are *Cladosporium*-like (IMI 44943).

and that *Cladosporium cellare* would be the most satisfactory disposition of *R. cellare* at this time.

## RHINOCLADIELLA Nannf.

Type Species: *Rhinocladiella atrovirens* Nannf.

**Generic Description:** Conidiophores more or less erect, simple or branched in a few species, septate, brown, main axis and branches terminate in solitary sporogenous cells; sporogenous cells sympodulae, cylindrical or sometimes slightly inflated and with prominent scars in the spore-bearing region; conidia produced terminally and in acropetal succession during sympodial extension of the sporogenous cell; conidia lightly pigmented, globose to ovoid or long-ellipsoid, continuous, rarely one-septate, smooth or sometimes minutely roughened, narrowly truncate at the attachment end, rounded at the distal end.

**Diagnostic Features:** The majority of *Rhinocladiella* species produce sympodulae with an elongate, scarred sporogenous apex bearing a succession of amerospores (Fig. 173).

**Notes:** *Rhinocladiella* was erected by Nannfeldt (see Melin and Nannfeldt, 1934) to include a single species, *R. atrovirens*. In this species the conidia secede to leave raised, dark scars. In most of the soil literature

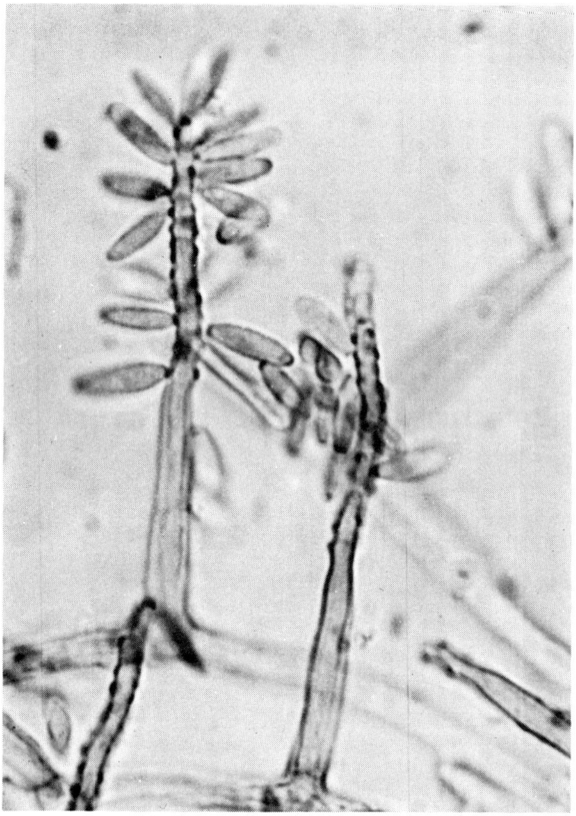

FIG. 173. *Rhinocladiella* sp. Conidia are produced in acropetal succession and secede to leave prominent, dark scars on the sporogenous cell (OAC 9002).

*Rhinocladiella* species have been listed and described under *Chloridium*. It was shown by Hughes (1958), however, that the type species of *Chloridium*, *C. viride*, produces phialospores. *Rhinocladiella* shows affinities to *Nodulisporium*, *Geniculisporium*, *Hansfordia* and *Pleurophragmium*. The conidial state of *Acrospermum gramineum*, as figured by Webster (1956), appears *Rhinocladiella*-like.

## RHINOCLADIUM Sacc. and March.

Type Species: *Rhinocladium coprogenum* Sacc. and March.

**Generic Description:** Vegetative hyphae stout, darkly pigmented, septate; conidiophores lacking, conidia borne directly on the vegetative hyphae, sessile, dark, thick-walled, with prominent germ pores, variable in size and shape, anvil-shaped, triangulate, ovate, or irregularly angular, nonseptate. (Description based on *R. coprogenum*.)

**Diagnostic Features:** The type species, *R. coprogenum*, has very dark, angular, thick-walled, sessile aleuriospores with prominent germ pores (Fig. 174).

**Notes:** This genus is rarely reported from soil. The only records are those of Kamyschko (1960), who described *R. nigrosporoides* and *R. sporotrichoides* as new from soil in the USSR. I have not seen these species.

## RHIZOCTONIA de Candolle

Type Species: *Rhizoctonia violacea* Tul.

**Generic Description:** No asexual spores present; sclerotia irregular in

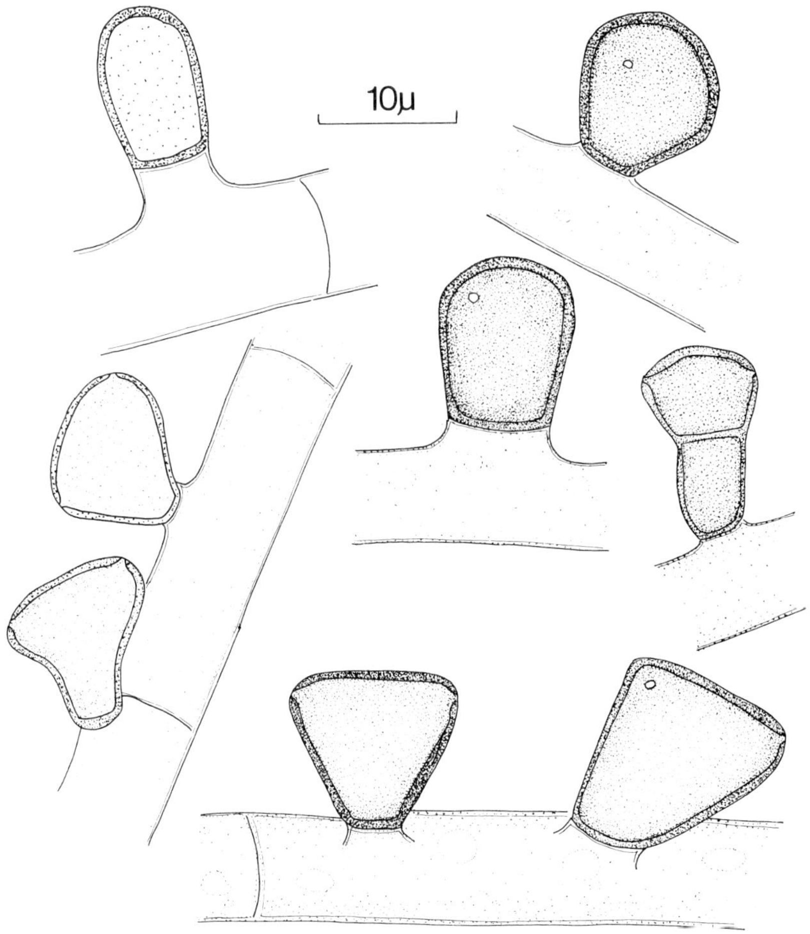


FIG. 174. *Rhinocladium coprogenum*. The large aleuriospores are sessile on the hyphae and have prominent germ pores (IMI 17311).

shape and size, sometimes confluent, brown or black, more or less loosely packed; hyphae stout, anastomosing frequently, branching more or less at right angles, pale brown to brown.

**Diagnostic Features:** The broad hyphae and irregular, somewhat spongy sclerotia serve to distinguish *R. solani*, the common species (Fig. 175).

**Notes:** *Ainsworth and Bisby's Dictionary of the Fungi* indicates that there are about 15 species in the genus *Rhizoctonia*. Because conidia are not produced, the taxonomy of this group of species is very difficult. Saksena and Vaartaja (1960) studied 60 representative strains of 700 isolates of *Rhizoctonia* recovered from forest nurseries in Saskatchewan and described four new species. These workers drew attention to the difficulties with this group, especially in the light of the incomplete and confusing descriptions in the literature. There is a wealth of material on *Rhizoctonia* as a pathogen in the literature of plant pathology.

Papavizas and Davey (1961) studied the saprophytic behavior of *Rhizoctonia* in unsterilized soil. Survival decreased with prolonged incubation, but the fungus was still recoverable after 120 days.

The nomenclature concerning the perfect state of *Rhizoctonia solani* has been satisfactorily resolved by Donk (1956) who recognized the Basidiomycete *Thanatephorus cucumeris* (Frank) Donk as the perfect state. Details on the culture and morphology of this latter species are also given by Warcup and Talbot (1962).

Parmeter *et al.* (1967) noted that the genus *Rhizoctonia* is a heterogeneous assemblage of mycelia of Basidiomycetes, Ascomycetes and

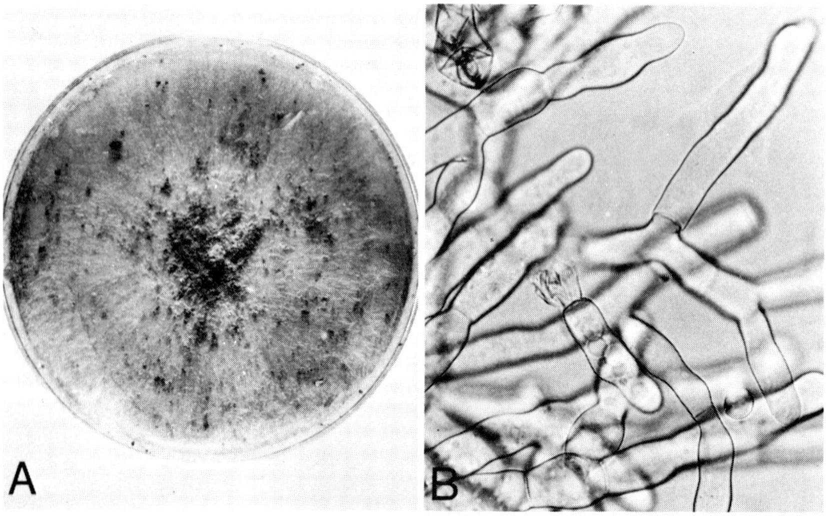


FIG. 175. *Rhizoctonia solani*. *A*, 3-week-old colony on potato dextrose agar. *B*, no conidia are produced. The hyphae are broad and pigmented (OAC 10279).

Fungi Imperfecti and that many of the mycelia in *Rhizoctonia* bear little resemblance to the mycelial state of *Thanatephorus cucumeris*. Comparative studies by Parmeter *et al.* showed that *Rhizoctonia solani*-like isolates could be separated into two groups: one with multinucleate hyphal cells and one with binucleate hyphal cells. Multinucleate isolates produced the *Thanatephorus cucumeris* perfect state, whereas the binucleate isolates produced a *Ceratobasidium* perfect state. They noted that *R. endophytica*, *R. callae* and *R. fragariae* are all mycelial states of a single *Ceratobasidium* species and pointed out that much of the previous work on *R. solani* might need reevaluation as frequently two distinct fungi may have been involved.

## RHODOTORULA Harrison

Type Species: *Rhodotorula glutinis* (Fres.) Harrison.

**Generic Description:** Cells round, oval, or elongate; reproduce by multilateral budding; sometimes produce pseudomycelium; distinct red or yellow pigments produced; do not ferment. (Description from Lodder and van Rij, 1952.)

**Diagnostic Features:** This genus includes red or yellow yeasts microscopically similar to *Cryptococcus* (Fig. 176).

**Notes:** Lodder and van Rij noted that the genera *Rhodotorula* and

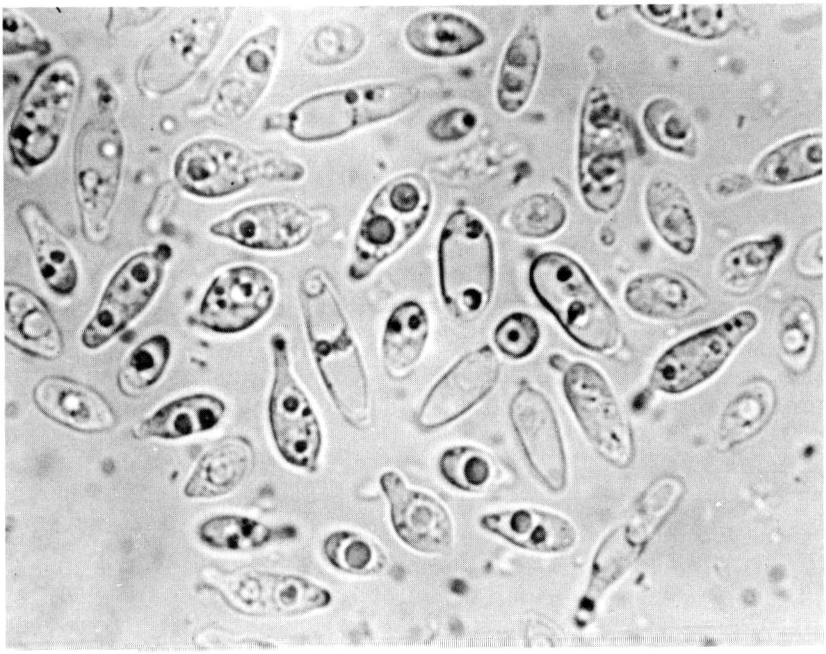

FIG. 176. *Rhodotorula aurantiaca*. No mycelium is produced. The blastospores are morphologically similar to *Cryptococcus*. (ATCC 10657).

*Chromotorula* were erected by Harrison for the asporogenous, pigment-forming yeasts, the former for the red yeasts and the latter for those other than red. They pointed out that it was hardly possible to separate red, yellow-red, orange and yellow yeasts, and also noted that the black yeasts placed in *Chromotorula* deviated markedly in morphological aspects from the other forms assigned to the group. The asporogenous yeasts with carotenoid pigments were therefore placed in *Rhodotorula*, and *Chromotorula* was rejected. For the taxonomy of *Rhodotorula* and related genera see Lodder and van Rij (1952). Members of this genus have been recorded from glaciated soils in Alaska by Cooke and Lawrence (1959) and from bat cave soil in Honduras by Orpurt (1964). Di Menna (1965) indicated that *Rhodotorula* was probably common in soil.

## RIESSIA Fres.

Type Species: *Riessia semiophora* Fres.

**Generic Description:** Synnemata erect, capitate-clavate; stalk composed of extremely thin, septate, hyaline hyphae, which diverge near the apex to form conidiophores; conidia of four, rarely five, cells radiating out from a small central cell, hyaline, produced singly, dry. (Description from Morris, 1963.)

**Diagnostic Features:** The unusual spore form, which has the aspect of a four-leaf clover in *R. semiophora*, makes this genus readily distinguished from all other coremial forms (Fig. 177).

**Notes:** Members of this genus are rarely reported from soil. The only record is that of Kamyschko (1961), who described *R. naumovii* as new from soil in the USSR. I have not seen this species.

In a study of an isolate of *Riessia semiophora*, isolated from decayed wood collected in Rhode Island, Goos (1967) found hyphae bearing clamp connections. He subsequently found clamp connections in the type material of *R. semiophora*. Conidia from the Rhode Island strain germinated to produce mycelia with clamp connections, confirming his conclusions that this species was a dicaryotic Basidiomycete. Goos also noted that the compound conidium of *R. semiophora* appeared to consist of several cells (usually four) which fused during development into a single morphological unit.

## SCLEROTIUM Tode

Type Species: *Sclerotium complanatum* Tode.

**Generic Description:** No conidia produced; sclerotia variable, sometimes regular and more or less globose; other times irregular in size and shape, compact, hard, brown to black, with pseudoparenchymatous rind; hyphae hyaline to subhyaline, clamp connections sometimes present.

**Diagnostic Features:** Hard, dark sclerotia produced on sterile, white mycelium are distinctive for *Sclerotium* (Fig. 178).

**Notes:** *Sclerotium* species are probably best known as plant pathogens and there is a wealth of information concerning the genus in the pathology literature. Two of the best known species are *Sclerotium rolfsii* and the sclerotial state of *Sclerotinia sclerotiorum.*

*S. rolfsii* attacks a wide range of host plants and was the subject of a symposium (see *Phytopathology, 51:* 107) which contained contributions from a number of authors on the history, taxonomy, host range, distribution and etiology of this fungus and the diseases it causes. *S. rolfsii* has been connected to its perfect state which belongs in the genus *Pellicularia* of the Basidiomycetes.

The taxonomy of *Sclerotinia sclerotiorum* and related species was studied by Purdy (1955). Purdy concluded that it was not possible to distinguish between *S. sclerotiorum*, *S. minor* or *S. trifoliorum* on the basis of asci or ascospores and that, because of the wide range in size of the sclerotia produced and the intergradation in size between species, this character was of little value in species determination. Purdy considered all three species as *S. sclerotiorum.*

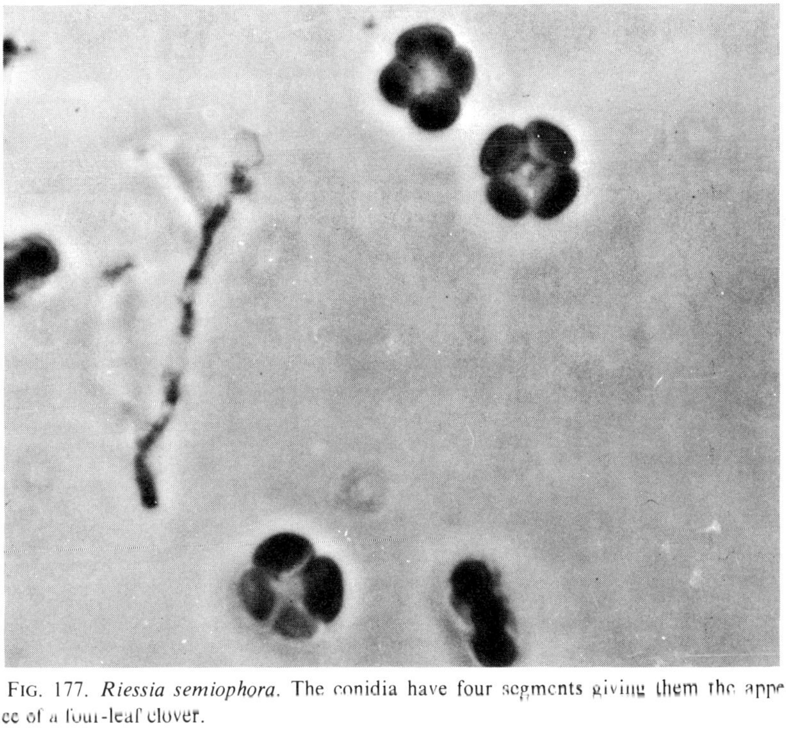

FIG. 177. *Riessia semiophora.* The conidia have four segments giving them the appearance of a four-leaf clover.

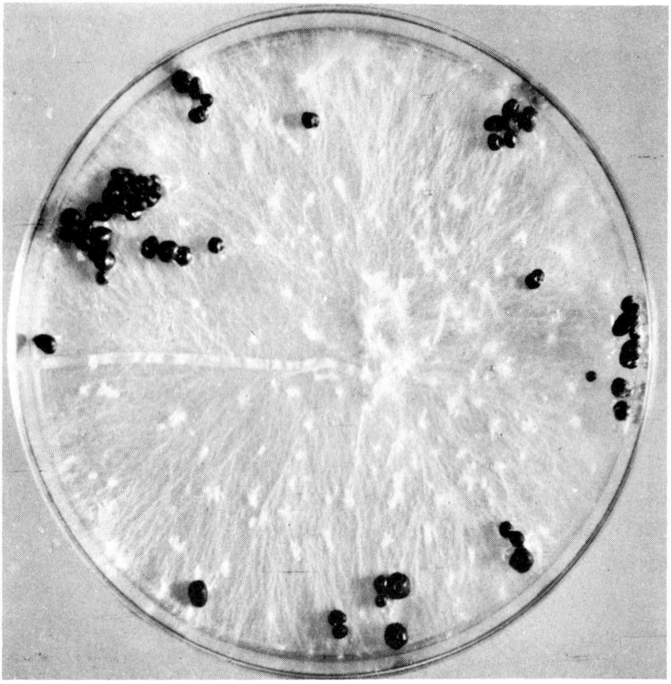

FIG. 178. *Sclerotium rolfsii*. No conidia are produced. The colonies produce hard, black sclerotia in concentric rings or irregular clusters (ATCC 16648).

*Sclerotium* is not recovered frequently in standard studies of the soil mycoflora (Gilman, 1957).

## SCOLECOBASIDIUM Abbott

Type Species: *Scolecobasidium terreum* Abbott.

**Generic Description:** Conidiophores arise from the aerial hyphae or hyphal ropes, continuous or septate, hyaline or pigmented, ovoid, clavate, wedge-shaped, cylindrical, or irregular; conidia produced in clusters or acropetal series from the ends of tubular extensions of the conidiophores; conidia Y-shaped, T-shaped, ovoid, or cylindric, continuous or more often one- to three-septate, olivaceous-brown, smooth or more often finely echinulate to verrucose.

**Diagnostic Features:** This genus is distinguished by the septate, pale brown conidia, developing in acropetal succession on narrow pedicels (Figs. 117*I*, 179 and 180). In most species the conidiophores are relatively short.

**Notes:** *Scolecobasidium* species are relatively common in soils, particularly those high in organic matter. The genus was reviewed by Barron and Busch (1962), but since this publication several species have been added to the genus. *S. verruculosum* and *S. macrosporum* have both been described from soil in India by Roy *et al.* (1962). In Italy, Graniti (1962) described *S. anellii* producing a superficial darkening on the surface of stalactites in the Castellara grottoes.

The ecology of *Scolecobasidium* species is of some interest in that they are frequently fairly numerous on dilution plates. According to Upadhyay (1966) of the Instituto de Mycologica, Recife, Brazil, *S. constrictum* is predaceous on nematodes.

## SCOPULARIOPSIS Bainier

Type Species: *Scopulariopsis brevicaulis* (Sacc.) Bain.

**Generic Description:** Conidiophores short and undistinguished; sporogenous cells anellophores; annellophores variable, short and swollen, flask-shaped or cylindrical, produced singly or in groups or in complex penicillate arrangements; spores globose, subglobose, ovate or pyriform, frequently truncate at the attachment point, papillate or rounded at the distal end, produced in long chains, smooth or coarsely roughened, non-septate, hyaline to darkly pigmented.

**Diagnostic Features:** Annellophores which arise singly or in groups and give rise to long chains of basally truncate amerospores serve to distinguish *Scopulariopsis* (Figs. 181 and 182).

**Notes:** Annellations on the sporogenous cells are not always prominent

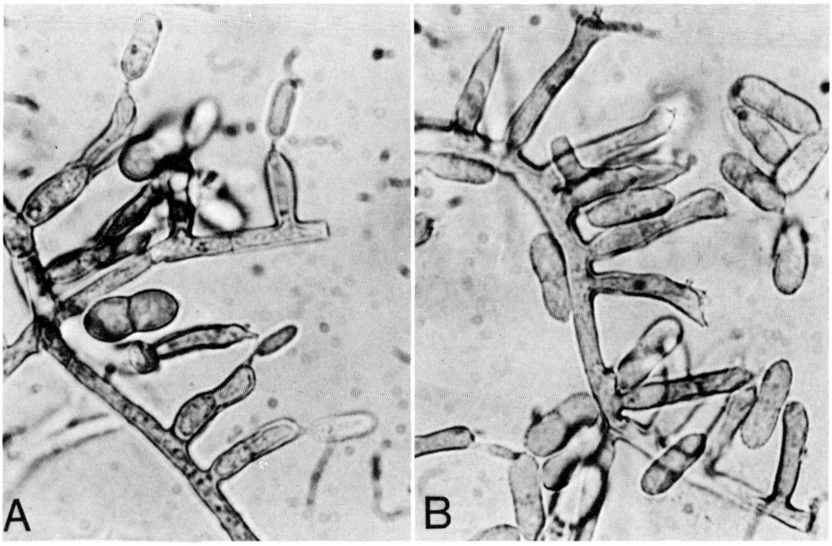


FIG. 179. *Scolecobasidium constrictum. A* and *B*, the two-celled conidia are blown out at the ends of slender pedicels arising from the sporogenous cells. The conidia secede to leave the pedicels as delicate tubular appendages on the sporogenous cell (OAC 10469).

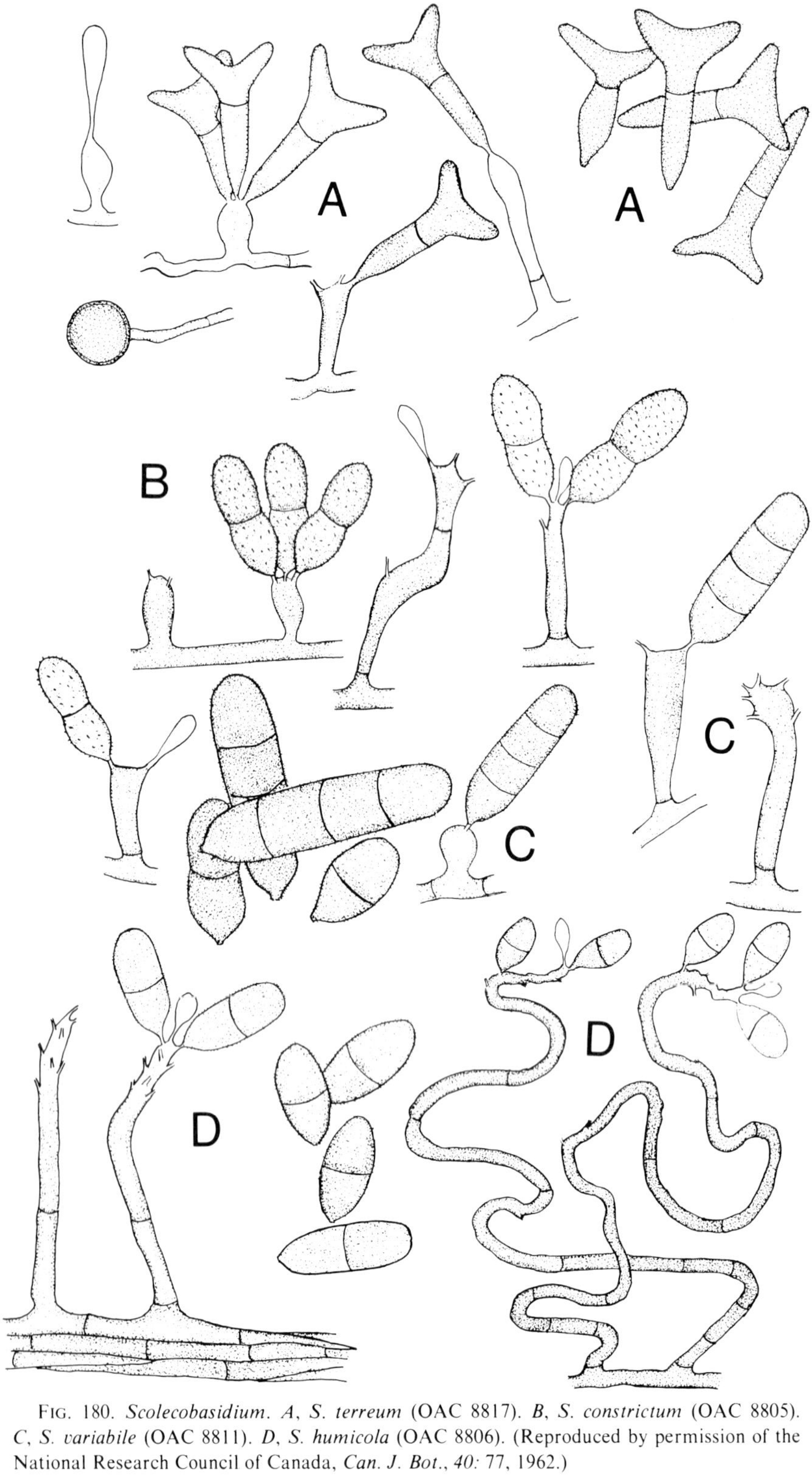


FIG. 180. *Scolecobasidium*. *A*, *S. terreum* (OAC 8817). *B*, *S. constrictum* (OAC 8805). *C*, *S. variabile* (OAC 8811). *D*, *S. humicola* (OAC 8806). (Reproduced by permission of the National Research Council of Canada, *Can. J. Bot.*, *40:* 77, 1962.)

despite the most careful search. In older cultures the long tubular (sometimes tapering) apex produced by successive proliferations appears much darker due to the closely packed annular scars (Fig. 139*D*). Shoemaker (1964) has suggested staining with 1 per cent aqueous azur-A solution to reveal annellations.

Culturally and morphologically, species of the genus *Scopulariopsis* show wide variation. Colonies may be white, tan, buff, brown or almost black in appearance; they may be smooth, mealy or strongly funiculose. Some species are slow growing and restricted, others rapidly spreading.

The genus *Scopulariopsis* was erected by Bainier (1907), with *S. brevicaulis* as the type species. Later, Ôta (1928) erected the genus *Phaeoscopulariopsis* to include two dematiaceous species of *Scopulariopsis*-like fungi. A new genus *Masoniella* was established by Smith (1952), with the type species *M. grisea.* Hughes (1953) regarded *Masoniella* as a synonym of *Phaeoscopulariopsis* and suggested that these should all be considered under one genus, *viz. Scopulariopsis.* This suggestion was supported by Barron *et al.* (1961) and more recently by Morton and Smith (1963), who published a monograph on the genus.

*Scopulariopsis* species are commonly found on all types of decaying organic materials; they are quite frequently isolated from soils, although not in large numbers.

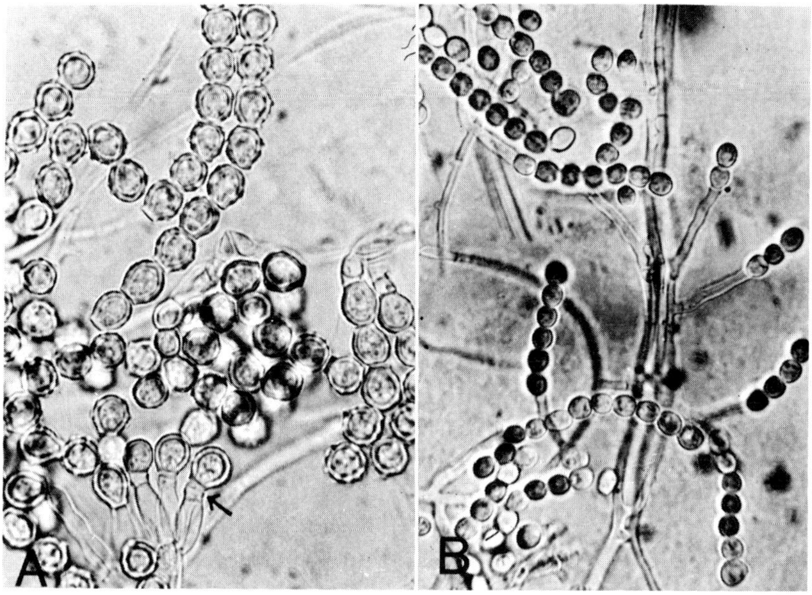


FIG. 181. *A*, *Scopulariopsis brevicaulis*. The conidia have a broadly truncate base. Annellophore (*arrow*) can be seen at the *bottom left* of the field (OAC 10161). *B*, *Scopulariopsis* state of *Microascus longirostris*. (Reproduced by permission of the National Research Council of Canada, *Can. J. Bot.*, *39:* 1609, 1961.)

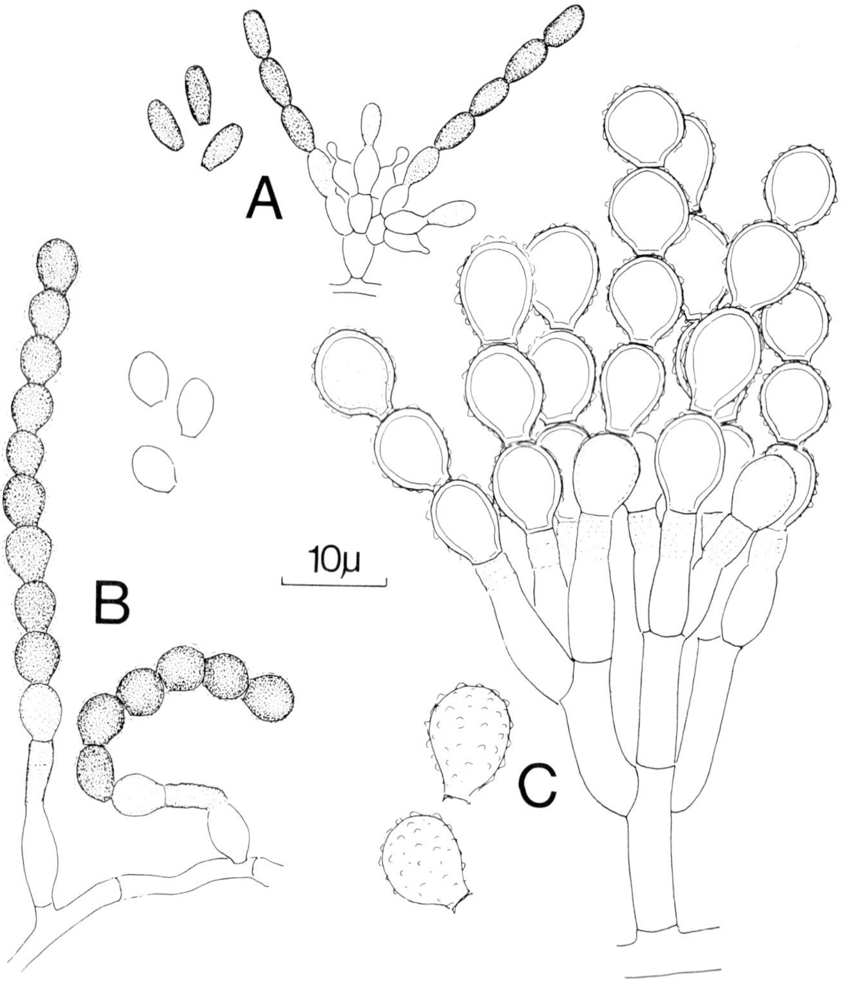


FIG. 182. *Scopulariopsis*. *A*, *S. humicola* (OAC 10260). *B*, *S. brumptii* (OAC 10261). *C*, *S. brevicaulis* (OAC 10171).

With a single exception, *Scopulariopsis* has always been associated with the Plectomycete genus *Microascus*. Corlett (1965), however, has reported that *Chaetomium trigonosporum* has a *Scopulariopsis* conidial state.

## SEPEDONIUM Link

Type Species: *Sepedonium chrysospermum* (Bull.) Link.

**Generic Description:** Conidiophores undistinguished, not sharply differentiated from the vegetative hyphae, simple or branched, hyaline, continuous or septate; conidia aleuriospores, developing singly as blown out ends of the conidiophores, sometimes produced in dense clusters, verrucose, nonseptate, large, globose to ovoid, hyaline to lightly pigmented; *Verticillium*-like phialospore state sometimes associated. (Description based on *S. chrysospermum*.)

**Diagnostic Features:** *Sepedonium* is typified by large, rough-walled, lightly colored aleuriospores borne singly on the conidiophore branches (Fig. 183).

**Notes:** *Sepedonium* species are commonly found as parasites on higher Basidiomycetes, particularly the Agaricaceae and Boletaceae.

The species commonly isolated from soil is *S. chrysospermum* (see Gilman, 1957; Miller *et al.*, 1957; McLennan and Ducker, 1954). We have isolated both *S. chrysospermum* and *S. ampullosporum* (Fig. 139*A*) from soils in Ontario. A paper on the morphology of *Sepedonium* species has been presented by Damon (1952a).

*S. chrysospermum* is the conidial state of *Apiocrea chrysosperma* (=*Hypomyces chrysospermus*). *Thielavia sepedonium*, as the name implies, has a *Sepedonium*-like conidial state, although Carmichael (1962) refers this state to *Chrysosporium*.

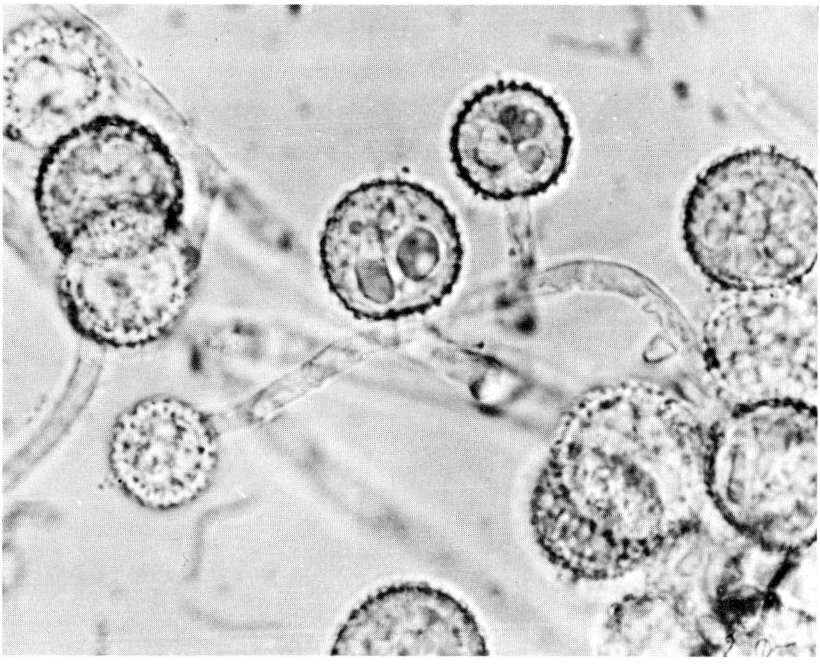

FIG. 183. *Sepedonium chrysospermum*. The aleuriospores are borne singly and terminally at the ends of short sporogenous branches. (OAC 10262).

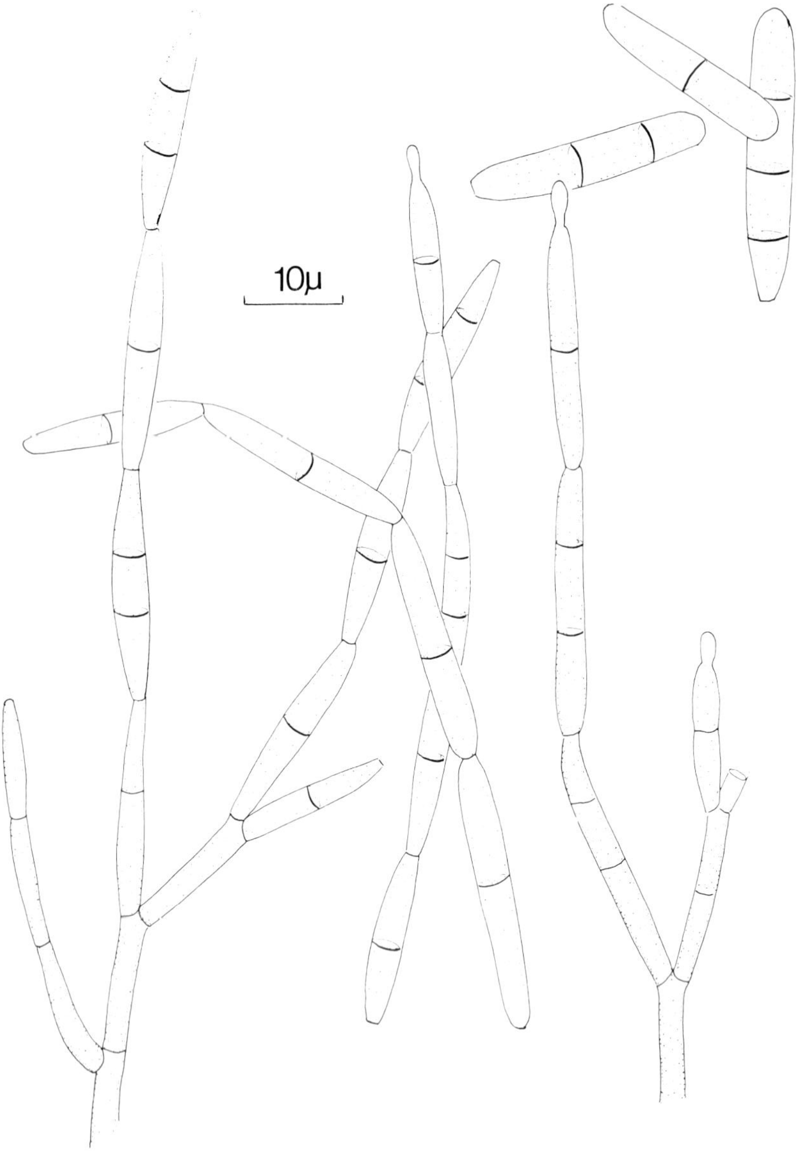


FIG. 184. *Septonema chaetospira*. The branching chains of blastospores have one or several highly refractive septa in each conidium (IMI 109928).

## **SEPTONEMA** Corda

Type Species: *Septonema secedens* Corda.

**Generic Description:** Conidiophores more or less erect, simple or irregularly branched, sometimes little distinguished from the vegetative hyphae, septate, pigmented; conidia blastospores, continuous or more often several septate, cylindric, lightly pigmented, truncate at the ends, borne in acropetal succession to form branching chains.

**Diagnostic Features:** The production of simple or branched chains of cylindric phragmospores developing acropetally as blown out ends of the conidiophores is characteristic of *Septonema* (Fig. 184).

**Notes:** Members of the genus *Septonema* are commonly found on the bark of deciduous or coniferous trees and frequently on other fungi, particularly Pyrenomycetes. The most extensive treatment is that of Hughes (1952b).

Members of this genus are rarely recorded from soil. *S. chaetospira* has been recorded from soil in England by Brown (1958) and Chesters (1960) isolated *S. hormiscium* from buried root pieces. This fungus is a member of the Blastosporae and shows relationships to *Cladosporium*.

## SPEGAZZINIA Sacc.

Type Species: *Spegazzinia tessarthra* (Berk. and Curt.) Sacc.

**Generic Description:** Conidiophore mother cells arise from the vegetative hyphae; conidiophores arise singly from mother cells, short or long, simple, flexuous, smooth or slightly roughened, brown-pigmented, somewhat broader at the apex; conidia produced singly as blown out ends of conidiophores; conidia composed usually of four dark brown cells adpressed in one plane and bearing large, irregular, spiny projections; four-celled smooth or slightly irregular spores may also be produced.

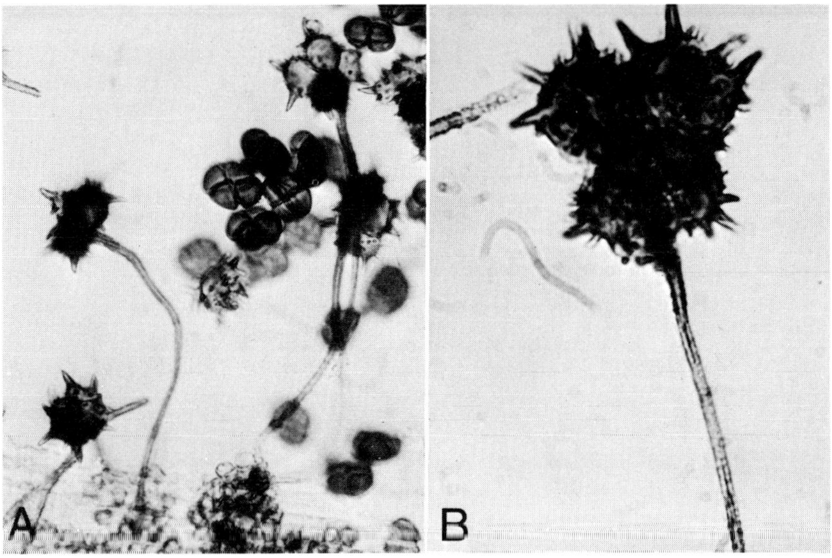


FIG. 185. *Spegazzinia tessarthra*. *A*, the two morphologically distinct types of spores produced. *B*, the "macroconidia" have prominent spiny outgrowths (IMI 93936).

**Diagnostic Features:** The large, usually four-celled spiny spores are distinctive for *Spegazzinia* (Fig. 185).

**Notes:** An interesting feature of this genus is that the conidiophores arise endogenously from conidiophore mother cells. Elongation of the conidiophore occurs by intercalary growth near the base. The process is described in detail by Hughes (1953), who discussed the origin and development of conidia in *S. tessarthra*, the type species. The above description is based on Hughes' findings.

*Spegazzinia* is rarely reported from soil. *S. lobulata* has been recorded by McLennan and Ducker (1954) from soil in Australia and a *Spegazzinia* species by Hodges (1962) from forest tree nursery soil in the United States and by Thrower (1954) from Australia.

## SPHAERIDIUM Fres

Type Species: *Sphaeridium vitellinum* Fres.

**Generic Description:** Reproductive hyphae gathered into sporodochia or coremia, darkly pigmented, ultimate branches sporogenous, giving rise to branching chains of blastospores; spores nonseptate, hyaline, long-ovoid to cylindric.

**Diagnostic Features:** Branching chains of hyaline blastospores borne on sporodochia or coremia are distinctive for *Sphaeridium* (Fig. 186).

FIG. 186. *Sphaeridium candidum*. The dark synnemata produce the branching chains of blastospores in cottony tufts (OAC 11901).

**Notes:** *Sphaeridium* is generally regarded as a sporodochial fungus. I have isolated *S. candidum* from peat soil in Ontario. In this species sporodochia are not produced in culture; the fungus produces branching synnemata from which cottony, white masses of blastospores are produced. The cottony spore masses stand out in sharp contrast to the dark, almost black, colonies. This is probably the only record of this genus from soil.

## SPOROBOLOMYCES Kluyver and van Niel

Type Species: *Sporobolomyces salmonicolor* Kluy. and v. Niel.

**Generic Description:** Organisms yeast-like; colonies salmon-pink to red; true mycelium and pseudomycelium may be present; hyphal cells oval to elongate; vegetative reproduction mainly by budding; ballistospores kidney-shaped, or sickle shaped, developing in an oblique position to the aerial "sterigmata," shot off violently. (Description based on Lodder and van Rij, 1952.)

**Diagnostic Features:** Yeast-like organisms which shoot off their spores violently characterize *Sporobolomyces* (Fig. 187).

**Notes:** The taxonomy of *Sporobolomyces* was considered in detail by Lodder and van Rij (1952), who reported that Kluyver and van Niel were the first to observe that the kidney-shaped or sickle-shaped conidium formed on a denticle was discharged violently by the "drop-excretion" method characteristic of the basidiospores of the hymenial Basidiomycetes.

Lodder and van Rij recognized seven species. Some of these form no mycelium, some pseudomycelium and some true mycelium. Irrespective of the presence or absence of mycelium, however, all of these species

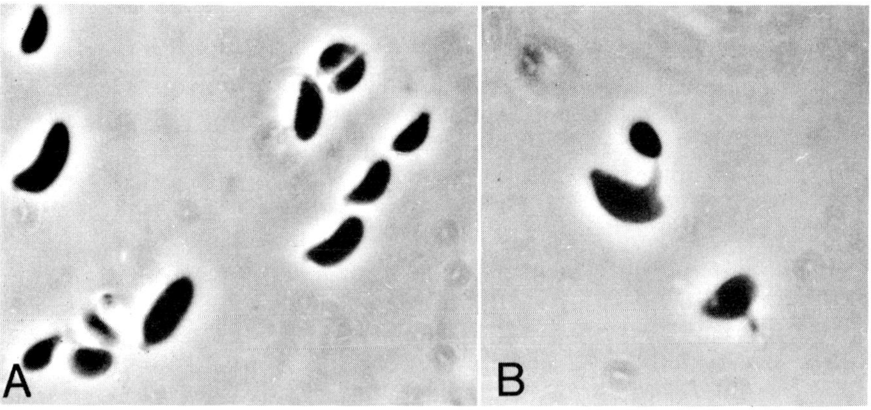


FIG. 187. *Sporobolomyces roseus*. *A*, typical asymmetric apiculate conidia. *B*, conidium germinating to produce secondary conidium on sterigma (OAC 10645).

were considered as *Sporobolomyces* on the basis of the spore discharge mechanism.

Members of this genus are commonly recovered as aerial contaminants or found growing superficially on decaying plant parts, woodpulp, leaves or fruits. They are rarely recorded from soil.

## SPOROTHRIX Hektoen and Perkins

Type Species: *Sporothrix schenckii* Hektoen and Perkins.

**Generic Description:** Conidiophores not distinguished; sporogenous cells arise more or less directly from the vegetative hyphae; continuous or septate, hyaline, denticulate, frequently broader at the base and tapering to a long, narrow apex; spores borne singly on denticles and in acropetal succession at the apex of the sporogenous cell, which may enlarge to form a denticulate vesicle or increase in length sympodially; sporogenous cell sometimes lacking and spores borne singly on short denticles directly from the vegetative hyphae; conidia nonseptate, hyaline, globose to ovoid, sometimes clavate, usually truncate at the attachment point.

**Diagnostic Features:** This genus is characterized by hyaline amerospores produced on short denticles acrogenously on a simple sporogenous cell or directly from the vegetative hyphae (Fig. 188).

**Notes:** There has been much confusion surrounding the two genera *Sporotrichum* and *Sporothrix*. The matter was largely resolved by Carmichael (1962), who compared *S. schenckii*, the type species of *Sporothrix*, with *Sporotrichum aureum*, the type species of *Sporotrichum*.

FIG. 188. *Sporothrix* sp. The sporogenous cells arise from the vegetative hyphae and a succession of spores results in a swollen denticulate apex or an elongate rachis (OAC 10180).

Carmichael noted that the method of conidium formation was quite different in the two species. In *Sporothrix*, as described above, the conidia are sympodulospores. In *Sporotrichum* the conidia are aleuriospores.

As pointed out by Carmichael, *Sporothrix* shows affinities to *Calcarisporium*. It was noted under *Calcarisporium* that *C. pallidum*, described by Tubaki (1955), could be contained in *Sporothrix*. The so-called "*Sporotrichum*" states described for certain *Ceratocystis* species should also be referred to *Sporothrix*.

## SPOROTRICHUM Link

Type Species: *Sporotrichum aureum* Link.

**Generic Description:** Hyphae broad, septate, hyaline to lightly pigmented; conidiophores little differentiated from the vegetative hyphae; conidia solitary, with broad attachment to the parental hypha, borne terminally or laterally on the sporogenous cells, nonseptate, hyaline to lightly pigmented, usually rough-walled, seceding by rupture of the subtending sporogenous hypha, globose or, more often, ovoid to clavate, rounded at the distal end, truncate, and often with an annular frill at the base.

**Diagnostic Features:** Nonseptate, rough-walled aleuriospores borne laterally or terminally on undistinguished sporogenous cells characterize the common species of *Sporotrichum* (Fig. 189).

**Notes:** The confusion surrounding the relationships between *Sporotrichum*, *Sporothrix* and *Chrysosporium* are discussed under the last two genera. As pointed out, the distinctions between *Sporotrichum* and *Sporothrix* are quite clear (see *Sporothrix*). The differences between *Sporotrichum* and *Chrysosporium* are not so obvious. If *Chrysosporium* were

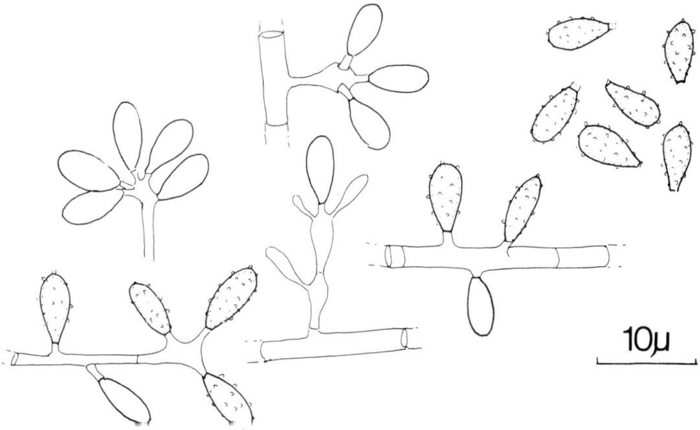


FIG. 189. *Sporotrichum thermophile*. The conidia are borne singly, apically and laterally on the hyphae or sporogenous cells.

circumscribed around *C. pannorum* and similar species, then we could remove the *Sporotrichum*-like forms into *Sporotrichum* proper.

## STACHYBOTRYS Corda

Type Species: *Stachybotrys atra* Corda.

**Generic Diagnosis:** Conidiophores erect or suberect, hyaline and smooth at first, becoming pigmented and roughened in age in some species, especially in the upper portions; conidiophores septate, frequently with a well developed foot cell; phialides borne apically in clusters of three to seven, inflated, hyaline or pigmented; phialospores sometimes in chains, more commonly sliming down to form gloeoid heads, dark or hyaline, rough or smooth, nonseptate, spherical, oval, long-cylindric or lemon-shaped.

**Diagnostic Features:** The apical cluster of several swollen phialides at the apex of a simple conidiophore is characteristic of *Stachybotrys* (Fig. 190).

**Notes:** In most of the common species of *Stachybotrys*, the dark amerospores slime down to form glistening heads. In *S. echinata*, and other species presently considered under *Memnoniella*, the spores are produced in chains.

Members of this genus are commonly isolated from soil with *S. atra*, *S. echinata*, *S. cylindrospora* and *S. bisbyi* the most frequently reported. Separation to species is principally on spore characters. No comprehensive treatment of the genus yet exists, but a comprehensive and detailed review of the earlier literature has been given by Bisby (1943).

The genus *Hyalostachybotrys* was erected by Srinivasan (1958), with *H. bisbyi* as the type. This genus is identical in all respects with *Stachybotrys*, except for the hyaline nature of the conidia and conidiophores. The spores of this species are pink to salmon-colored in the mass. Barron (1964) suggested that, despite the lack of dark pigment, *Hyalostachybotrys* and similar forms should be considered under *Stachybotrys*. In *Memnoniella* the spores do not slime down but are held together in long chains. Smith (1962) relegated *Memnoniella* to synonymy with *Stachybotrys*, considering that the distinction between the two genera was not sufficient to warrant separation. It had previously been reported by Zuck (1946) that there were isolates intermediate between *Stachybotrys* and *Memnoniella*.

## STACHYLIDIUM Link ex Fr.

Type Species: *Stachylidium bicolor* Link.

**Generic Description:** Main stalk of conidiophore unbranched, brown, straight, simple or bearing solitary or paired primary, secondary, and tertiary lateral branches below septa towards the distal end; phialides oval or cylindric-oval, produced terminally on main stalk and lateral branches and in verticils below the septa; collarette inconspicuous; phialospores oval, slimy, one-celled, pale brown to brown. (Description from Hughes, 1951.)

**Diagnostic Features:** The tall, slender, pigmented conidiophores bearing verticils of cylindrical phialides are diagnostic for *Stachylidium* (Fig. 192).

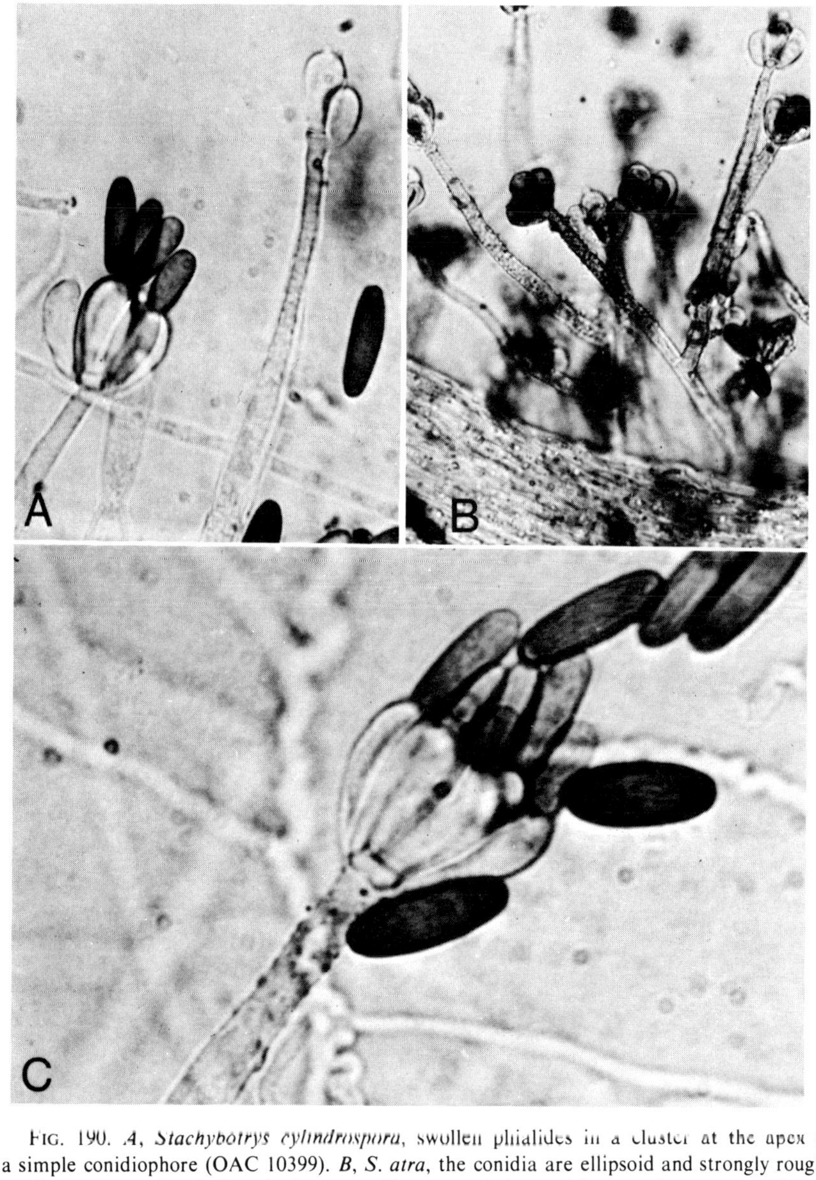


FIG. 190. *A*, *Stachybotrys cylindrospora*, swollen phialides in a cluster at the apex of a simple conidiophore (OAC 10399). *B*, *S. atra*, the conidia are ellipsoid and strongly roughened (OAC 10217). *C*, *S. cylindrospora*. The apex of the conidiophore becomes roughened and the conidia have delicate striations running longitudinally down the spore.

**Notes:** *Stachylidium* is commonly recorded from herbaceous or woody substrata. *S. bicolor* is the common species, and both the conidiophores and phialides are uniformly roughened. The genus approaches *Verticillium* in its diagnostic characters. In the literature some brown-pigmented *Verticillium* species such as *V. tenuissimum* have been figured and described under *Stachylidium*. The genus is probably monotypic. A detailed account of a large number of collections of *S. bicolor* from Britain and West Africa is given by Hughes (1951c). The genus is rarely reported from soil but is listed in the recent literature by Johnson and Osborne (1964).

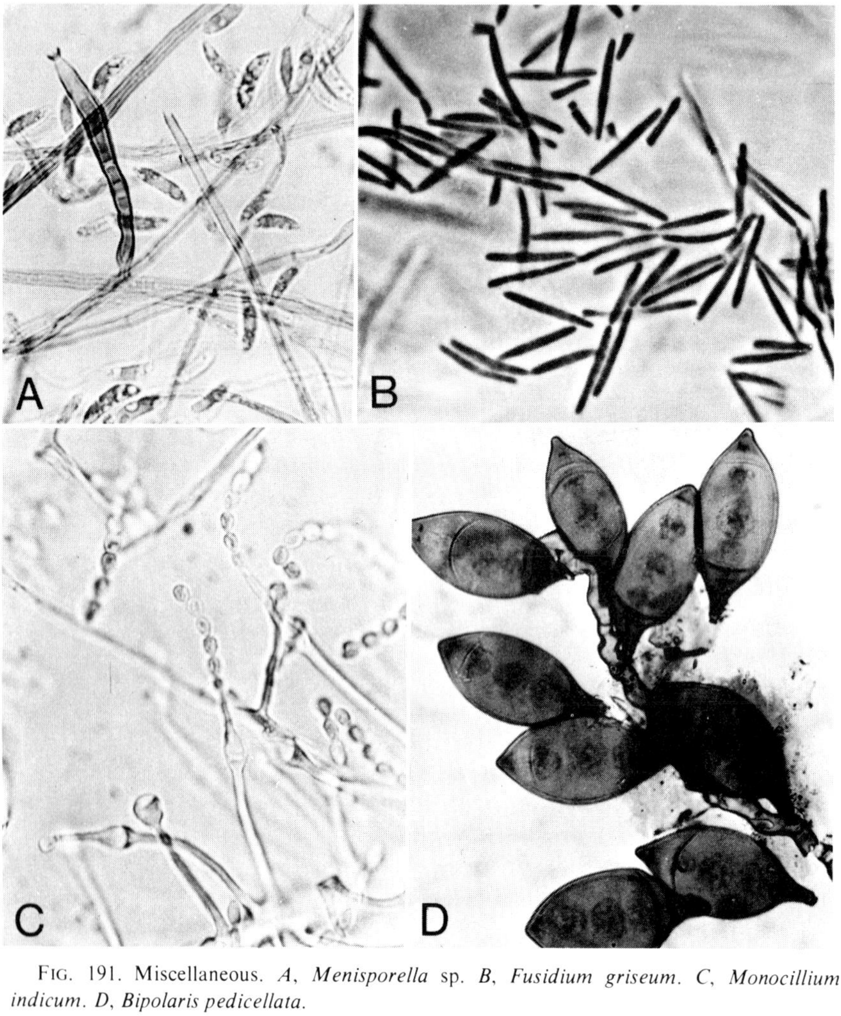


FIG. 191. Miscellaneous. *A*, *Menisporella* sp. *B*, *Fusidium griseum*. *C*, *Monocillium indicum*. *D*, *Bipolaris pedicellata*.

FIG. 192. *Stachylidium bicolor*. The conidia are produced in clusters at the apex of the phialides. Phialides and conidiophores are roughened in this species (IMI 79363).

### STAPHYLOTRICHUM Meyer and Nicot

Type Species: *Staphylotrichum coccosporum* Meyer and Nicot.

**Generic Description:** Mycelium septate, branching, hyaline to lightly pigmented; conidiophores erect, septate, stout, thick-walled, dark brown below, becoming paler to hyaline above, branching irregularly at the apex; conidia aleuriospores, formed at the ends of the branches and branchlets of the conidiophore or directly from short branches on the aerial hyphae; aleuriospores spherical to ovoid, nonseptate, thick-walled, light brown, smooth; no accessory phialospore state known. (Description based on *S. coccosporum*.)

**Diagnostic Features:** The large, mostly spherical, pale brown aleuriospores borne singly or in clusters on the branches of tall, stout, dark conidiophores make *Staphylotrichum* easily recognized. Colonies are frequently yellow-orange in color (Fig. 193).

**Notes:** Only one species, *S. coccosporum*, has been described in this genus and that originally from soil in tropical Africa (Nicot and Meyer, 1956). What appears to be the same fungus was described from soil of the Botanic Gardens, Turin, Italy by Badura (1963) as *Botrydiella*.

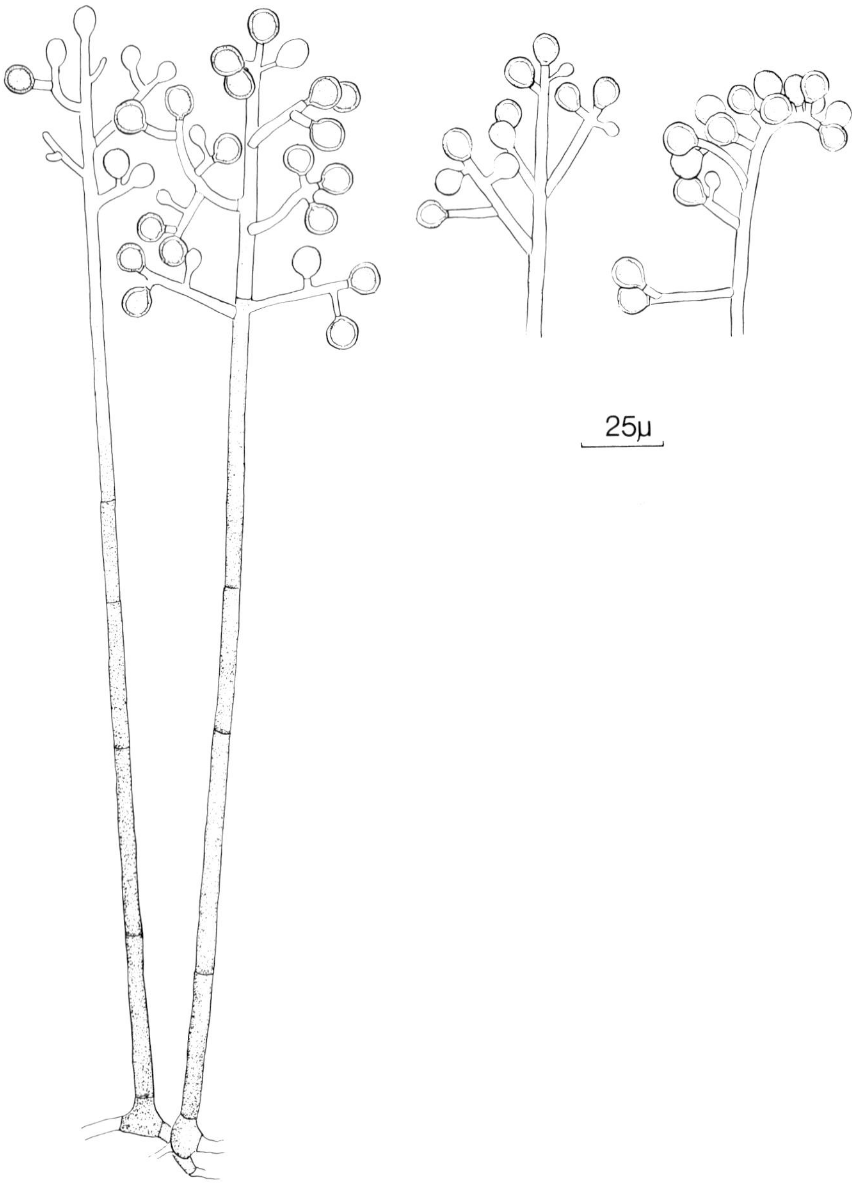


FIG. 193. *Staphylotrichum coccosporum*. The aleuriospores are borne on short branches at the apex of the stout, tall conidiophores (10093).

The morphology of a number of isolates of *S. coccosporum* has been studied in some detail by Maciejowska and Williams (1963). In Ontario we have found this fungus to be a dominant in the cultivated organic soils (muck soils) of Bradford Marsh. It has also been recorded from virgin forest soil in Panama (Goos, 1962) and is probably common in organic soils throughout the world. This genus is very close morphologically to *Botryotrichum*. It bears striking similarities to *B. piluliferum*, and an isolate of *Staphylotrichum* which fails to produce the conidiophores (as some do) might be misclassified as *Botryotrichum*.

## STARKEYOMYCES Agnihot.

Type Species: *Starkeyomyces koorchalomoides* Agnihot.

**Generic Description:** Sporodochia bright-colored, superficial, entire, conidiophores irregularly ramose, forming a fertile layer; conidia hyaline, continuous, acrogenous, noncatenate, with an apical membranous appendage. (Description from Agnihothrudu, 1956.)

**Diagnostic Features:** Bright-colored sporodochia producing conidia each with a flaring, delicate appendage characterize *Starkeyomyces*.

**Notes:** The genus is monotypic. It was erected by Agnihothrudu (1956) on *S. koorchalomoides*, isolated from the rhizosphere of pigeon pea (*Cajanus cajana*) at Madras University. Agnihothrudu's figures show the conidiophores irregularly branched and bearing the sporogenous cells (apparently phialides) singly or more often in pairs or groups of three terminally on the conidiophore branches. As pointed out by Agnihothrudu, the genus is close morphologically to *Koorchalomella* described by Subramanian (1953), but differs from this in lacking setae.

## STEMPHYLIUM Wallr.

Type Species: *Stemphylium botryosum* Wallr.

**Generic Description:** Conidiophores pigmented, septate, simple or sparingly branched, with darker terminal swelling through which a solitary, terminal/porospore is blown out; conidiophore proliferates through the conidial scar and forms a secondary swelling bearing a second conidium; successive proliferations give the conidiophore a noded appearance; conidia porospores, darkly pigmented, muriform, large, smooth or roughened, frequently with marked constriction at the central septum, variable in shape. (Description based on *S. botryosum*.)

**Diagnostic Features:** *Stemphylium* is typified by conidiophores with dark, intercalary swellings bearing solitary, pigmented, muriform porospores (Fig. 194).

**Notes:** Members of this genus are commonly found as parasites of higher plants (Neergaard, 1945) but are also relatively frequently recovered from soil (Gilman, 1957). The genus has been studied in some detail by Wiltshire (1938) and by Neergaard (1945). Wiltshire appreciated that, in the popular concept of the genus, two morphologically distinct types of conidiophore development were present. In the first of these, typified by *S. botryosum*, the conidiophore proliferated through the conidial scar at the apex of the conidiophore in the manner described above. Wiltshire segregated these forms into the subgenus *Eu-Stemphylium*. In the second type, exemplified by *S. lanuginosum*, after spore formation the conidiophore resumed growth sympodially from a point just behind the apex of the conidiophore and bipassed the terminal spore. These forms, with a geniculate conidiophore were separated into the subgenus *Pseudostemphylium* by Subramanian (1961), but more recently they have been considered under *Ulocladium* by Simmons (1967).

## STEPHANOMA Wallr.

Type Species: *Stephanoma strigosum* Wallr.

**Generic Description:** Conidiophores inconspicuous; conidia aleuriospores, borne as spherical enlargements form more or less elongate pedicels arising from the vegetative hyphae; three to six globose, bullate swellings protrude from the cell wall; phialospore state present; phialophores erect or suberect, with phialides arranged in verticillate fashion, *Verticillium*-like; phialospores hyaline, continuous or septate, elongate-ovate.

**Diagnostic Features:** The unusual aleuriospore consisting of a central living cell with globose "bullate" outgrowths protruding from the surface make *Stephanoma* very distinctive (Fig. 195).

**Notes:** This genus is rarely reported from soil. The only record from this substratum is that of Williams and Schmitthenner (1956), *Stephanoma* is the conidial state of certain *Hypomyces* species and is considered by

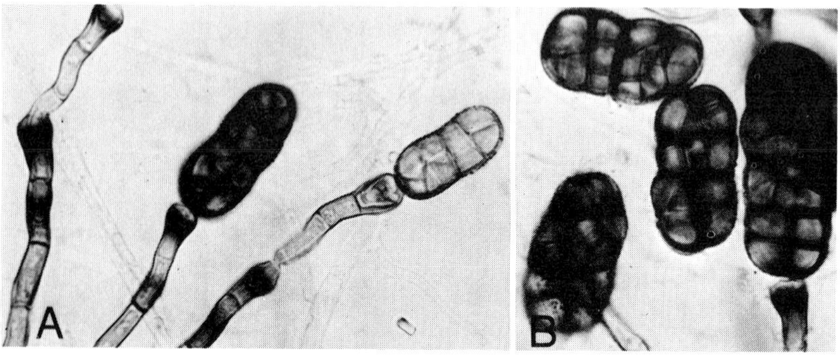


FIG. 194. *Stemphylium botryosum*. *A* and *B*, the conidia are produced singly and terminally. The apex of the conidiophore is darker and slightly swollen and the conidiophore proliferates through the apex to produce a spore at a higher level (OAC 10280).

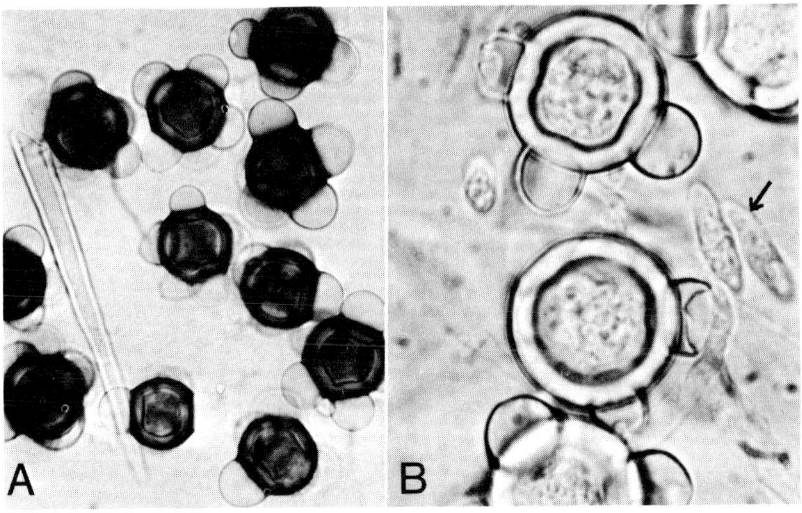


FIG. 195. *Stephanoma strigosum*. *A*, mature aleuriospores stained with aniline blue. *B*, higher magnification showing the hemispherical vesicular outgrowths from the spore. Conidia of phialospore state (*arrow*) can also be seen (IMI 96750).

Howell (1939) to be close to *Sepedonium*. *S. strigosum*, the most common species of the genus, is usually found on the fructifications of Discomycetes (see Tubaki, 1963a).

## STEPHANOSPORIUM Dal Vesco

Type Species: *Stephanosporium cerealis* (Thum.) Swart.

One species of the genus *Oidiodendron*, *O. cerealis* (Fig. 196) remains controversial. This species was for a long time disposed as *Trichosporium cerealis* and was the species described by Robak (1932) as *O. nigrum*. Barron (1962), agreeing with Robak that the species was an *Oidiodendron*, made the transfer. This same fungus was described by van Beyma (1933) as *Haplographium fuliginea* and was also the basis of a new genus of the Hyphomycetes erected by Dal Vesco (1961) called *Stephanosporium*, with *S. atrum* as the type. Swart (1965) pointed out the relationships between *H. fuliginea*, *S. atrum* and *O. cerealis*, but agreed with Dal Vesco that it was not congeneric with *Oidiodendron*, and made the combination *Stephanosporium cerealis*. I am not convinced that the separation of this species into a new genus is justified. The method of formation of the conidium is exactly as found in *O. echinulatum*. The only significant difference between *Oidiodendron* and *Stephanosporium* is that the spores in the latter are flattened and have a conspicuous darkened band round the perimeter.

## **STILBUM** Tode ex Fr.

Type Species: None designated.

**Generic Description:** Conidiophores aggregated into hyaline or bright-colored synnemata; diverging at the tip and terminating in solitary phialides; phialospores nonseptate, hyaline, gathering in light- or bright-colored mucoid heads.

**Diagnostic Features:** This genus is characterized by light synnemata terminating in simple conidiophores bearing amerospores in mucoid heads (Fig. 197).

**Notes:** The fungi included in this genus have in other treatments been included in *Stilbella* Lindau, which is regarded by several workers as a synonym of *Stilbum*. The synonymy was proposed by Wakefield and Bisby (1941) and developed by Mains (1948), who stated the case as follows.

> Mycologists are not in agreement concerning the application of the generic name *Stilbum*. Tode proposed the name in 1790 for a genus, describing six species, the first being *Stilbum vulgare*. Fries in 1832 in his treatment of *Stilbum* in his *Systema Mycologicum* included twenty-two species. The first species described is *Stilbum hirsutum*. *S. vulgare* is the nineteenth. Fries placed the genus in his third class, Hyphomycetes. For many years *Stilbum* has been used for a genus of the Hyphomycetes and more than 100 species have been described (140 included by Saccardo according to Ainsworth and Bisby). In 1900, Lindau placed *Stilbum* in the Basidiomycetes near *Pilacre* including only one species *S. vulgare*. For the remaining species in the Hyphomycetes, he proposed the name *Stilbella*. *Stilbum* as pub-

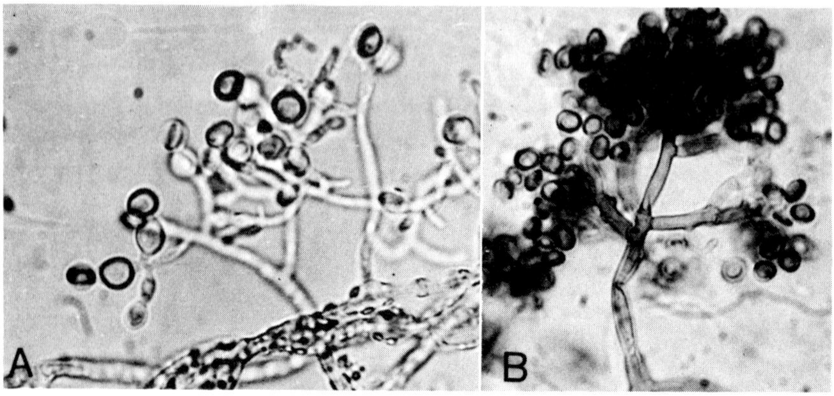


FIG. 196. *Oidiodendron cerealis* (=*Stephanosporium cerealis*). *A*, arthrospores develop in basipetal succession from the sporogenous hyphae. *B*, the mature conidiophore is pigmented and the conidia are in branching chains (OAC 10222).

FIG. 197. *Stilbum* sp. The conidia gather in slimy balls on hyaline synnemata (OAC 10282).

lished by Fries was unquestionably a genus of the Hyphomycetes, apparently with *S. vulgare* as the only exception among the species. It continued to be treated as such without question until 1900 with the addition of many species. Lindau's assumption that *S. vulgare* should be the type of Stilbum, resulting in the application of the name to a monotypic genus of the Basidiomycetes, does not appear to be justified if recommendation VI of the International Rules of Botanical Nomenclature (Cambridge revision) is followed.

*Stilbum* or *Stilbum*-like fungi are not uncommon in soil. In our studies on cultivated peat soils (muck soils) in Ontario, we have frequently isolated a *Stilbum* species which compares very well with the description of *Stilbella bulbicola* Hennings. Mathur and Thirmulachar (1960) have described a new *Emericellopsis* species from soil with a *Stilbella*-like conidium state. This raises an interesting problem on the relationships between *Stilbum* and *Cephalosporium*.

Stilbum is a large genus with over 100 species described, and so authoritative identification to species would at present be difficult or impossible. The species of the genus are commonly found as entomogenous fungi (see Mains, 1948).

## SYMPODIELLA Kendrick

Type Species: *Sympodiella acicola* Kendrick.

**Generic Description:** Conidiophores more or less erect, solitary, simple, septate, brown, increasing in length by sympodial growth; conidia acro-pleurogenous, cylindrical with truncate ends, continuous, hyaline, produced in dry, unbranched chains. (Description from Kendrick, 1958.)

**Diagnostic Features:** Short chains of hyaline, truncate conidia produced in acropetal succession from a simple, darkly pigmented conidiophore characterize *Sympodiella* (Fig. 198).

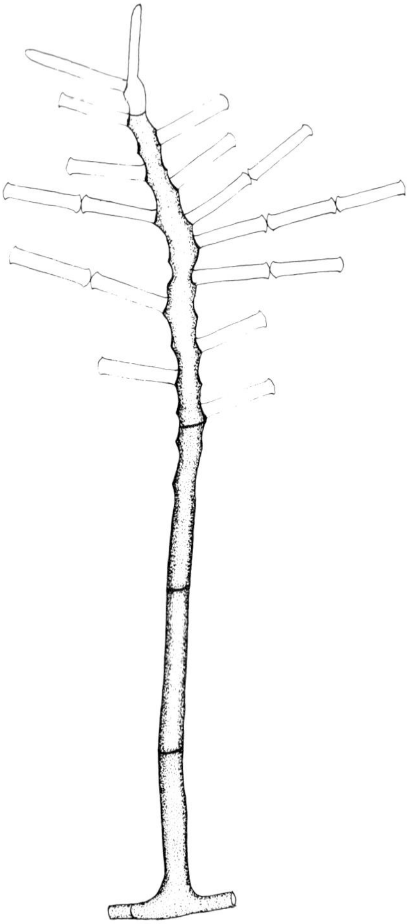

FIG. 198. *Sympodiella acicola*. The pigmented conidiophore increases in length sympodially producing short lateral chains of conidia. (Reproduced courtesy of Kendrick, *Trans. Brit. Mycol. Soc.*, *41*: 519–521, 1958. Fig. 2.)

**Notes:** This unusually interesting Hyphomycete was described by Kendrick (1958) on a single species, *S. acicola*, isolated from decaying needles of *Pinus sylvestris*. As described by Kendrick, the first conidium is produced from the conidiophore apex from which a new growing point develops just below and to one side of the original apex, but almost immediately forms another conidium. This process is repeated so that the conidiophore elongates considerably by sympodial growth. Secondary conidia develop in acropetal chains from the primary conidia by apical extension and abstriction; growth is restricted to the terminal conidium.

This genus shows some relationship to *Polyscytalum*. It has rarely been isolated from soil, and the only record of which I am aware is in the herbarium of the Commonwealth Mycological Institute, Kew, a *Sympodiella* species isolated by F. T. Last from soil in Sitka spruce nurseries in England.

## THERMOMYCES Tsiklinsky

Type Species: *Thermomyces lanuginosus* Tsiklinsky.

**Generic Description:** Conidiophores inconspicuous; conidia borne on short sporogenous cells arising more or less at right angles from the vegetative hyphae; sporogenous cells cylindrical or slightly inflated, almost flask-shaped; conidia aleuriospores, borne singly at the apex of each sporogenous cell, smooth or more often roughened with warty outgrowths, brown, nonseptate, thick-walled.

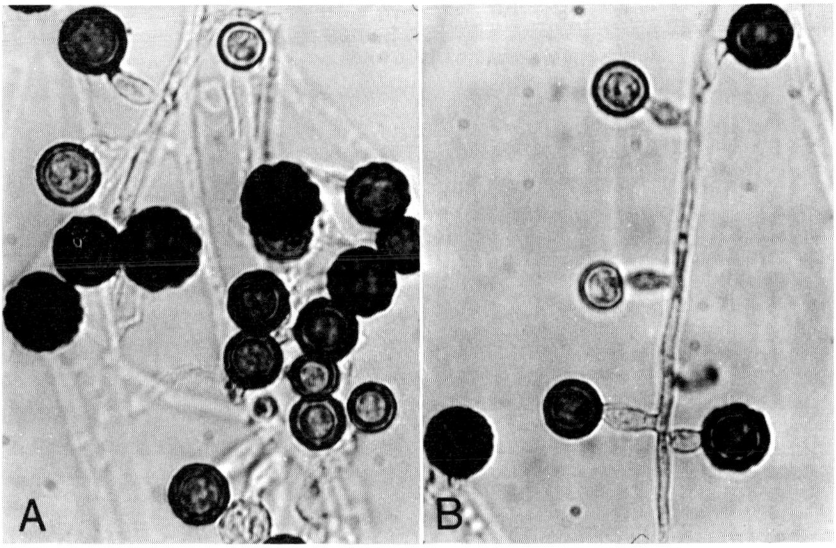


FIG. 199 *Thermomyces lanuginosus*. *A* and *B*, the roughened aleuriospores are borne on short lateral branches from the vegetative hyphae (OAC 10560).

**Diagnostic Features:** Rough-walled, spherical aleuriospores borne on short sporogenous cells are characteristic of *Thermomyces* (Fig. 199).

**Notes:** *Thermomyces* was erected by Tsiklinsky (1899) on *T. lanuginosus* isolated from garden soil. This genus has had a somewhat confused taxonomic history which was reviewed by Pugh *et al.* (1964), who redescribed *T. lanuginosus* and described a new species, *T. verrucosus*, isolated from sand dune soils in England, and also from soils in Chile and Peru. *T. lanuginosus* and *T. verrucosus* resemble each other closely in morphology but can be distinguished on spore characteristics. *T. lanuginosus* is a thermophilic fungus, while *T. verrucosus* is mesophilic. Apinis (1963) transferred *Humicola stellatus* to *Thermomyces*. The biology of *Thermomyces* and other thermophilic Hyphomycetes is considered in detail by Cooney and Emerson (1964). *Thermomyces* is morphologically close to *Humicola* and, indeed, there is little difference between these two genera.

## THIELAVIOPSIS Went

Type Species: *Thielaviopsis ethaceticus* Went.

**Generic Description:** Conidiophores undistinguished, with sporogenous cells arising more or less directly from the vegetative hyphae; sporogenous cells phialides, slightly inflated below, tapering to a long cylindrical apex, pale brown; phialospores endoconidia, produced inside sporogenous cells in basipetal succession and forming long fragile chains; conidia variable in length and width, cylindrical, truncate at both ends, continuous, lightly pigmented; aleuriospores produced abundantly, large, very dark, terminal on short branches, mostly three- to seven-celled, fragmenting at maturity into unicellular, arthrospore-like units.

**Diagnostic Features:** *Thielaviopsis* is characterized by cylindrical endospores with associated dark-brown, septate aleuriospores (Fig. 200).

**Notes:** *Thielaviopsis basicola* is the only species of the genus recorded from soil (Gilman, 1957). This fungus is an aggressive pathogen of tobacco and is known to have a very large host range (Johnson, 1916). *T. basicola* is difficult to isolate from soil by conventional procedures, but use of special techniques, such as use of live carrot tissue as bait (Yarwood, 1946), has facilitated detection and isolation of this fungus (Bateman, 1963).

Taxonomically, the genus *Thielaviopsis* has little to recommend it. It belongs to the *Chalara* complex, which would also include the genus *Chalaropsis*. These genera have a phialospore state which is essentially the same, and differ in presence or nature of the aleuriospore state. If one subscribes to the concept that a form genus should be circumscribed around one state only, then these genera must be regarded as the same. *Thielaviopsis* and *Chalaropsis* must eventually, therefore, be considered as synonyms of *Chalara*.

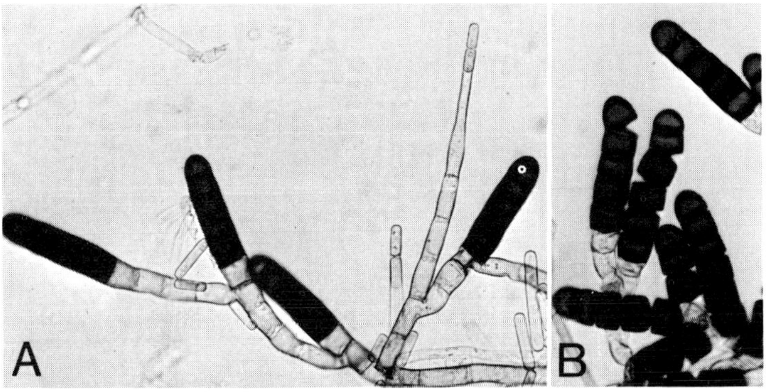


FIG. 200. *Thielaviopsis basicola. A*, the cylindrical endospores can be seen developing in the phialides. The septate aleuriospores are very dark. *B*, aleuriospores fragment at maturity into arthrospore-like units (OAC 10211).

## THOZETELLOPSIS Agnihothrudu

Type Species: *Thozetellopsis tocklaiensis* Agnihot.

**Generic Description:** Sporodochia cushion-shaped, superficial from the first, typically sessile, glistening, moist, glabrous; sterile mycelium hyaline, forming a byssoid stroma at the base of the sporodochium; conidiophores well formed, ramose, septate, hyaline, closely packed to form a superficial fertile layer; intermingled with the conidiophores are somewhat awl-shaped sterile setae; conidia produced singly, acrogenously and successively at the tip of the conidiophores, fusoid-curved to falcate, smooth-walled, hyaline, continuous, with a terminal setula at each end. (Description from Agnihothrudu, 1958.)

**Diagnostic Features:** This genus is distinguished by the production of sporodochia, setulate hyaline amerospores and awl-shaped, sterile setae interspersed with the conidiophores (Fig. 201).

**Notes:** This unusual Hyphomycete was described by Agnihothrudu (1958) and originally isolated from the decaying floral parts of *Camellia sinensis*. It was later recorded from soil by Agnihothrudu (1962a). There are no other records of the fungus known.

## THYSANOPHORA Kendrick

Type Species: *Thysanophora penicillioides* (Roum.) Kendrick.

**Generic Description:** Conidiophores dark brown, clustered or solitary, erect or nearly so, simple, septate, bearing apical or apical and lateral penicillate heads; penicilli initially apical, may become lateral after subapical proliferation of stipe, subhyaline to pale brown, becoming darker in age; ultimate sporogenous cells phialides; phialides bottle-shaped; phialospores hyaline, nonseptate, in long dry chains, smooth to slightly

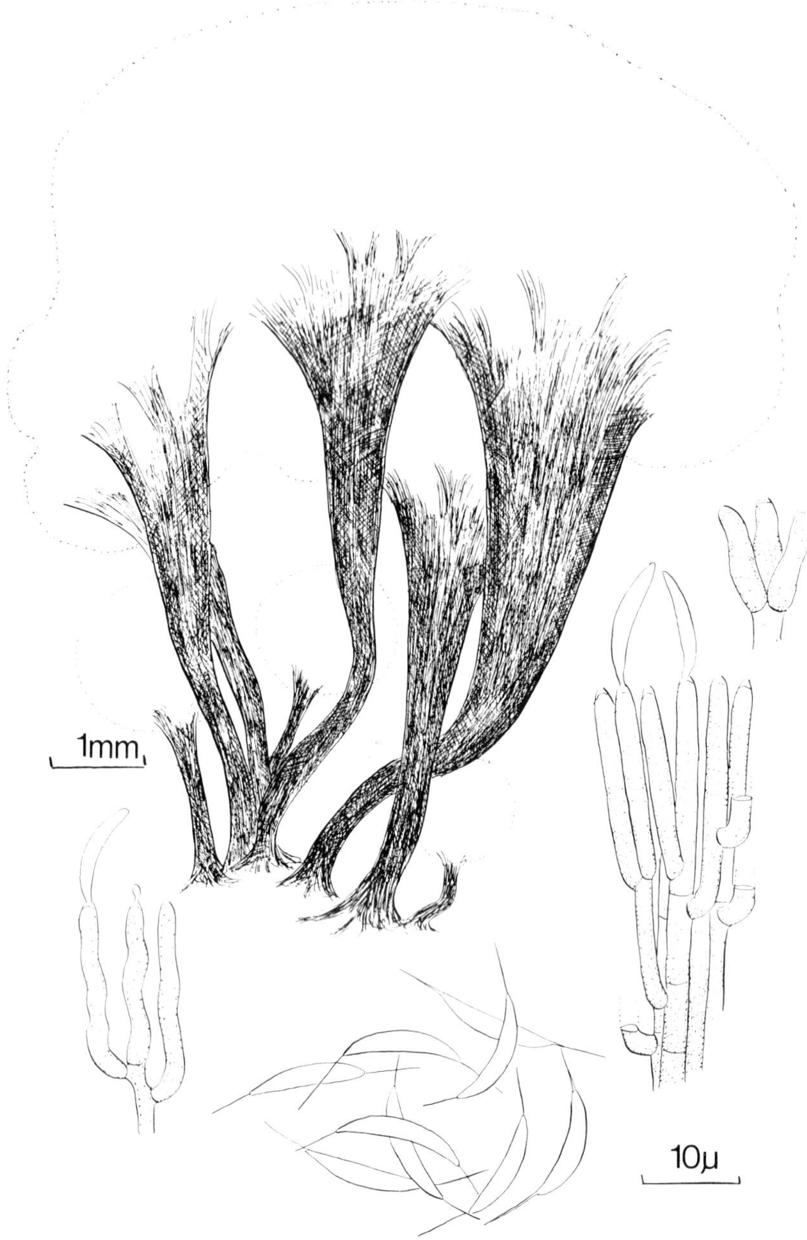


FIG. 201. *Thozetellopsis toklaiensis*. In this species the "synnemata" flare at the apex trumpet-like. The setulate conidia gather in slimy masses (IMI 74806, ex type).

roughened; sclerotia sometimes produced, translucent to creamy-white becoming pale brown in age. (Description based on Kendrick, 1961.)

**Diagnostic Features:** *Thysanophora* is characterized by producing stout, dark conidiophores, with penicillate heads of phialides giving rise to long chains of dry amerospores (Figs. 125*C* and 202).

**Notes:** The genus *Thysanophora* was erected by Kendrick (1961) based on *Haplographium penicillioides.* Two species were described, *viz. T. penicillioides* and *T. longispora.* The genus is associated with the decay of conifer needles and, in our experience, *T. penicillioides* is not uncommon in soils from coniferous woods. As described by Kendrick, after the production of an apical penicillus, secondary growth of the stipe often occurs by means of a subapical proliferation. A succession of such penicilli and proliferations is often produced by a single conidiophore. *T. longispora* has been recorded from soil only by Christensen and Whittingham (1965).

## TORULA Pers.

Type Species: *Torula herbarum* Pers.

**Generic Description:** Conidiophores short or lacking, not readily distinguished, with conidia arising more or less directly from the vegetative hyphae; conidia porospores, produced in acropetal succession to form simple or branching chains, darkly pigmented, continuous or septate, smooth or roughened. (Description based on *T. herbarum.*)

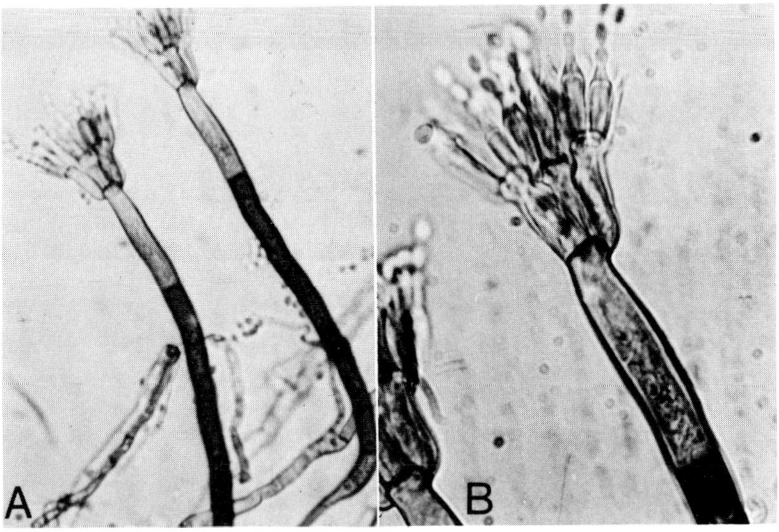


FIG. 202. *Thysanophora penicillioides. A*, the dark conidiophores have an apical penicillus. *B*, the flask-shaped phialides produce the conidia in chains (UAC 10063).

**Diagnostic Features:** *Torula* is characterized by simple or branching chains of dark porospores which break up readily into amerospores or phragmospores and which arise more or less directly from the vegetative hyphae (Fig. 203).

**Notes:** *Torula* species are reported infrequently from soil (Gilman, 1957). The genus by definition produces porospores and it is probable that some of the previous records of *Torula* from soil do not belong in this genus. The only *Torula* we have isolated from soil in our laboratory is *T. herbarum* and this was found on only two occasions. A *Torula* species was listed by Chesters and Thornton (1956) from soil in England. Research on the nature and method of spore formation in *Torula* has been reported by Joly (1964a) who described spore formation in *T. herbarum*, *T. graminis* and *T. dicoccum*. *T. terrestris* has been described from forest soil in India by Misra (1967).

## TORULOMYCES Delitsch

Type Species: *Torulomyces lagena* Delitsch.

**Generic Description:** Conidiophores undistinguished, short, simple, cylindrical, borne more or less at right angles to the vegetative hyphae, bearing a single, terminal, usually inflated phialide at the apex; phialospores hyaline to lightly pigmented, continuous, smooth or slightly roughened, borne in more or less long chains, globose to subglobose in shape. (Description based on *T. lagena*.)

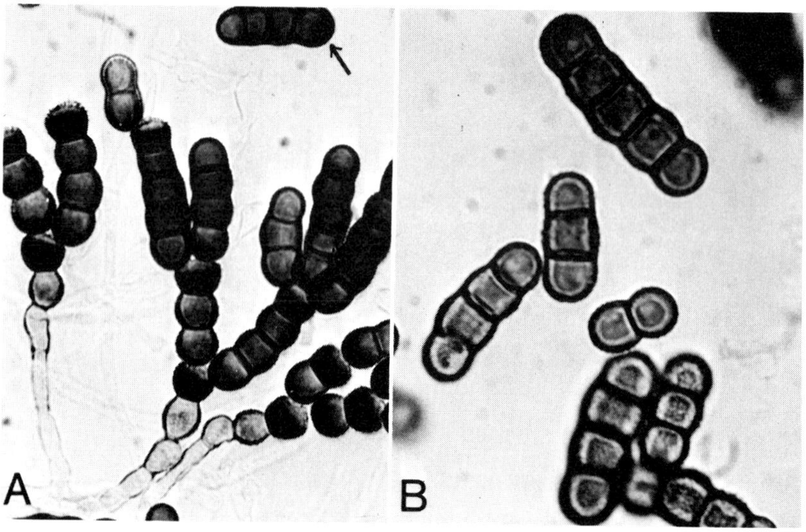


FIG. 203. *Torula herbarum*. *A* and *B*, the porospores are produced in branching chains which break up readily into phragmospores. The pore can be seen (*arrow*) at the base of one of the spores (OAC 10031).

**Diagnostic Features:** The solitary, inflated phialide borne on a short, cylindrical stalk and producing chains of spherical, nonseptate spores is typical of the common species *T. lagena* (Figs. 59*F* and 204).

**Notes:** The genus *Torulomyces* was erected by Delitsch (1943), with *T. lagena* as the type species. A second species *T. viscosus* was also described and both species were originally recovered from soil. Due apparently to poor wartime communications, this genus remained more or less in obscurity. Saksena (1955) created the genus *Monocillium* to accommodate *M. indicum* recovered from soil in India. A second species of this genus, *M. humicola*, was described by Barron (1961) from Ontario soils, and Christensen and Backus (1964) described a new variety, *M. humicola* var. *brunneum*, from soils in Wisconsin. Barron (1967) reported that *M. humicola* was conspecific with *T. lagena*. He suggested that the method of conidium production in *Monocillium* was not clear and that, until it could be determined whether or not *Monocillium* was a phialospore-forming fungus, the genus should remain distinct from *Torulomyces*. *T. lagena* is a relatively small, delicate Hyphomycete. The phialides and subtending stalk (conidiophore) measure about 20 $\mu$. The conidia are produced in long, delicate chains and are spherical, slightly roughened and very small (2 $\mu$ in diameter). This species has been recorded in Europe,

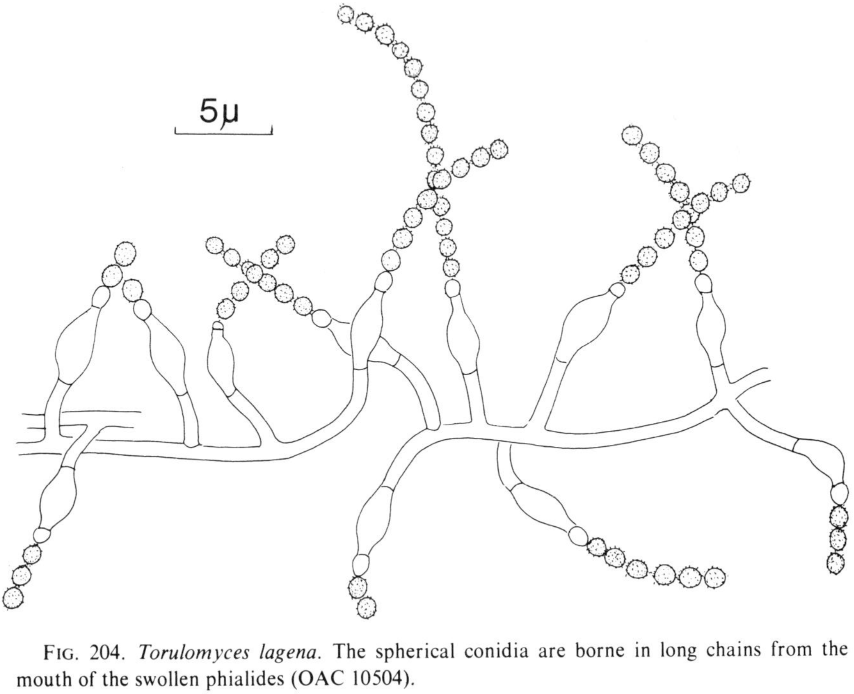


FIG. 204. *Torulomyces lagena*. The spherical conidia are borne in long chains from the mouth of the swollen phialides (OAC 10504).

Canada and the United States and is probably of widespread occurrence in soils. Christensen and Backus (1964) noted that their variety occurred in 26 out of 38 forests studied.

## TORULOPSIS Berlese

Type Species: *Torulopsis colliculosa* (Hartmann) Sacc.

**Generic Description:** Cells round or oval, or very rarely somewhat elongate; reproduction by multilateral budding; cells only exceptionally capsulated; no production of a starch-like compound as in the genus *Cryptococcus*; pseudomycelium lacking or very primitive; sediment and ring, but rarely a pellicle, produced in liquid media; pellicle if present never dry, dull or creeping; fermentative ability usually present, seldom absent; no red or yellow carotinoid pigments produced. (Description from Lodder and van Rij, 1952.)

**Diagnostic Features:** *Torulopsis* is similar morphologically to *Cryptococcus* but usually lacks a capsule and does not form starch (Fig. 205).

**Notes:** *Torulopsis* is probably common in soils. Jensen (1963) found that species of this genus were second only to *Candida* in their frequency of occurrence in Danish beech forest soils.

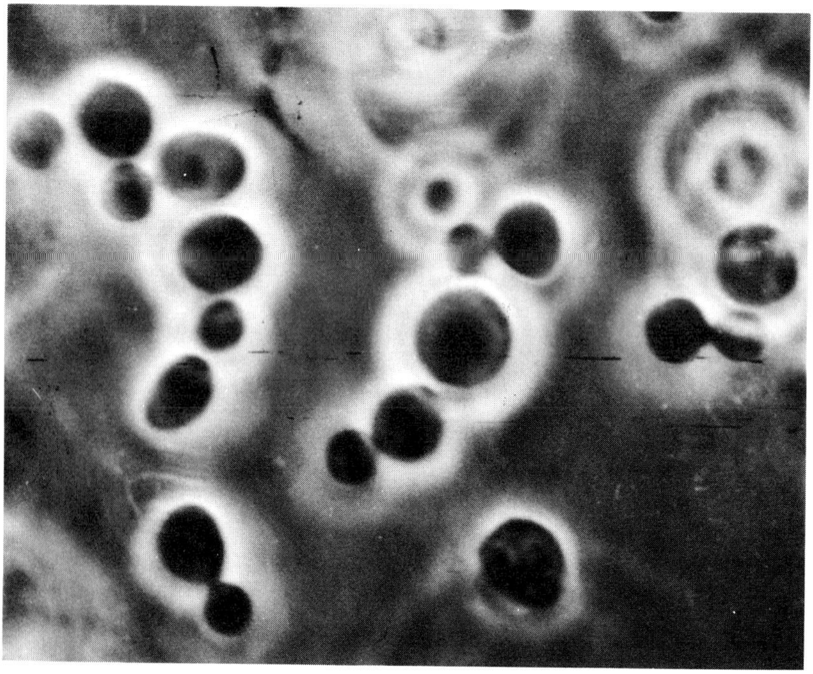

FIG. 205. *Torulopsis candida*. The blastospores are very similar morphologically to those of *Cryptococcus* (ATCC 2560).

This is a large genus comprising at least 22 species. The taxonomy and morphology of these are discussed in detail by Lodder and van Rij (1952). Many species have been isolated from a variety of infections on humans, but the pathogenicity of *Torulopsis* is generally regarded as limited and their presence in mycotic or other infections is incidental.

## TRICHOCLADIUM Harz

Type Species: *Trichocladium asperum* Harz.

**Generic Description:** Conidiophores poorly differentiated and inconspicuous, appearing as more or less short pedicels, consisting of from one to four cells, simple, less often branched, hyaline to subhyaline; conidia aleuriospores developing singly as blown out ends of the sporogenous cells, subglobose to ovoid or clavate, sometimes cylindrical, darkly pigmented, continuous or more often one- or several-septate, flattened at the basal scar, which usually bears a frill of the parental cell, smooth or roughened, sometimes with a pronounced germ pore.

**Diagnostic Features:** This genus is characterized by terminal, solitary aleuriospores which are usually septate, dark-brown, thick-walled and ovoid to short-cylindric in shape (Fig. 206).

**Notes:** *T. asperum* (=*Dicoccum asperum*) is the most common species of the genus recovered from soil (Gilman, 1957). The taxonomy and morphology of this species are given in some detail by Hughes (1952c) who also redescribed *T. opacum*. He noted that an isolate of *T. opacum* has

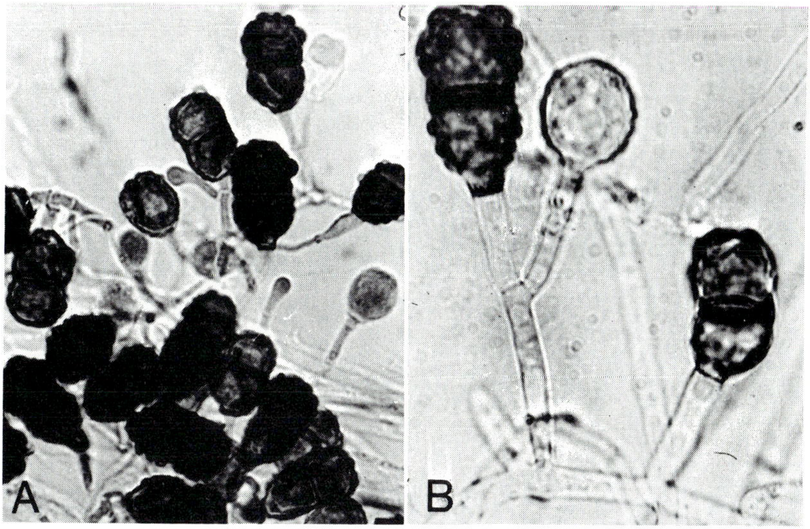


FIG. 206. *Trichocladium asperum*. *A*, various stages in the development of the aleuriospores. *B*, higher magnification showing the broad attachment of the aleuriospore to the sporogenous cell (OAC 10214).

been recorded by Bisby *et al.* (1935) under the name *Torula allii.* Both of these species and also *T. canadense* (see Hughes, 1959) have been isolated from Ontario soils in our laboratory. The normal habitat for *Trichocladium* is on wood and Hughes cites 20 collections of *T. canadense* from the wood, of *Betula, Acer, Fagus, Populus* and *Thuja.* It is not unexpected, then, that this species would be isolated from time to time in forest soils. Kendrick and Bhatt (1966) draw attention to a number of records of *Trichocladium* from soil and note that they recovered the three recognized species from organic soils in Ontario.

## TRICHODERMA Pers.

Type Species: *Trichoderma viride* Pers.

**Generic Descriptions:** Conidiophores erect or straggling, solitary or frequently aggregated into floccose tufts, hyaline, septate, branching irregularly, or weakly or strongly verticillate, frequently branching more or less at right angles to the main axis; sporogenous cells phialides, borne singly or in clusters, hyaline, ovate to flask-shaped; when aggregated into tufts, conidiophores frequently intermixed with sinuous sterile hyphae; sterile hyphae sinuous, smooth or encrusted with wart-like protuberances, anastomosing below; phialospores hyaline or green, nonseptate, gathering in balls at the mouths of the phialides.

**Diagnostic Features:** *Trichoderma* is very distinctive in culture, producing rapidly growing, floccose colonies which are white, yellow-green or bright green in the common species (Fig. 207).

**Notes:** This genus is one of the most ubiquitous of soil fungi and is rarely excluded from floristic listings. It has been extensively studied from the aspects of its saprophytic potential and because of its antagonism and parasitism against other fungi. The taxonomy of this genus has nevertheless been somewhat confused and is still not clear. In cultural studies of *Hypocrea* and *Trichoderma*, Webster (1964) noted that it was unwise at present to press for a precise nomenclature of the conidial states of *Hypocrea.* He quotes Mr. J. J. Elphick of the Commonwealth Mycological Institute as saying that "The only distinction made here between the green Trichodermas is to call the globose or somewhat ovate-spored species *T. viride* and the ellipsoid to cylindrical-spored ones *T. koningi.*"

Hughes (1958) regarded *Pachybasium* as congeneric with *Trichoderma.* In this genus the colonies are white and the spores hyaline. The floccose masses of conidiophores are intermingled with stout, sinuous, sterile hyphae, which frequently bear warty encrustations.

Regarding the ability of *Trichoderma* to produce an antibiotic, Webster and Lomas (1964) tested a number of *Trichoderma* isolates and failed to show that any of these could produce an antibiotic. A re-examination of

Weindling's strain which produced gliotoxin and Brian's isolates (see Brian, 1944, and Brian and McGowan, 1945) which produced viridin showed that these were not *Trichoderma viride* and that they matched the type of *Gliocladium virens*.

## TRICHOPHYTON Malmsten

Type Species: *Trichophyton tonsurans* Malmsten.

**Generic Description:** Colonies granular, powdery or velvety and white to light brown or red to purple in color; microconidia (aleuriospores) prominent spore form, small, continuous, thin-walled, hyaline, subspherical to clavate, borne singly or in grape-like clusters; macroconidia and abundant spherical or clavate microconidia also produced (Fig. 208). Both spore forms are aleuriospores.

**Notes:** *Trichophyton* is a pathogenic organism commonly listed under Dermatophytes in the nomenclature of medical mycology. Capitalizing on their ability to utilize keratin, species of the genus attack the hair, skin and nails in animals and man. As pointed out by Emmons *et al.* (1963), most of the Dermatophytes produce two types of conidia in culture. Both types are aleuriospores and attached by a broad base to the parent hypha or conidiophore, seceding eventually by rupture of the connective wall. The two spore types are usually designated in the literature as either microspores (microaleuriospores) or macrospores (macroaleuriospores) according to size.

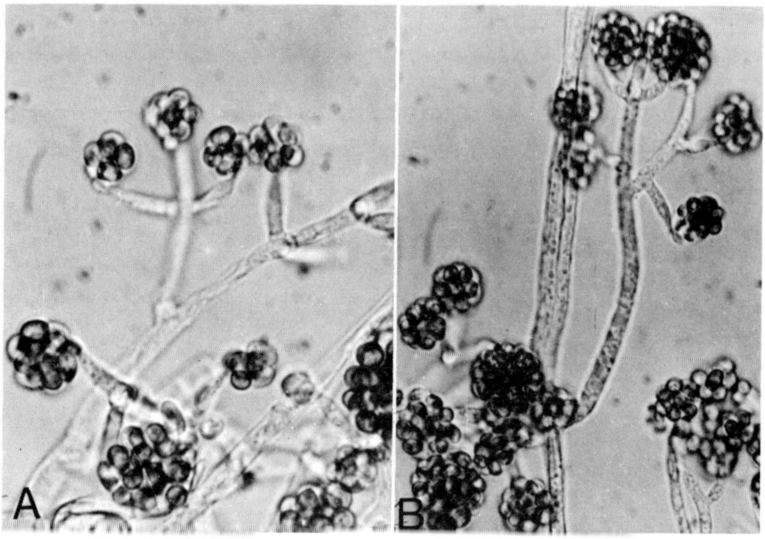


FIG. 207. *Trichoderma viride*. The phialides arise alternately or in pairs or verticils and are frequently borne at right angles to the parent conidiophore or branch (OAC 10239).

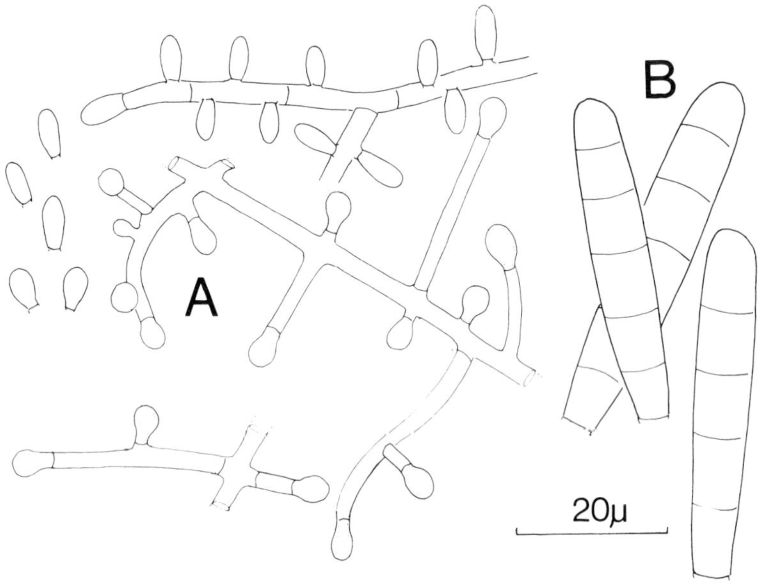


FIG. 208. *Trichophyton* sp. *A*, *Microconidia*. *B*, *Macroconidia*.

*Trichophyton* species are rarely reported from soil in the usual floristic listings of soil fungi. When selective techniques such as baiting with hair are used, the results indicate that *Trichophyton* is common in soil, *viz.* Frey (1965), Durie (1964) and Ajello *et al.* (1965).

*Trichophyton terrestre* is the conidial state of *Arthroderma quadrifidum* (see Pugh, 1964).

### TRICHOSPORON Behrand

Type Species: *Trichosporon cutaneum* (de Beurmann *et al.*) Ôta.

**Generic Description:** Pseudomycelium and true mycelium develop abundantly; both blastospores and arthrospores produced; chlamydospores sometimes occur; mostly oxidative, rarely fermentative. (Description from Lodder and van Rij.)

**Diagnostic Features:** *Trichosporon* includes the asporogenous yeasts which reproduce by forming both blastospores and arthrospores (Fig. 209).

**Notes:** *Trichosporon* is a yeast organism of the family Cryptococcaceae in the asporogenous yeasts. Eight species of the genus are recognized (Lodder and van Rij, 1952), of which two are parasitic on man. *T. beigelii* is the causal organism of white piedra (Emmons *et al.*, 1963). The genus is infrequently recorded in listings of soil fungi but is probably common in soil (Di Menna, 1965; Maciejowska, 1964).

## TRICHOTHECIUM Link

Type Species: *Trichothecium roseum* Link.

**Generic Description:** Conidiophores erect or suberect, produced singly or in groups, simple or sparingly branched, hyaline, septate; conidia produced in short, fragile chains in basipetal succession, held together by mucus; conidia large, two-celled, hyaline, with well marked, truncate attachment point. (Description based on *T. roseum.*)

**Diagnostic Features:** The short chains of two-celled conidia at the apex of a hyaline, simple conidiophore are diagnostic for the common *T. roseum* (Figs. 16*A* and 210).

**Notes:** The only species of this genus reported from soil with any frequency is *T. roseum.* In this species the conidiophores are produced in a dense stand and the colony is a pale rose color. The conidia are large (12 to 18 by 8 to 10 $\mu$ ) and the upper cell is somewhat larger than the lower one. A *Trichothecium* species has been connected to an ascigerous state *Hypomyces* (*H. trichothecioides*) by Tubaki (1960).

The confusion surrounding the method of spore formation in *Trichothecium* was largely clarified by Ingold (1956). Because of the unusual

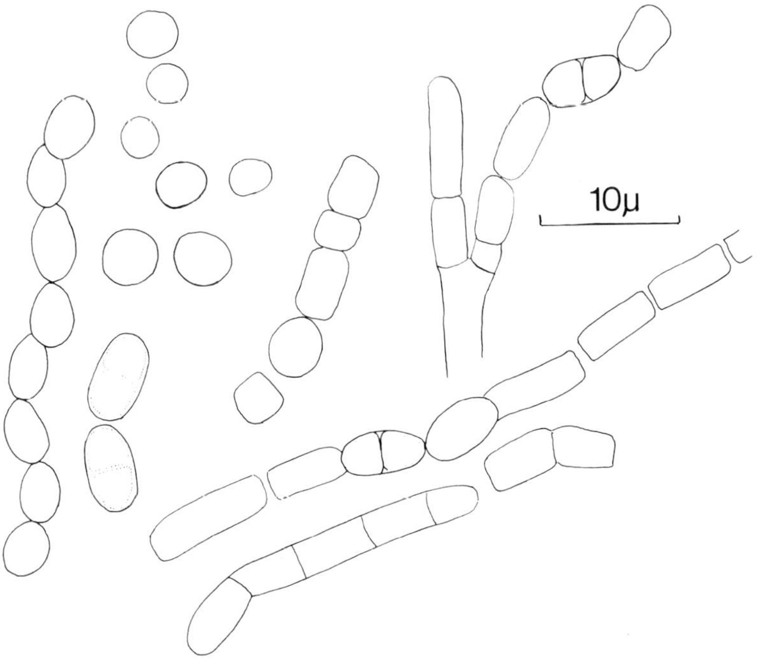


FIG. 209. *Trichosporon cutaneum.* The mycelium may bud yeast-like to produce a pseudomycelium or fragment into arthrospores (IMI 95036).

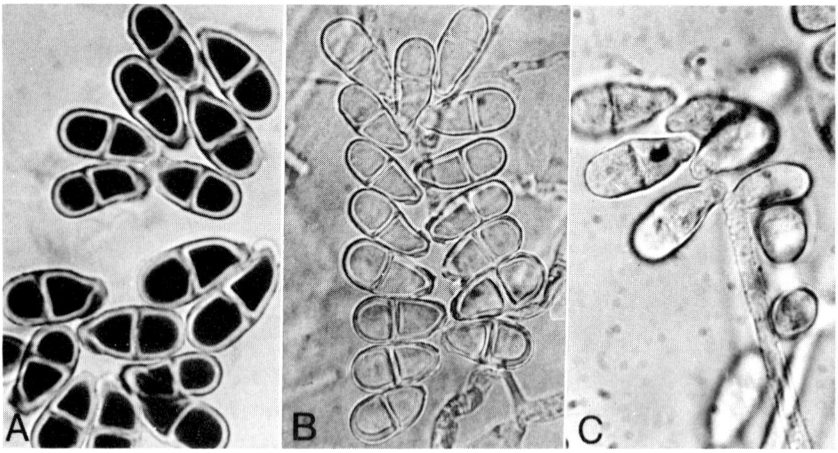


FIG. 210. *Trichothecium roseum*. *A*, conidia stained with aniline blue. *B*, unstained spores. Note material at base of spores which "cements" them in short chains. *C*, simple conidiophores produces conidia in basipetal succession (OAC 10170).

method of spore origin, the systematic position of *Trichothecium* has caused difficulties and it does not fit well into the system as proposed by Hughes. Tubaki (1958) referred to the conidia as meristem thallospores and placed them in a new group (Section IX) in his classification of the Hyphomycetes. From my own observations it seems that part of the conidiophore is incorporated into each spore. If the conidiophore continues growth from below, the conidia might be regarded as meristem spores. On the other hand, if the conidiophore becomes shorter with each spore, then the conidia could possibly be modified arthrospores.

In the past there has been confusion surrounding the relationships between *Trichothecium* and *Arthrobotrys*. This has been elucidated by Sidorova *et al.* (1964) and Rifai and Cooke (1966). The latter workers reappraised the genus and gave descriptions of four species.

### **TRICHURUS** Clements and Shear

The genus *Trichurus* was erected by Clements and Shear to include *Stysanus*-like fungi which bore sterile setae in the head. It is unfortunate that Morton and Smith (1963), in their treatment of *Doratomyces* (=*Stysanus*), did not consider the validity of this closely related taxon.

Swart (1964) studied the production of the coremia in three species of the genus *Trichurus* and concluded that the separation of *Trichurus* and *Doratomyces* may not be a natural one. He did not, however, make the necessary transfers to validate this opinion. The method of conidium production in *Trichurus* and *Doratomyces* is identical, and a generic description for *Trichurus* would be exactly as for *Doratomyces*, with the addition of a note on the occurrence of the sterile setae. *Trichurus* is not uncommon in soil. *T. terrophilus* is reported by Swift (1929) and the Commonwealth Mycological Institute records list an isolate from Ontario by R. F. Cain. *T. gorgonifer* has been isolated by Guillemat and Montégut (1957) and *T. spiralis* by Peyronel and Dal Vesco (1955), Miller *et al.* (1957) and Rai *et al.* (1961). We have isolated *T. spiralis* on two occasions from soil in Ontario. As can be seen (Fig. 211), this fungus has a striking appearance.

## TRIMMATOSTROMA Corda

Type Species: *Trimmatostroma salicis* Corda.

**Generic Description:** Conidiophores lacking; conidia originate from more or less undifferentiated hyphae; conidia dictyospores, darkly pigmented, produced in simple chains in basipetal succession from an inter-

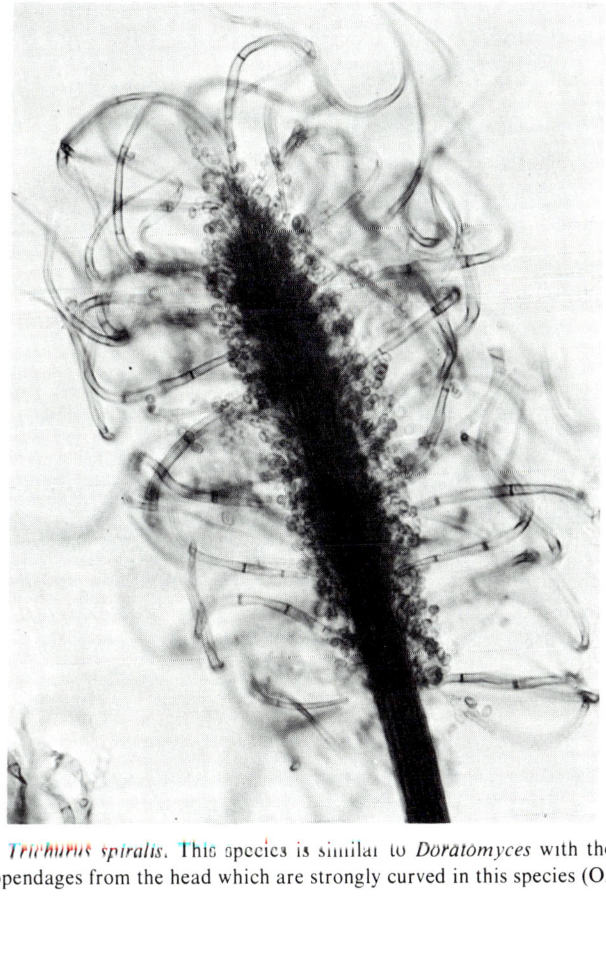

FIG. 211. *Trichurus spiralis*. This species is similar to *Doratomyces* with the addition of the sterile appendages from the head which are strongly curved in this species (OAC 10136).

calary meristem, with spores merging imperceptibly with the "conidiophore."

**Diagnostic Features:** In this genus the conidia are dictyospores which develop basipetally in simple chains from more or less undifferentiated hyphae of the stroma (Fig. 9).

**Notes:** There is a little confusion surrounding the taxonomy of *Trimmatostroma*, *Coniothecium* and *Melanconium*. It was pointed out by Hughes (1953) that *C. atrum*, the type species of *Coniothecium*, was described as having the conidia embedded in slime and had little in common with 11 species subsequently classified by Corda as *Coniothecium*. Hughes transferred these, including *C. betulinum*, to *Trimmatostroma*. Later (Hughes, 1958), on the basis of an examination of type material, he regarded *Trimmatostroma* as congeneric with *Melanconium*. Sutton (1964) re-evaluated *Melanconium* (see *Melanconium*) and, if we accept his reappraisal of the situation, then we fall back on *Trimmatospora* as the valid generic name for *C. betulinum* and similar forms. *C. betulinum* has been recovered from soil in Ireland by C. Dickinson (private communication).

## TRIPOSPERMUM Speg.

Type Species: *Tripospermum acerinum* (Syd.) Speg.

**Generic Description:** Conidiophores lacking or inconspicuous and little differentiated from the vegetative hyphae; conidia subhyaline to dark brown, staurospores, consisting of two pairs of divergent, septate, subulate arms, each pair arising from two connected basal cells, one of which bears a stalk cell by which the conidiophore is attached to the parent hyphae; conidia originate by blowing out a bulb from the inner wall through the pellicle of a parent cell; phialide-like cells with open ends have been observed in some collections. (Description based on Hughes, 1951.)

**Diagnostic Features:** Staurospores formed from two pairs of divergent arms arising from two connected basal cells are characteristic of *Tripospermum* (Figs. 212 and 213).

**Notes:** *Tripospermum* is rarely reported from soil. The only collections I am aware of from this source are our own from soil in Ontario. We have isolated *T. myrti* and an unidentified *Tripospermum* species.

The morphology and taxonomy of *Tripospermum* has been treated in some detail by Hughes (1951b), but since this publication a number of species have been added to the genus (see the *Index of Fungi of the Commonwealth Mycological Institute*). *Tripospermum* species are usually found on leaves, sometimes twigs, of deciduous plants in both tropical and temperate regions. Spore formation in *Tripospermum* was described from culture material by Ingold and Cox (1957).

## **TRITIRACHIUM** Limber

Type Species: *Tritirachium dependens* Limber.

**Generic Description:** Conidiophores long, erect or decumbent, septate, simple or verticillately branched, sometimes secondary or tertiary verticils produced; ultimate branchlets sympodulae, slightly swollen near the

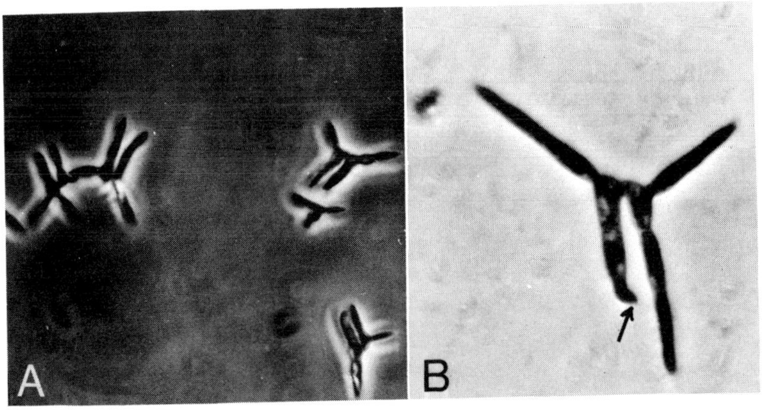


FIG. 212. *Tripospermum myrti*. *A* and *B*, the staurospores have usually five arms disposed in three planes. Attachment point to the parent hyphae is indicated by the *arrow* (OAC 10324).

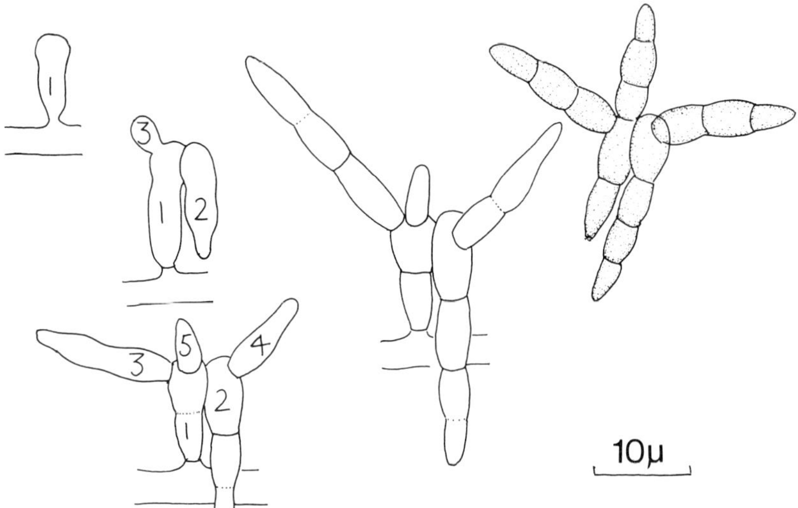


FIG. 213. *Tripospermum myrti*. *Numbers* indicate chronological sequence in the formation of the arms of the conidium.

base and tapering to a rachis-like, conidium-bearing portion; conidia produced in acropetal succession, globose to ovate, hyaline to dilutely colored, often conglutinate.

**Diagnostic Features:** The rachis-like sympodulae, produced in verticillate arrangements, are characteristic of most species of *Tritirachium* (Fig. 214).

**Notes:** *Tritirachium* is one of the nicer examples of sympodulospore formation in the Hyphomycetes. The verticillately arranged sporogenous cells are frequently slightly swollen at the base and taper gradually to a long, narrow rachis. For descriptions of species the reader is referred to Limber (1940), van Beyma (1942) and MacLeod (1954). The genus is close to *Beauveria* in its morphology.

Species of *Tritirachium* are infrequently reported from soil. Sewell (1959) reported a *Tritirachium* from heathland soil in England, and we have isolated *T. roseum* and an unassigned species from soil in Ontario. Mosca (1960) has recovered *T. cinnamoneum* from alpine soil in Italy. *T. roseum* has been reported from the former Belgian Congo by Meyer (1959).

## TUBERCULARIA Tode

Type Species: *Tubercularia vulgaris* Tode.

**Generic Description:** Fructification a sporodochium; sporodochium more or less cushion-shaped, sessile or short-stalked, smooth or wrinkled,

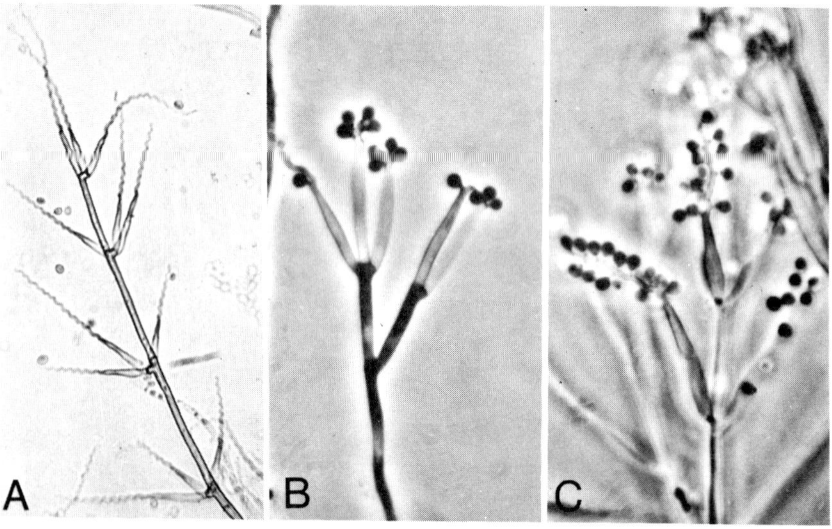


FIG. 214. *A*, *Tritirachium roseum*. The conidia have been dispersed to reveal the whorls of sporogenous cells each with a rachis-like, spore-bearing apex. *B* and *C*, *Tritirachium* sp. showing young (*B*) and older (*C*) sporogenous cells with the conidia attached (OAC 10587).

sometimes with marginal hairs; conidiophores erect, forming a compact layer over the surface of the sporodochium, simple or sparingly branched; conidia phialospores, ovate to elongate-cylindric, sometimes globose, rarely naviculate, gathering in masses.

**Diagnostic Features:** *Tubercularia* is typified by bright-colored sporodochia bearing a compact layer of simple or branched conidiophores, producing masses of small, continuous phialospores (Fig. 215).

**Notes:** This genus is rarely reported from soil (Gilman, 1957). *T. vulgaris*, the conidial state of *Nectria cinnabarina*, has been recorded from soil in Canada by Bisby *et al.* (1935). *N. cinnabarina* causes the "coral spot" disease of deciduous woody plants and is very common in some forests as a saprophyte or wound parasite. It would not be surprising, therefore, to recover it from stray spores in forest soil samples.

### ULOCLADIUM Preuss

Type Species: *Ulocladium botrytis* Preuss.

**Generic Description:** Conidiophores darkly pigmented, arising as branches from the vegetative hyphae, simple, sometimes branched, septate, with tips not markedly swollen as in *Stemphylium;* conidia poro-

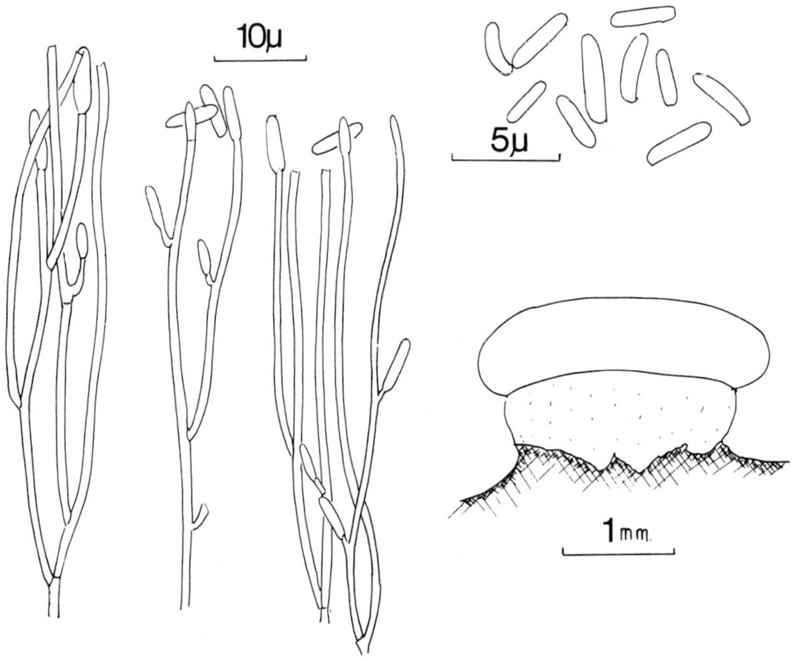


FIG. 215. *Tubercularia vulgaris*. Conidiophores are produced in a dense stand from the stromatic sporodochium. (Drawing from mount from infected maple.)

spores, variable in shape, mostly ovate, not beaked, dark muriform, usually without constriction at the major median septum, borne singly or in botryose clusters on conidiophores, acrogenous, displaced laterally due to renewed growth of the conidiophore from a point below apex; succession of conidia produced acrogenously by sympodial growth of the conidiophores. (Description from Subramanian, 1961.)

**Diagnostic Features:** According to Simmons (1967), the distinguishing character of *Ulocladium* is the fundamentally obovoid, nonbeaked form of the conidium as contrasted with the ovoid, distally tapered or beaked form of *Alternaria* (Fig. 216).

**Notes:** Simmons (1967) illustrated and contrasted the type species of *Alternaria*, *Stemphylium* and *Ulocladium*. He noted that *Ulocladium* resembled *Alternaria* in the production of dictyoporospores on well differentiated conidiophores which commonly become geniculate through successive, lateral renewals of growth. Conidia of *Ulocladium*, however, are fundamentally obovoid (sometimes ellipsoidal or subspherical) and do not taper distally into true beaks. The residual pore in the conidium of *Ulocladium* is in the narrowed base which often terminates in a small, rounded, subhyaline apiculus.

Simmons pointed out that in some species of *Ulocladium*, *e.g. U. lanuginosum*, the conidia may germinate *in situ*, giving rise apically or laterally to secondary conidiophores. When apical, such secondary conidiophores resemble the true conidial beaks of *Alternaria* and were referred to by Simmons as "false beaks." *Ulocladium* species with false beaks may be easily confused with *Alternaria*. (See also Subramanian, 1961.)

## UMBELOPSIS Amos and Barnett

Type Species: *Umbelopsis versiformis* Amos and Barnett.

**Generic Description:** Conidiophores hyaline, often septate, bearing a swollen head with sporogenous branches at the apex; conidia one-celled, hyaline, globose. (Description from Amos and Barnett, 1966.)

**Diagnostic Features:** *Umbelopsis versiformis*, the only species, is characterized by spherical, hyaline spores (4 to 5 $\mu$ in diameter) produced singly and terminally on slightly tapered sporogenous cells radiating out from a vesicle (Fig. 217).

**Notes:** This genus was erected by Amos and Barnett (1966) to include a single species, *U. versiformis*, recovered from the roots of various deciduous trees and from the soil in the vicinity of the roots. We have recovered the same fungus from forest soils in Ontario on two occasions.

Colonies of the fungus are white at first but become pale buff in age. The color is apparently due to a slight pigmentation taken up by the mature conidia. Amos and Barnett do not commit themselves as to the type

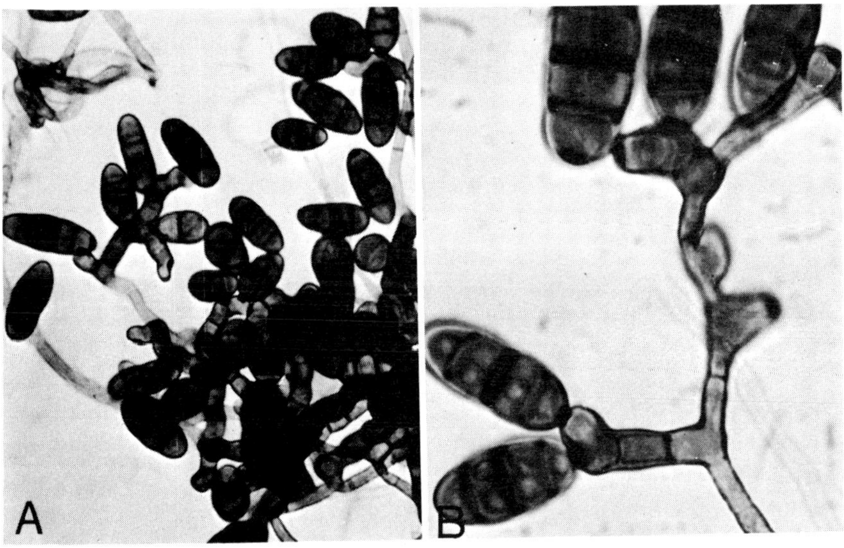


FIG. 216. *Ulocladium* sp. Solitary, muriform porospores are produced in acropetal succession from sympodially extending conidiophores.

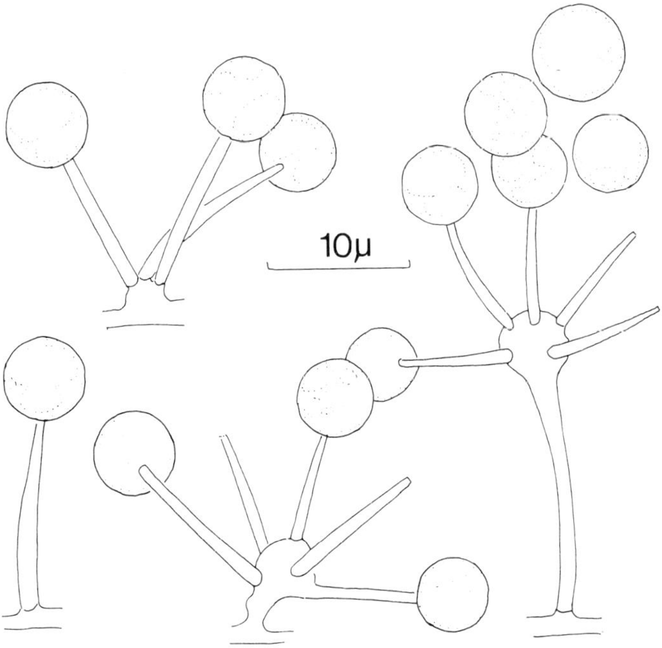


FIG. 217. *Umbelopsis versiformis*. The spherical, subhyaline spores are borne singly on short sporogenous cells which frequently radiate from a stalked vesicle (OAC 10349, ex type).

of spore produced in this genus. The conidia secede readily and there is no sign of a basal frill indicative of rupture of the supporting sporogenous cell. It seems possible that they might be blastospores.

## VARICOSPORIUM Kegel

Type Species: *Varicosporium elodeae* Kegel.

**Generic Description:** Conidiophores poorly differentiated from the vegetative hyphae, bearing conidia apically; conidia hyaline, consisting of an elongated main axis which bears several short laterals on the sides; laterals are septate and may show further branching. (Description based on *V. elodeae.*)

**Diagnostic Features:** *Varicosporium* has an unusual spore form. The elongate main axis with simple or branching laterals make it distinctive (Fig. 218).

**Notes:** This genus has been reported from soil infrequently. It has been recorded from this source by Bessey (1939), who recovered it from soil in water baited with hemp seed. It has also been recorded by Williams and Schmitthenner (1956) from Ohio. The genus, in its normal ecological habitat, is considered a parasite or saprophyte on aquatic plants (see Ingold, 1942).

Ingold (1942) studied *V. elodeae* on decaying alder leaves and in pure culture. He noted that the conidiophores were simple and 100 to 200 $\mu$ long by 2 to 3 $\mu$ broad, bearing a number of branched conidia, of which one was usually terminal and the remainder lateral. Ingold found the conidia to consist of a man axis (60 to 120 $\mu$ long by 3 $\mu$ wide) with one to three laterals of the same width developed on one side only of the main axis. Each of the laterals sometimes branched again in the same one-sided manner. Ingold drew attention to the narrow constriction at the point of origin of the spore with the conidiophore and of each branch and noted that the conidia could fragment, to some extent, by part of the spore breaking off at this isthmus.

Ingold regarded the first-formed terminal spore as a phialospore and the lateral spores as radulaspores. I have not seen this species in culture, but the conidia would appear to be blastospores.

## VERTICICLADIELLA Hughes

Type Species: *Verticicladiella abietina* (Peck) Hughes.

**Generic Description:** Conidiophores stout, more or less erect, solitary or aggregated into fascicles, septate, darkly pigmented, branching at the top to produce a penicillate head, with three to five series of metulae; metulae brown (lower) to hyaline (upper) terminating in sporogenous cells (sympodulae); conidia sympodulospores, borne in acropetal succession, hyaline, nonseptate, ovoid to clavate, straight or slightly curved, with

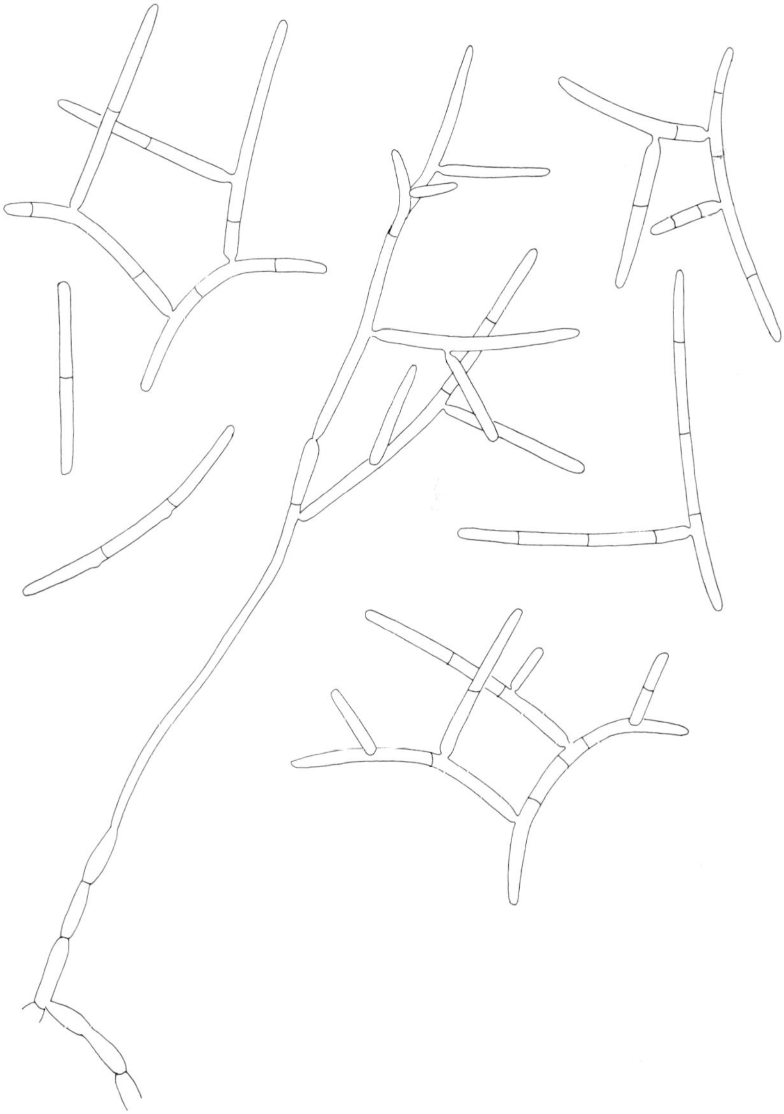

FIG. 218. *Varicosporium elodeae.* Conidiophore and conidia. Each conidium consists of a main axis from which one to three branches grow out laterally, usually to one side of the axis. Primary branches may produce secondary or even tertiary branches. (Redrawn from Ingold, *Trans. Brit. Mycol. Soc.*, *25:* 339–417, 1942. Fig. 39.)

truncate basal scars, gathering in mucus. (Description based on Kendrick, 1962.)

**Diagnostic Features:** *Verticicladiella* is characterized by stout, darkly pigmented conidiophores with an apical penicillus bearing a mucoid head of hyaline amerospores produced by sympodulae (Fig. 219).

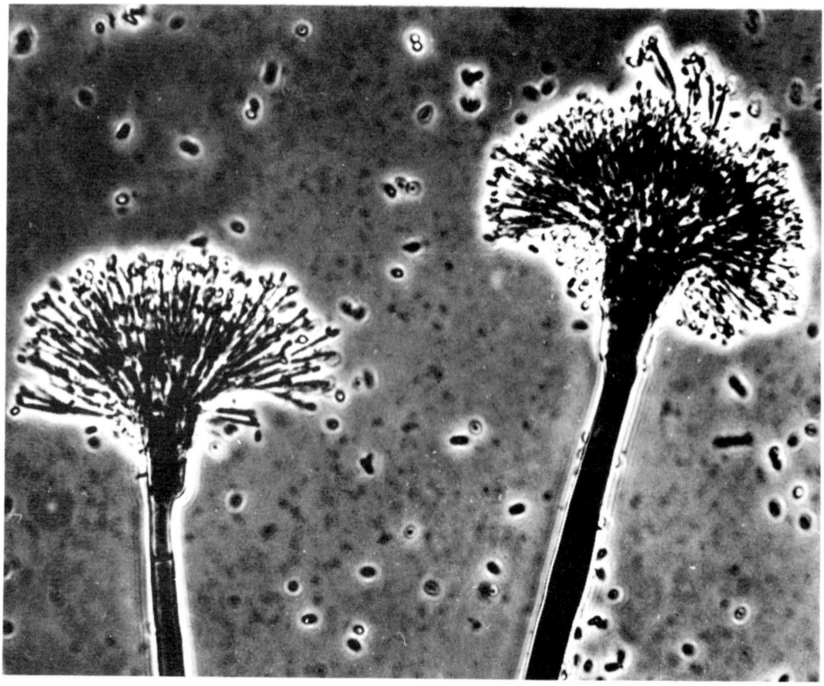

FIG. 219. *Verticicladiella abietina.* Conidiophores, sporogenous cells and conidia. The spores are produced from a sympodially extending sporogenous cell and slime down to form gloeoid masses at the conidiophore apex (DAOM 63700). (Reproduced by the courtesy of W. B. Kendrick.)

**Notes:** *Verticicladiella* is a member of the so-called *Leptographium* complex of Hyphomycetes. This complex contains several genera and is characterized by darkly pigmented conidiophores bearing penicillate heads in which amerospores gather in mucus. They are superficially very similar, but, as has been pointed out by Kendrick in a series of publications, the method of spore formation of the different genera involved may be quite different. The morphology and taxonomy of *Verticicladiella* has been described in some detail by Kendrick (1962), who recognized seven species.

In theory this genus should be rather easy to recognize from the morphologically similar *Phialocephala* and *Leptographium*, but in practice there may be some confusion. A careful study of the sporogenous cells in old and young heads is the key to genus identification.

Undoubtedly many of the records of this fungus have been listed under other genera and it is not known how common it is in soil. We have isolated a *Verticicladiella* species on a number of occasions from organic soils of peat bogs in Ontario. The specific identity of these isolates is not yet established (see *Graphium*).

## VERTICICLADIUM Preuss

Type Species: *Verticicladium trifidum* Preuss.

**Generic Description:** Conidiophores stout, erect, septate, solitary or in groups, pigmented, bearing branches singly or more often in primary, secondary or tertiary verticils, with ultimate branchlets sympodulae; conidia borne apically in acropetal succession with sympodial growth of the sporogenous cell, continuous, dry, hyaline or lightly pigmented.

**Diagnostic Features:** *Verticicladium* is characterized by stout conidiophores with verticillate branching at the top and dry conidia borne acropetally on sympodulae (Fig. 220).

**Notes:** *Verticicladium* is rarely reported from soil. *V. trifidum* has been recorded by McClennan and Ducker (1954) from soil in Australia and by Chesters from England (1960).

The genus is normally recorded as a saprophyte on woody substrates and conifer leaf litter. A detailed account of its morphology and taxonomy given by Hughes (1951) is the basis for the description above.

## VERTICILLIUM Nees

Type Species: *Verticillium lateritium* Berk.

**Generic Description:** Conidiophores erect or nearly so, septate, hya-

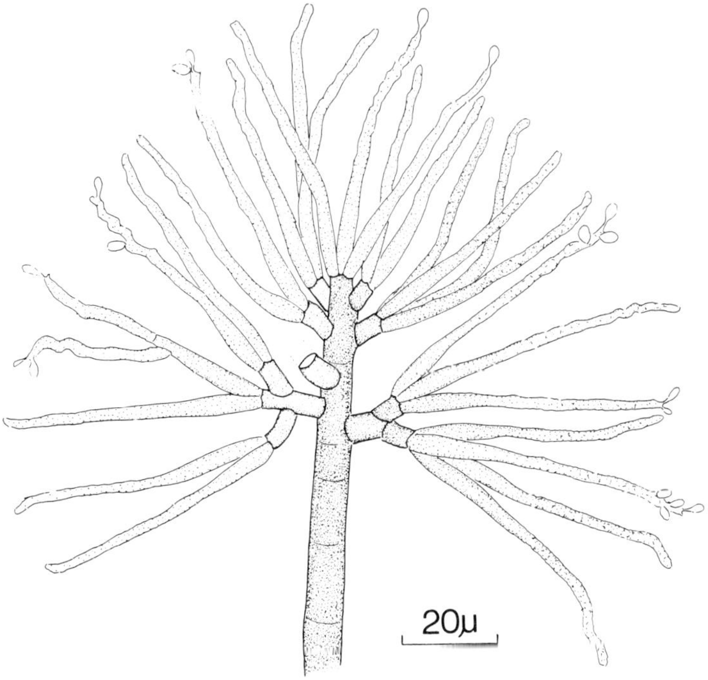


FIG. 220. *Verticicladium trifidum*. Conidiophores, sporogenous cells and conidia. The conidia are borne dry on the sympodially extending sporogenous cells.

line or pigmented, simple or branched; branching sometimes irregular, more often arranged in primary, secondary or higher order verticils; sporogenous cells phialides; phialides borne in whorls on the primary conidiophore axis or branches; phialospores hyaline or subhyaline, nonseptate, produced in mucus and sliming down to form balls at the apex of the phialides; spores ovoid or short-cylindric, sometimes flattened on one side or even allantoid; chlamydospores and aleuriospores associated with some species.

**Diagnostic Features:** The production of balls of amerospores on verticillately arranged phialides is characteristic of *Verticillium* (Fig. 221).

**Notes:** *Verticillium* species are extremely common in soil. One of the most common species, the conidial state of *Nectria inventa*, has been variously referred to as *Acrostalagmus cinnabarinus*, *Verticillium cinnabarinum* and *Verticillium lateritium*. The latter name is probably the best taxonomic disposition for conidial *Nectria inventa*. This is a striking species readily identified by its bright orange-red coloration. The pathogen *V. albo-atrum* is infrequently recorded from soil. Although probably common enough in this environment, it is more readily recovered by special techniques (Nadakavukaren and Horner, 1959).

*Verticillium* species are frequently misidentified because the conidiophore development is not always typical. In many cases the main conidiophore axis with its verticils of phialides is hard to find and the isolates appear to all intents and purposes as *Cephalosporium*. A careful search,

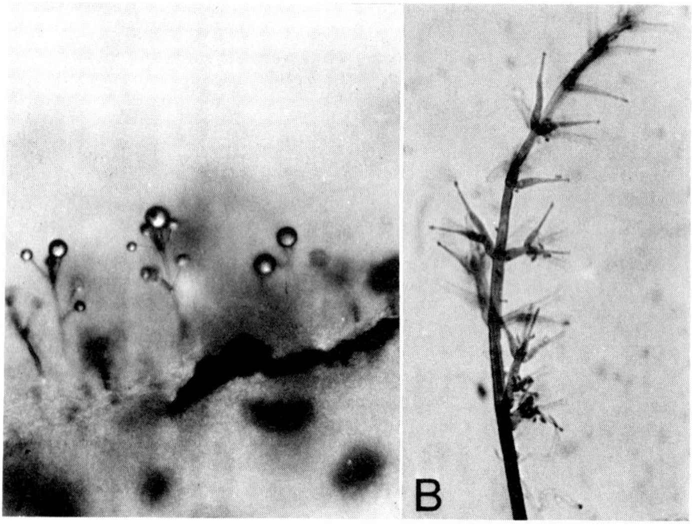


FIG. 221. *A*, *Verticillium albo-atrum*, verticils of phialides with conidia gathering in balls (OAC 10098). *B*, *Verticillium tenuissimum*, in this species the conidiophore and phialides are pigmented (OAC 10290).

however, reveals typical verticils in most cases. The aleuriospore genera *Mycogone*, *Sepedonium* and *Diheterospora* may have accessory phialospore states of the *Verticillium* type.

### **VIRGARIA** Nees ex Sacc.

Type Species: *Virgaria nigra* (Link) Nees.

**Generic Description:** Conidiophores erect or suberect, simple or sparingly branched, darkly pigmented, septate; conidia produced in acropetal succession over the upper part of the conidiophores and branches; conidia pigmented, nonseptate, asymmetric, with well marked basal scar indicating attachment point. (Description based on *V. nigra.*)

**Diagnostic Features:** The dark, asymmetric amerospores (sympodulospores), produced in acropetal succession from sparingly branched conidiophores serve to distinguish the common *Virgaria nigra* (Fig. 222).

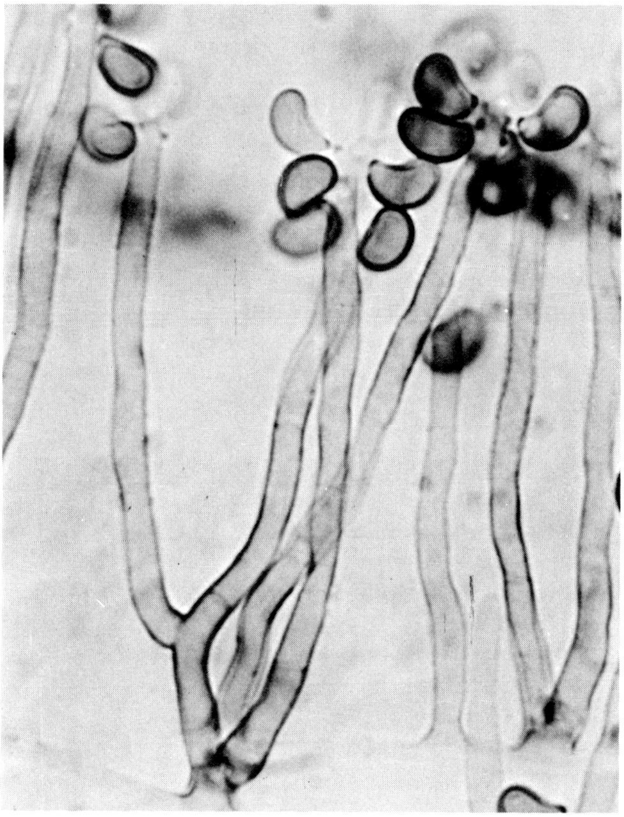

FIG. 222. *Virgaria nigra*. The bean shaped conidia secede to leave conspicuous denticles on the sympodulae (OAC 10245).

**Notes:** *Virgaria* is commonly found on dead wood (Commonwealth Mycological Institute records) and is rarely isolated from soil. Christensen and Whittingham (1965) record a *Virgaria* species from cedar-fir swamps in Wisconsin. We have isolated *V. nigra* on one occasion from peat soils in Ontario.

### VOLUCRISPORA Haskins

Type Species: *Volucrispora aurantiaca* Haskins.

**Generic Description:** Conidiophores simple or branched; conidia blastospores, produced in slime, consisting of a basal cell bearing one to five cells at or near its distal end, originally connected to the basal cell by narrow protoplasmic isthmuses or constrictions, but eventually separated from it by deposition of wall material. (Description from Petersen, 1962.)

**Diagnostic Features:** Blastospores which arise from short conidiophores consisting of a basal cell with one to several apical "wings" is characteristic of *Volucrispora* (Fig. 223).

**Notes:** *Volucrispora* was erected by Haskins (1958) based on *V. aurantiaca* isolated from a water culture containing soil collected near Saskatoon, Canada. Petersen (1962) studied the origin and development of the spores in *Volucrispora* and *Tricellula* with slide-culture techniques and phase-contrast microscopy. He concluded that the spores were blastospores, and presented emended diagnoses for both genera. From Petersen's results on the developmental morphology, there seems little to choose between the two genera. They are the same except for the number of lateral cells which develop from the basal cells, and this hardly seems to be a generic distinction. Consideration of *Volucrispora* under *Tricellula* seems inevitable.

### VOLUTELLA Tode ex Fr.

**Generic Description:** Fructification a sporodochium, sessile or on a short stalk, with a fringe of long, bristle-like setae; conidiophores in a dense stand covering the sporodochium, simple or more often irregularly branched; conidia phialospores, produced successively and abundantly, oval to elliptical or short-cylindric, hyaline, continuous, gathering in masses.

**Diagnostic Features:** *Volutella* is characterized by bright-colored, setose sporodochia on which nonseptate phialospores gather in mucoid masses (Fig. 224).

**Notes:** According to Hughes (1958), *Volutella* is a *nomen illegitimum.* Unfortunately, the species previously considered under this name have not been disposed of in any other genus. As a consequence, *Volutella* is retained for the moment to accomodate this group of species. It seems possible that *Volutina* Penzig and Sacc. could, with some modification,

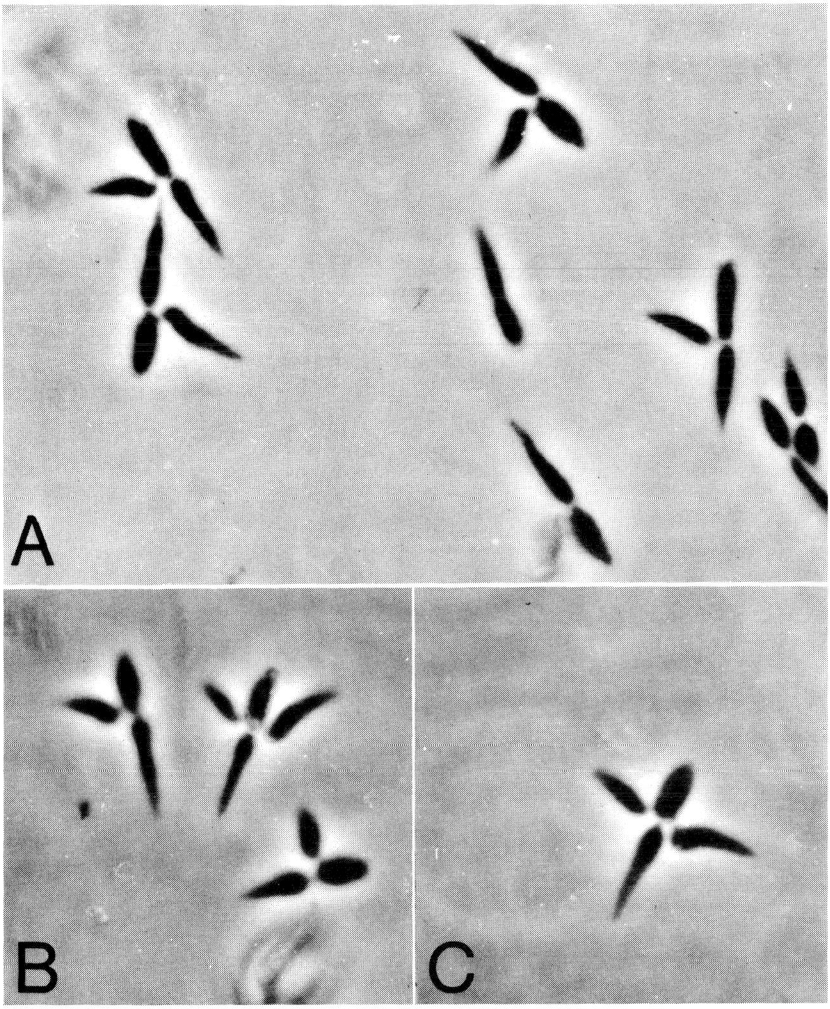

FIG. 223. *Volucrispora aurantiaca. A–C*, the blastospores usually have three or four arms connected by narrow isthmi (OAC 10618, ex type).

satisfactorily include *Volutella* in its popular concept. As it stands at present, *Volutina*, based on *V. concentrica*, is described as having the conidia in chains. It would then bear the same relationship to *Volutella* as *Metarrhizium* does to *Myrothecium*.

*Volutella* species are relatively common in soil (Gilman, 1957). We have isolated two species of this genus from soil. One of these agrees with the description of *V. ciliata* and is very common in organic and forest soils in Ontario.

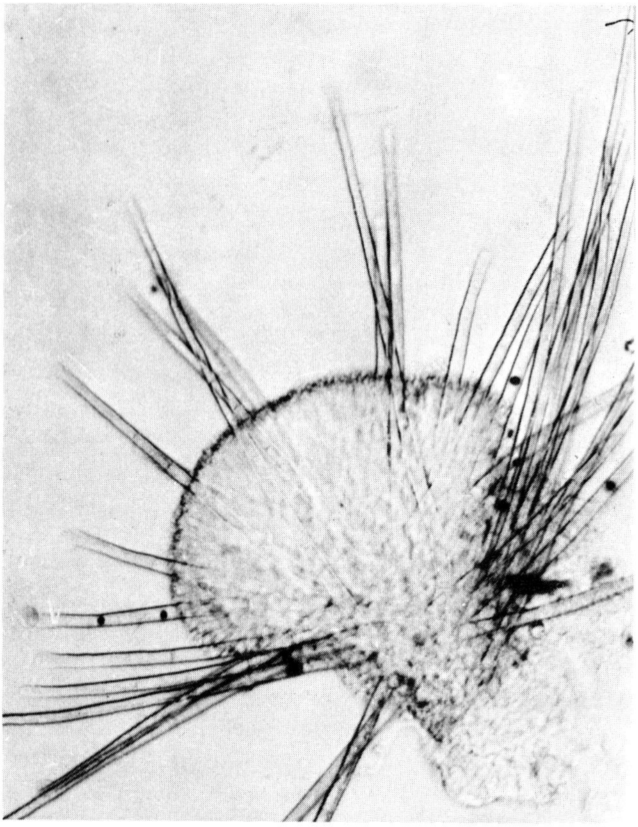

FIG. 224. *Volutella ciliata.* The sporodochium has a fringe of stout, blunt setae (OAC 10561).

## WALLEMIA Johan-Olson

Type Species: *Wallemia ichthyophaga* Johan-Olson.

**Generic Description:** Conidiophores hyaline to subhyaline, more or less erect in a dense stand, cylindrical, terminating in a single phialide-like structure with a dark cup; each "phialide" produces a filamentous sporogenous hypha; sporogenous hypha continuous at first, becoming septate basipetally to form a simple chain of arthrospores; arthrospores more or less cubical at first, rounding off to produce spherical, pigmented, non-septate spores. (Description based on *W. ichthyophaga.*)

**Diagnostic Features:** Arthrospores which develop from a sporogenous hyphae originating from a phialide-like structure are very unusual and a distinctive feature of *Wallemia* (Figs. 16*B* and 59*E*).

**Notes:** This genus is monotypic. *W. ichthyophaga* is perhaps better known under its synonyms *Sporendonema sebi* or *S. epizoum.*

*Sporendonema* is numerically a small genus, containing two or three species. *S. purpurascens* occurs as the "lipstick mould" or "red *Geotrichum*" of commercial mushroom beds. *S. casei*, the type species, is commonly found on cheese rind, where it forms small, orange-colored colonies (see Smith, 1960). The species known as *S. sebi* (*i.e. W. ichthyophaga*) is found chiefly on substrates with high sugar or salt concentrations. This species was given an extensive treatment by Redaelli and Ciferri (1934), and later Ciferri (1958) published an extensive list of synonyms, including *W. ichthyophaga.* He concluded that *S. epizoum* was the correct name for this fungus.

On the basis of the method of spore formation, however, it is clear that *S. epizoum* is not congeneric with *S. casei*, the type species of the genus, and *Wallemia* seems the proper disposition. *W. ichthyophaga* was isolated on several occasions by Wang (1965) from samples of paper.

### **WARDOMYCES** Brooks and Hansford

Type Species: *Wardomyces anomala* Brooks and Hansford.

**Generic Description:** Conidiophores short, arising singly on the sides of hyphae, hyaline, branched, septate; primary and secondary branches arise successively from cells of the main stalk or its branches; sporogenous cells terminal, inflated; conidia borne successively in small numbers on the sporogenous cells, the first being apical and remaining so, the others arising laterally; conidia continuous or one-septate, ovoid or ellipsoid, papillate in some species, brown to blackish, thick-walled and marked with a single longitudinal germ slit, seceding by rupture of the wall of the sporogenous cell and retaining a minute, hyaline basal frill.

**Diagnostic Features:** *Wardomyces* typically produces short, somewhat inflated conidiophores and sporogenous cells which bear apical clusters of dark aleuriospores with delicate, longitudinal germ slits (Fig. 225).

**Notes:** The genus *Wardomyces* was erected by Brooks and Hansford (1922) to include a single species, *W. anomala*, isolated from rabbit flesh in cold storage. The genus has been characterized more fully by Hennebert (1962), who added two species, *W. hughesii* and *W. humicola.* Dickinson (1964) isolated a fourth species, *W. papillata*, from salt marsh muds in England and wheat fields in Ireland. Dickinson (1966) found that *Echinobotryum pulvinatum* Matruchot was an earlier valid name for *W. papillata* and recombined this as *W. pulvinata.* Dickinson reported that this same species was reported from beechwood soil in Poland by Krzemieniewska and Badura (1954) under the name *Stachybotrys humilis.* A second record of *W. humicola* from soil was reported by Dal Vesco (1963), who also drew attention to the figures of Delitsch (1943) as a possible earlier record of *W. humicola.* This photograph, labeled as *Stachybotrys* species, may well be *W. pulvinata.* We have isolated all three species of *Wardomyces* described by Hennebert from Ontario soils.

The ecological significance of *Wardomyces* is not clear. As evidenced

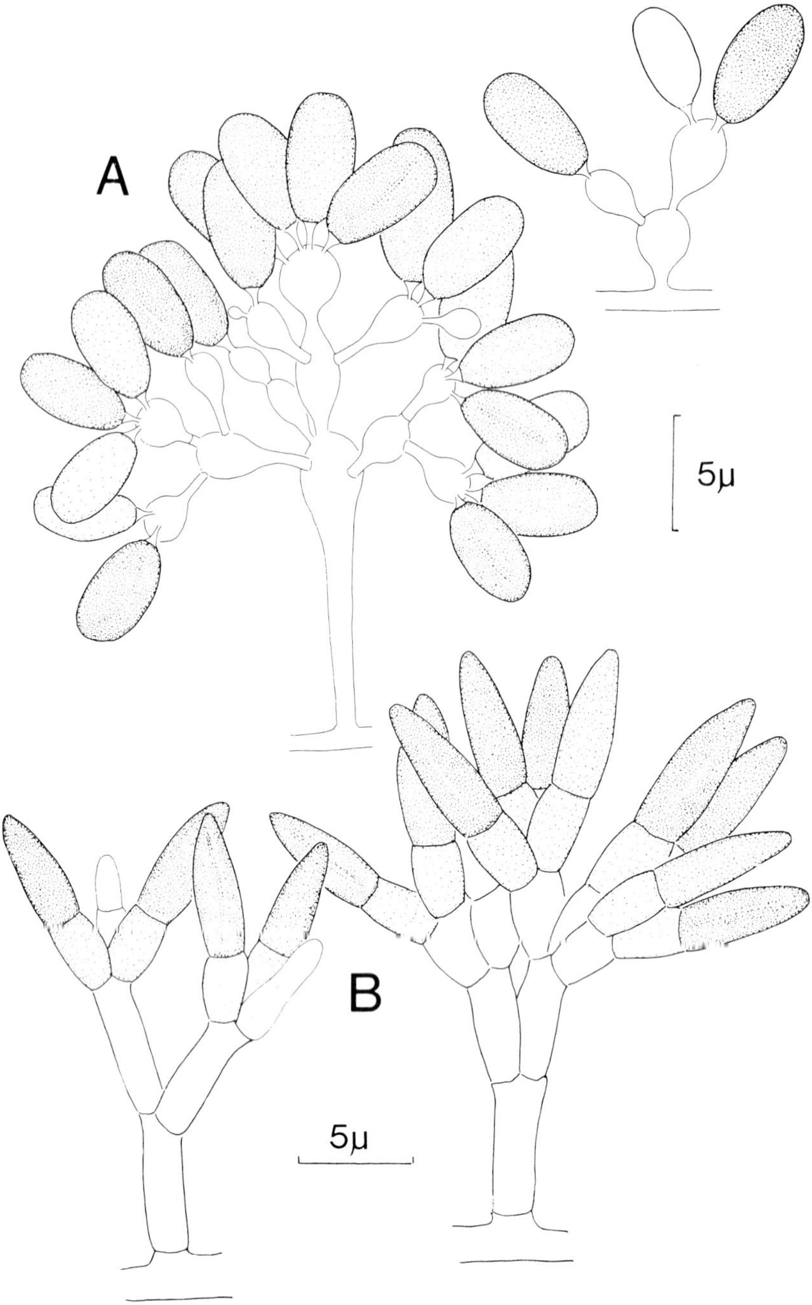


FIG. 225. *Wardomyces*. *A*, *W. hughesii* (OAC 10402). *B*, *W. humicola* (OAC 10403).

by the initial isolation of *W. anomala*, this species grows well at low temperatures. Hennebert also recorded isolates from eggs in cold storage. One of the Ontario isolates of this species was recovered from soil which had been stored several weeks in a refrigerator.

## ZYGOSPORIUM Mont.

Type Species: *Zygosporium oscheoides* Mont.

**Generic Description:** Conidiophore consisting of a more or less upright simple or branched main axis (falciphore) bearing, singly and laterally, specialized branches (falces) with two or more sporogenous cells at the distal end; falciphores more or less erect, simple or branched, septate,

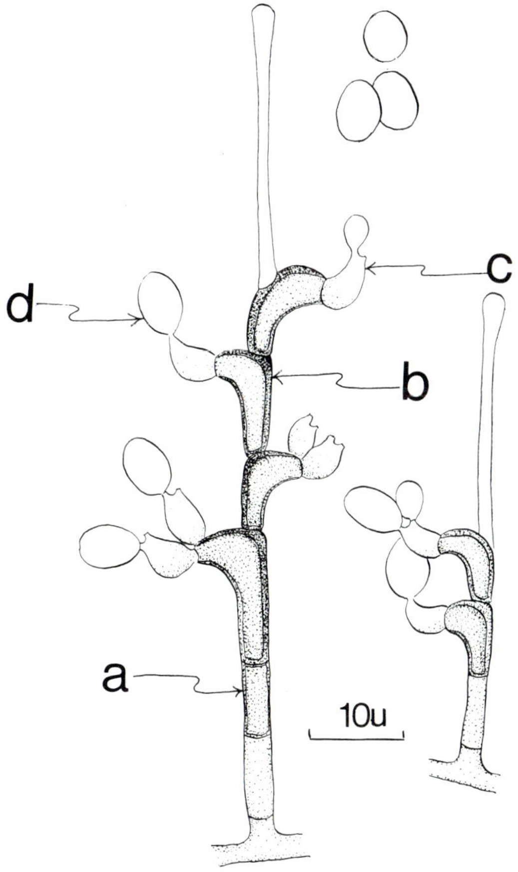


FIG. 226. *Zygosporium masonii*. In this species the falciphore (*a*) produces reflexed branches called falces (*b*). Each falx bears one or several sporogenous cells (*c*) with the conidia (*d*) borne singly on short truncate denticles (IMI 111971)

brown to dark brown, smooth or slightly roughened, sometimes terminating in a hyaline subglobose vesicle; falx shaped like a bill-hook, more or less stalked, septate, darkly pigmented, swollen at the apex; sporogenous cells hyaline to subhyaline; conidia nonseptate, produced as blown out ends of the sporogenous cells, sometimes enveloped in mucilage, hyaline to pale brown, smooth or warted, globose to ellipsoid. (Description based on Hughes, 1951.)

**Diagnostic Features:** The dark-brown falx, shaped like the blade of a bill-hook and bearing several sporogenous cells apically, is distinctive for *Zygosporium* (Figs. 59*A* and 226).

**Notes:** The terms falciphore and falx were proposed by Mason (1941) to describe the unusual conidiophore apparatus produced in this genus. A review of the morphology and taxonomy of *Zygosporium* is given by Hughes (1951a), who described and illustrated seven species. It was pointed out by Hughes that the term "prophialide" proposed by Vuillemin for the falx was not desirable, since the sporogenous cells were not phialides.

The falciphore is regarded as the upright branch of a repent hypha which bears one or more lateral falces, the whole comprising the conidiophore. In some species the falciphore is lacking and the falces originate directly from the repent hyphae.

*Zygosporium* is rarely reported from soil. Agnihothrudu (1962) recovered *Z. echinosporum* from tea soils in Assam, and in the herbarium of the Commonwealth Mycological Institute there is a record of *Z. masonii* isolated from cabbage soil in Hong Kong by M. Chu. Meredith (1962) has presented evidence that the conidia in *Zygosporium* are violently discharged.

CHAPTER

# VII

# Excluded Genera

The following genera have been recorded from soil but have been excluded from the present treatment for a variety of reasons. Many are synonyms of established genera, some are homonyms, others are illegitimate, ambiguous or confused names. In a few cases the genera have been excluded from detailed consideration because of a lack of culture or herbarium material or sufficient information to fully characterize these genera for comparison with similar genera described herein.

***Acremonium* Link ex Fr. = ?

*Acrocylindrium* Bon. = *Verticillium*

*Acrophialophora* Edwards = *Paecilomyces*

*Acrostalagmus* Corda = *Verticillium*

*Acrostaphylus* Arnaud = *Nodulisporium*

*Acrotheca* Fuckel = *Ramularia*

*Acrothecium* Preuss—*nomen ambiguum* (see the text)

*Aleurisma* Link = *Chrysosporium*

*Alphitomyces* Reissek—single soil record probably a coremial *Paecilomyces*

*Bactridiopsis* Hennings—see the text

*Beltraniella* Subram.—only soil record *B. humicola* Rao = *Beltrania rhombica*

*Bisporomyces* van Beyma = *Chloridium*

*Blodgettia* Wright—fungus component of lichen (see the text)

* Excluded because of insufficient information and material.

*Botrydiella* Badura = *Staphylotrichum*

*Cephalodiplosporium* Kamyschko = *Cephalosporiopsis*
*Cephalotrichum* Link—see *Doratomyces*
*Cephalothecium* Corda = *Trichothecium*
**Chlamydorubra* Deshpande and Deshpande
*Chlamydosporium* Peyronel—*Mycelia Sterilia*
*Cirrhomyces* Höhn. = *Chloridium*
**Cladorrhinum* Sacc. and March.
*Clonostachys* Corda = *Gliocladium*
*Coniothecium* Corda = *Trimmatostroma*
*Coremium* Link = *Penicillium*
*Corethropsis* Corda—only record *C. hominis* = *Chrysosporium pannorum*
**Cryptomela* Sacc.
*Cylindrium* Bon. = *Fusidium*
*Cylindrocephalum* Bon. = *Chalara*

*Dactylium* Nees = *nomen dubium* (see the text)
*Dematium* Pers. ex Fr.—*D. pullulans* = *Aureobasidium pullulans*
*Dendryphiella* Bubak and Ranoj. = *Dendryphion*
*Dicoccum* Corda—*D. asperum* = *Trichocladium asperum*
*Didymocladium* Sacc. = *Cladobotryum*
*Diplocladium* Bon. = *Cladobotryum*
**Discocolla* Prill. and Delacr.

**Fumago* Pers.

*Geomyces* Traaen = *Chrysosporium*
*Gliobotrys* Höhn. = *Stachybotrys*
*Gliocladiopsis* Saksena = *Cylindrocladium*

*Haplaria* Link = *Botrytis*
*Haplochalara* Linder—*H. pidoplitschkoi* Zhdanova = ? *Chloridium chlamydosporis*
*Heterosporium* Klotzsch = *Cladosporium*
*Hormiscium* Kunze = *Torula*
*Hormodendrum* Bon. = *Cladosporium*
*Hyalobotrys* Pido. = *Stachybotrys*
*Hyalopus* Corda = ? *Cephalosporium*
*Hyalostachybotrys* Srin. = *Stachybotrys*
**Hymenella* Fr.
*Hymenula* Fr. = *Hymenella*
*Hyphoderma* = *nomen illegitimum*
**Hyphosoma* Sydow

**Illosporium* Mart.

*Keratinomyces* Vanbreus. = *Microsporum*

*Macrosporium* Fr. = *Alternaria*
*Margarinomyces* Laxa = *Phialophora*
*Masoniella* Smith = *Scopulariopsis*
*Mastigosporium* Riess—only record *M. heterosporum* = *Cylindrocarpon* sp.
*Melanconium* Link
*Memnoniella* Höhn. = *Stachybotrys*
*Mesobotrys* Sacc. = *Gonytrichum*
*Moeszia* Bubak = *Cylindrocarpon*

*Monacrosporium* Oud. = *Dactylella*
*Monosporium* Höhn. = *nomen illegitimum*
*Monotospora* Sacc. = *Humicola*
*Monotospora* Corda — ?
**Mucrosporium* Preuss
**Multicladium* Deshpande and Deshpande
*Myrotheciella* Speg. —only isolates examined proved to be *Metarrhizium*

*Oospora* Wallr. = *nomen illegitimum*
*Oothecium* Speg. = *Capnodiastrum*

*Pachybasium* Sacc. = *Trichoderma*
*Papularia* Fr. = *Arthrinium*
**Patouillardiella* Speg.
*Pestalotiopsis* Stey.—see *Pestalotia*
*Phaeoscopulariopsis* Ôta = *Scopulariopsis*
*Pochonia* Batista and Fonseca = *Diheterospora*
**Polyschema* Upadhyay
*Pseudostemphylium* (Wilts.) Sub. = *Ulocladium*
*Pullularia* Berk. = *Aureobasidium*

**Rhinocephalum* Kamyschko
**Rhinocladiopsis* Kamyschko
*Rhinocladium* Kamyschko = homonym
*Rhinotrichum* Corda—see *Olpitrichum*
*Rhinocladiella* Kamyschko = homonym
*Rhopalomyces* Corda = Mucorales

*Saprophragma* Deshpande and Deshpande
**Sarcinella* Sacc.
*Schelleobrachea* Hughes = *Pithomyces*
**Schizoblastosporon* Ciferri
*Septogloeum* Sacc.
*Septomyxa* Sacc.
*Spicaria* Harting = *Paecilomyces*
*Spicularia*—*S. terrestris* = *Pseudobotrytis terrestris*
*Spondylocladium* Mart. = *Stachylidium*
*Sporocybe* Fr. = *Periconia*
*Sporodum* Corda = *Periconia*
**Stemmaria* Preuss
*Stilbella* Lindau = *Stilbum*
*Stysanus* Corda = *Doratomyces*
*Synsporium* Preuss = *Stachybotrys*

**Tawdiella* Deshpande and Deshpande
*Tetracoccosporium* Szabo = *Stemphylium*
*Tilachlidium* Preuss = *Cephalosporium*
*Trichobotrys* Penzig and Sacc. = *Periconia*
**Trichosporiella* Kamyschko
*Trichosporium* Fr. = *nomen confusum*
*Truncatella* Stey.—see *Pestalotia*

*Umbellula* Morris = *Pseudobotrytis*

*Verticilliastrum* Daszewska = *Trichoderma*
*Volutella* = *nomen illegitimum* (see the text)

*Zygodesmus* Corda = *nomen dubium*

# Glossary

**acervulus:** a flat or saucer-shaped stroma or hyphal mat bearing a stand of closely-packed conidiophores, as in *Colletotrichum*; fructification type characteristic of the Melanconiales.

**acropetal:** produced successively with the youngest at the apex, as the spores of *Tritirachium* or *Cladosporium*.

**acropleurogenous:** produced apically and on the sides, as the spores in *Diplococcium*.

**aleuriospore:** spore form characteristic of the Aleuriosporae; a thick-walled terminal spore formed from the blown out end of a sporogenous cell, as in *Humicola*.

**amerospore:** a nonseptate spore, as in *Penicillium*.

**ampulla:** the swollen sporogenous cell of the Botryoblastosporae on which all the blastospores develop simultaneously, as in *Botrytis*.

**ampulliform:** swollen below with a narrow neck, as the phialide of *Chaetopsina*.

**anastomosis:** fusion between hyphal elements to form a bridge, as in *Rhizoctonia*.

**annellophore:** the conidiophore or sporogenous cell characteristic of the Annellosporae which bears a succession of ring-like scars or annellations around the sporogenous cell, as in *Scopulariopsis*.

**annellospore:** spore form characteristic of the Annellosporae, usually truncate at the base.

**arthrospore:** spore form characteristic of the Arthrosporae; formed by septation and eventual fragmentation of the vegetative hyphae or specialized sporogenous hyphae, as in *Geotrichum* or *Oidiodendron*.

**aspergilliform:** name given to the *Aspergillus*-like accessory spore states of certain Hyphomycetes such as *Acremoniella*.

**asymmetric:** of spore shape—flattened or concave on one side, as in *Harposporium*.

**basipetal:** of spores—youngest at the base, as from a phialide or annellophore.

**biconic:** of spore shape—two cones attached base to base, as in *Beltrania*.

**bivalvate:** of spore shape—lens-shaped spores with a hyaline rim, as in *Arthrinium*.

**blastospore:** spore form characteristic of the Blastosporae; an asexual spore produced by budding from the sporogenous cell or a pre-existing spore, as in *Cladosporium*.

**botryoaleuriospore:** aleuriospores which develop successively and basipetally from the sporogenous cell to form an apical cluster, as in *Echinobotryum*.

**botryoblastospore:** blastospores which develop more or less simultaneously from a swollen ampulla, as in *Oedocephalum*.

**byssoid:** cottony, floccose; composed of delicate threads, as the mycelium of *Thozetellopsis*.

**catenate:** in chains, as the conidia of *Aspergillus*.

**chlamydospore:** thick-walled resting spore, frequently intercalary, which is formed by modification of a pre-existing cell, as in hyphae of *Geotrichum candidum* or spore of *Fusarium*.

**clavate:** club-shaped, as the vesicle of *Aspergillus clavatus*.

**collarette:** cup-shaped structure found at the apex of the sporogenous cell in certain Phialosporae such as *Phialophora*.

**conidiophore:** main axis or branch bearing sporogenous cells; frequently applied to the sporogenous cell itself, as in *Cephalosporium*.

**conidium:** asexual, nonmotile spore.

**coremium:** = synnema.

**dematiaceous:** belonging to the Dematiaceae in which the conidia or conidiophores are pigmented.

**dendritic:** irregularly branched, tree-like, as the conidiophore of *Cladosporium*.

**denticle:** small, tooth-like prolongation on which spore is borne, as on the ampulla of *Oedocephalum*.

**determinate:** applied to conidiophores where the growth of the conidiophore ceases with production of terminal conidia, as in *Dichotomophthora*.

**dichotomous:** branching into two more or less equal arms, as the conidiophore of *Botryosporium*.

**dictyospore:** spore which is longitudinally and transversely septate, as in *Alternaria*.

**didymospore:** two-celled spore, as in *Leptodiscus*.

**echinulate:** of conidia and conidiophores; with spiny ornamentation.
**ellipsoid:** of spores more or less elliptical in optical section.
**endoconidium:** name given to the phialospore produced within an elongate tubular phialide, as in *Chalara*; frequently also applied to a conidium produced within a collarette.

**falcate:** curved like the blade of a sickle.
**falciphore:** upright branch of a repent hypha bearing one or more lateral falces, as in *Zygosporium*.
**falx:** the fertile hypha shaped like a hook in *Zygosporium*.
**fascicle:** bundle or tuft, as the conidia of *Menispora* or the conidiophores of *Cercospora*.
**foot cell:** the basal cell sometimes found supporting the conidiophores of *Aspergillus*; also used to describe the basal cell of the macrospore of *Fusarium*.
**funiculose:** hyphae aggregated into straggling ropes or bundles.
**fusiform:** spindle-shaped, narrowing towards both ends, as the conidia of many *Paecilomyces* species.

**gangliospore:** aleuriospore but not chlamydospore.
**geniculate:** bent knee-like, as the conidiophore of *Curvularia*
**germ slit:** thin area in spore wall running the length of the spore and through which germination may occur; frequently found in aleuriospores such as those of *Mammaria*.
**globose:** spherical.
**gloeosporae:** spores which slime down in gloeoid masses, as in *Gliocladium*.

**hyaline:** nonpigmented.

**indeterminate:** conidiophores which continue growth more or less indefinitely either monopodially or sympodially.
**intercalary:** between two cells, as with chlamydospores of *Geotrichum*.

**macroconidium:** some fungi such as *Fusarium* (phialospores) and *Microsporum* (aleuriospores) produce two spore states which are similar in type but differ in size. The large septate spores are frequently referred to as macrospores and the smaller, nonseptate spores as microspores. The term macrospore is also used infrequently as synonymous with chlamydospore, as in *Paecilomyces*. This latter useage seems undesirable.
**metula:** name given to the specialized branches of the conidiophore bearing sporogenous cells, as in *Penicillium*.
**microconidium:** see macroconidium; also used to describe spermatium in Ascomycetes.
**moniliaceous:** belonging to the Moniliaceae; conidia and conidiophores hyaline or brightly colored.

**mononematous:** conidiophores solitary or in tufts or fascicles, but not aggregated to form synnemata or borne on sporodochia or acervuli.
**monopodial:** growth of main axis continues from the same apex.
**monoverticillate:** sporogenous cells arranged in a solitary cluster at the apex of the conidiophore, as in *Stachybotrys*
**mucedinaceous:** = moniliaceous.
**mucilaginous:** slimy.
**muriform:** longitudinally and transversely septate, as in *Stemphylium*.

**nomen ambiguum:** name of a taxon used with different meanings.
**nomen confusum:** name of a taxon based on two or more different elements.
**nomen illegitimum:** name of a taxon that is contrary to the rules of nomenclature.
**nomen nudum:** name published since 1934 without a Latin diagnosis.

**obclavate:** inversely clavate, as the spores of *Alternaria*.
**ovoid:** egg-shaped, as the spores of *Acremoniella*; name implies spores are broader at one end; frequently, however, the term has been used erroneously as a synonym of ellipsoid.

**palmate:** hand-like, usually with finger-like extensions, as the conidia of *Dactylosporium*.
**pedicel:** a slender stalk subtending a spore; sometimes it is either equivalent to a sporogenous cell, as in *Humicola*, or to a denticle, as in *Scolecobasidium*.
**penicillus:** a complex arrangement of sporogenous cells and metulae on a conidiophore having a brush-like appearance, as in *Penicillium*. The sporogenous cells need not be phialides before this term can be applied (see *Leptographium*).
**phialide:** sporogenous cell characteristic of the Phialosporae; sporogenous cell in which the spores are produced in basipetal succession from an open growing point.
**phialophore:** conidiophore bearing phialides.
**phialospore:** spore form characteristic of the Phialosporac.
**phragmospore:** multi-celled spore with transverse septa, as in *Curvularia*.
**polymorphic:** having more than one spore form, as in *Doratomyces stemonitis*, which has accessory spore states belonging to the genera *Echinobotryum* and *Scopulariopsis*.
**porospore:** spore form characteristic of the Porosporae; a spore produced through a minute pore in the wall of the sporogenous cell or hypha.
**prophialide:** more or less synonymous with metula; name given to cell which bears phialides; may be used in place of primary sterigma to describe the cells which arise from the vesicle of certain *Aspergillus* species such as *A. niger*.
**pyriform:** pear-shaped, as the spore of *Pyricularia*.

**rachis:** elongate, spore-bearing extension of the sporogenous cell of certain Sympodulosporae such as *Tritirachium*.

**radulaspore:** term used by Mason (1937) for the conidia borne over the surface of the ascospores of *Nectria coryli* while still within the ascus. Also used by Mason to describe conidia of *Botrytis*. Incorrectly used by many authors as equivalent to sympodulospore.

**sarciniform:** box-shaped, as the spores of *Stemphylium botryosum*.

**sclerotium:** a hard, resistant mass of hyphae or pseudoparenchyma, usually for survival under adverse conditions, as in *Sclerotium*.

**sessile:** without a stalk; borne directly from the vegetative hyphae, as the spores in *Mammaria*.

**seta:** stiff, bristle-like hair composed of one or many cells and found in association with the vegetative hyphae, as in *Botryotrichum*, or in association with a fruit body, as in *Colletotrichum*.

**setula:** a delicate, hair-like, nonmotile appendage arising from the ends of conidia, as in *Menisporopsis*; frequently associated with spore dispersal.

**sinuous:** wavy or serpentine, as the conidiophores of *Acremoniella serpentina*.

**sporodochium:** a cushion-shaped, stromatic mass bearing closely packed conidiophores on its upper surface; the fruit-body characteristic of the Tuberculariaceae; also applied to the fruit-body of *Myrothecium* or *Metarrhizium*, which are frequently nonstromatic.

**sporogenous cell:** cell which gives rise to one or several spores.

**staurospore:** conidium with arms extending in several directions, as in *Tripospermum*.

**stellate:** star-shaped; refers to staurospores.

**sterigma:** frequently applied to the phialide or prophialide of *Aspergillus*; properly applied to the denticle on which the basidiospore is produced in Basidiomycetes.

**stilbaceous:** belonging to the Stilbaceae; producing a synnema.

**subglobose:** subspherical.

**subulate:** tapering to a slender point; awl-shaped, as the conidiophore in *Gonytrichum*.

**sympodial:** main axis extending by growth of a succession of apices, each of which develops behind and to one side of the previous apex, which ceases growth and produces a spore or spores.

**sympodioconidium:** conidium produced from a sympodula.

**sympodula:** sporogenous cell of the Sympodulosporae which continues growth sympodially after each terminal spore is produced.

**sympodulospore:** = sympodioconidium.

**synnema:** a more or less compact association of erect conidiophores bearing spores along their length or in a sporiferous head. Conidiophores may be fused along their length, as in *Doratomyces*, or merely compacted, as in *Isaria cretacea*.

**verticil:** a whorl of three or more branches or sporogenous cells.

**vesicle:** swollen apex of a conidiophore, as in *Aspergillus*, or of a sporogenous cell, as in *Arthrobotrys*.

# Bibliography

Agnihothrudu, V. 1956. Fungi isolated from rhizosphere. II. *Starkeomyces*, a new genus of the Tuberculariaceae. *J. Ind. Bot. Soc.*, *35:* 38–42.

———. 1958. Notes on fungi from North-East India. I. A new genus of the Tuberculariaceae. *Mycologia*, *50:* 570–579.

———. 1961. Rhizosphere microflora of tea (*Camellia sinensis* (L) O. Kunze) in relation to the root rot caused by *Ustulina zonata* (Lév.) Sacc. *Soil Sci.*, *91:* 133–137.

———. 1962. Notes on fungi from North-East India. XVII. *Menisporella assamica* gen. et sp. nov. *Proc. Ind. Acad. Sci.*, Sect. B, *56:* 97–101.

———. 1962a. A comparison of some techniques for the isolation of fungi from tea soils. *Mycopathologia*, *16:* 235–242.

———. 1962b. Notes on fungi from North-East India. IX. Hyphomycetes. *J. Ind. Bot. Soc.*, *41:* 465–477.

———. 1964. The occurrence of *Leptodiscus terrestris* Gerdemann in the tea gardens of Assam. *Current Sci. 33:* 25–26.

———, and G. C. S. Barua. 1957. A new record of a rare fungus from tea rhizosphere. *Sci. Culture*, *22:* 568–569.

Ainsworth, G. C. 1963. *Ainsworth and Bisby's Dictionary of the Fungi*, Ed. 5. Commonwealth Mycological Institute, Kew, England.

Ajello, L. 1959. A new *Microsporum* and its occurrence in soil and on animals. *Mycologia*, *51:* 69–76.

Ajello, L., and S. Cheng. 1967. Sexual reproduction in *Histoplasma capsulatum*. *Science*, *155:* 1696.

———, E. Varsavsky, G. Sotgiu, A. Mazzoni, and A. Mantovani. 1965. Survey of soils for human pathogenic fungi from the Emilia-Romagna region of Italy. *Mycopathologia*, *26:* 64–71.

Alexopoulos, C. J. 1962. *Introductory Mycology*, Ed. 2. John Wiley and Son, New York.

Ames, L. M. 1963. *A Monograph of the Chaetomiaceae.* U. S. Army Res. Dev. Series, No. 2.

Amos, R. E., and H. L. Barnett. 1966. *Umbelopsis versiformis*, a new genus and species of the Imperfects. *Mycologia*, *58:* 805–808.

Apinis, A. E. 1963. Occurrence of thermophilous microfungi in certain alluvial soils near Nottingham. *Nova Hedwigia*, *5:* 57–78.

——— 1964. Revision of British Gymnoascaceae. Mycol. Paper No. 96, Commonwealth Mycological Institute, Kew, England.

Arnaud, G. 1953. Mycologie concrète: Genera II (suite et fin). *Bull. Soc. Mycol. Fr.*, *69:* 265–306.

Von Arx, J. A. 1957. Die arten der gattung *Colletotrichum* Cda. *Phytopath. Zeits.*, *29:* 413–508.

Badura, L. 1963. Richerche sulla micoflora del suolo sotto i Faggi dell'Orto Botanico di Torino. *Allionia, 9:* 65–74.

———. 1963a. Fungilli nuovi rari o critici isolati dal suolo i Faggi dell'Orto Botanico di Torino. *Allionia, 9:* 175–185.

Bainier, G. 1907. Mycothèque de l'École de Pharmacie. XIV. *Scopulariopsis* (*Penicillium* pro parte) genre nouveau de Mucédinées. *Bull. Soc. Mycol. Fr., 23:* 98–105.

Bakshi, B. K. 1950. Fungi associated with Ambrosia beetles in Great Britain. *Trans. Brit. Mycol. Soc., 33:* 111–120.

———. 1952. *Oedocephalum lineatum* is a conidial stage of *Fomes annosus*. *Trans. Brit. Mycol. Soc., 35:* 195.

Barnett, H. L. 1958. A new *Calcarisporium* parasitic on other fungi. *Mycologia, 50:* 497–500.

———. 1960. *Illustrated Genera of Imperfect Fungi.* Burgess Publishing Company, Minneapolis, Minnesota.

———. 1963. The physiology of mycoparasitism. In *The Physiology of Fungi and Fungus Diseases*, West Virginia University Agric. Expt. St. Bull. 488T, pp. 65–90.

———, and V. G. Lilly. 1962. A destructive mycoparasite, *Gliocladium roseum. Mycologia, 54:* 72–77.

Barron, G. L. 1961. Studies on species of *Oidiodendron*, *Helicodendron*, and *Stachybotrys* from soil. *Can. J. Bot., 39:* 1563–1571.

———. 1961a. *Monocillium humicola* sp. nov. and *Paecilomyces variabilis* sp. nov. from soil. *Can. J. Bot., 39:* 1573–1578.

———. 1962. New species and new records of *Oidiodendron*. *Can. J. Bot., 40:* 589–607.

———. 1964. A note on the relationship between *Stachybotrys* and *Hyalostachybotrys*. *Mycologia, 56:* 313–316.

———. 1964a. A new genus of the Hyphomycetes from soil. *Mycologia, 56:* 514–518.

———. 1967. *Torulomyces* and *Monocillium*. *Mycologia, 59:* 716–718.

———, and G. C. Bhatt. 1967. A new species of *Gonytrichum* from soil. *Mycopathologia, 32:* 126–128.

———, and C. Booth. 1966. A new species of *Arachniotus* with an *Oidiodendron* conidial state. *Can. J. Bot., 44:* 1057–1061.

———, and L. V. Busch. 1961. Studies on the soil Hyphomycete *Scolecobasidium*. *Can. J. Bot., 40:* 77–84.

———, R. F. Cain, and J. C. Gilman. 1961. The genus *Microascus*. *Can. J. Bot., 39:* 1609–1631.

———, ———, and ———. 1961. A revision of the genus *Petriella*. *Can. J. Bot., 39:* 837–845.

———, and A. H. S. Onions. 1966. *Verticillium chlamydosporium* and its relationships to *Diheterospora*, *Stemphyliopsis* and *Paecilomyces*. *Can. J. Bot., 44:* 861–869.

Bateman, D. F. 1963. Influence of host and non-host plants upon populations of *Thielaviopsis basicola* in soil. *Phytopathology, 53:* 1174–1177.

Batista, A. C., and O. M. Fonseca. 1965. *Pochonia humicola* n. gen. e. n. sp. una curiosa entidade fungica dos solos do Nordeste do Brasil. Inst. Micol. Univ. Recife. Brasil. Publ. No. 462.

———, and Heine, J. W. 1965. A new species of the genus *Monocillium* Saksena isolated from Brazilian soil. Inst. Micol. Univ. Recife. Publ. No. 457.

———, and H. P. Upadhyay. 1965. Soil fungi from north-east Brazil. Inst. Micol. Univ. Recife. Brasil. Publ. No. 442.

Bayliss Elliott, J. S. 1918. Some new species of fungi imperfecti. *Trans. Brit. Mycol. Soc., 6:* 56–61.

———. 1930. The soil fungi of the Dovey Salt Marshes. *Ann. Appl. Biol., 17:* 284–305.

Benedek, T. 1963. On *Anixiopsis stercoraria* (Hansen) Hansen, 1897, and its imperfect stage: *Keratinomyces ajelloi* Vanbreuseghem, 1952. *Mycopathologia, 21:* 179–203.

———. 1963a. *Microsporon* vs. *Microsporum*. *Mycopathologia, 19:* 269–270.

Benham, R. W., and J. L. Miranda. 1953. The genus *Beauveria*, morphological and taxonomical studies of several species and of two strains isolated from wharf-piling borers. *Mycologia, 45:* 727–746.

Berkeley, M. J., and C. E. Broome. 1861. Notices of British fungi, 952, *Acrospeira mirabilis*. *Ann. Mag. Nat. Hist. Ser. III*, VII, 449.

Bessey, E. A. 1939. *Varicosporium elodeae* Kegel, an uncommon soil fungus. *Papers Mich. Acad. Sci., 35:* 15–17.

Van Beyma, T. K. F. H. 1933. Beschreibung einiger neuer Pilzarten aus dem Centraalbureau voor Schimmelcultures—Baarn (Holland). *Zentralb. Bakt. Parasit. Inf., 88:* 132–141.

———. 1942. Beschreibung einiger neuer Pilzarten aus dem Centraalbureau voor Schimmelcultures, Baarn, Holland. *Antonie van Leeuwenhoek, 8:* 116–120.

Bisby, G. R. 1943. *Stachybotrys. Trans. Brit. Mycol. Soc., 26:* 133–143.

———, M. Timonin, and N. James. 1935. Fungi isolated from soil profiles in Manitoba. *Can. J. Res., Sect. C, 13:* 47–65.

Boedijn, K. B., and J. Reitsma. 1950. Notes on the genus *Cylindrocladium* (Fungi: Mucedinaceae). *Reinwartia, 1:* 51–60.

Booth, C. 1957. Studies of Pyrenomycetes. I. Four species of *Chaetosphaeria*, two with *Catenularia* conidia. Mycol. Paper No. 68, Commonwealth Mycological Institute, Kew, England.

———. 1959. Studies of Pyrenomycetes. IV. *Nectria* (Part I). Mycol. Paper No. 73. Commonwealth Mycological Institute, Kew, England.

———. 1959a. Studies of Pyrenomycetes. V. Nomenclature of some Fusaria in relation to their Nectrioid perithecial states. Mycol. Paper No. 74, Commonwealth Mycological Institute, Kew, England.

———. 1966. The genus *Cylindrocarpon*. Mycol. Paper No. 104, Commonwealth Mycological Institute, Kew, England.

———, and J. S. Murray. 1960. *Calonectria hederae* and its *Cylindrocladium* conidial state. *Trans. Brit. Mycol. Soc., 43:* 69–72.

Brian, P. W. 1944. Production of gliotoxin by *Trichoderma viride. Nature, 154:* 667–668.

———, and J. C. McGowan. 1945. Viridin: a highly fungistatic substance produced by *Trichoderma viride. Nature, 156:* 144–145.

Brooks, F. T. 1944. Notes on the pathogenicity of *Myrothecium roridum* Tode ex Fr. *Trans. Brit. Mycol. Soc., 27:* 155–157.

———, and C. G. Hansford. 1922. Mould growth upon cold-store meat. *Trans. Brit. Mycol. Soc., 8:* 113–142.

Brown, A. H. S., and G. Smith. 1957. The genus *Paecilomyces* Bainier and its perfect stage *Byssochlamys* Westling. *Trans. Brit. Mycol. Soc., 40:* 17–89.

Brown, J. C. 1958. Soil fungi of some British sand dunes in relation to soil type and succession. *J. Ecol., 46:* 641–664.

———, and W. B. Kendrick. 1958. *Gliomastix guttuliformis* sp. nov. *Trans. Brit. Mycol. Soc., 41:* 499–500.

Bunce, M. E. 1961. *Humicola stellatus* sp. nov., a thermophilic mould from hay. *Trans. Brit. Mycol. Soc., 44:* 372–376.

Butler, E. E., R. K. Webster, and J. W. Eckert. 1965. Taxonomy, pathogenicity, and physiological properties of the fungus causing sour rot of citrus. *Phytopathology, 55:* 1262–1268.

Cain, R. F., and N. A. Hastings. 1956. Studies of soil fungi. II. A new species of *Sphaerospora* with a *Botrytis*-like stage. *Can. J. Bot., 34:* 360–376.

Caldwell, R. 1963. Observations on the fungal flora of decomposing beech litter in soil. *Trans. Brit. Mycol. Soc., 46:* 249–261.

Capriotti, A. 1961. *Torulopsis castellii* sp. nov., a yeast isolated from a Finnish soil. *J. Gen. Microbiol., 26:* 41–43.

Carlile, M. J., B. G. Lewis, E. M. Mordue, and J. Northover. 1961. The development of coremia. I. *Penicillium claviforme. Trans. Brit. Mycol. Soc., 44:* 129–133.

Carmichael, J. W. 1957. *Geotrichum candidum. Mycologia, 49:* 820–830.

———. 1962. *Chrysosporium* and some other aleuriosporic Hyphomycetes. *Can. J. Bot., 40:* 1137–1173.

Chesters, C. G. C. 1960. Certain problems associated with the decomposition of soil organic matter by fungi. In *Ecology of Soil Fungi*, University of Liverpool Press, pp. 223–238.

———, and G. N. Greenhalgh. 1964. *Geniculisporium serpens.* gen. et sp. nov., the imperfect state of *Hypoxylon serpens. Trans. Brit. Mycol. Soc., 47:* 393–401.

———, and R. H. Thornton. 1956. A comparison of techniques for isolating soil fungi. *Trans. Brit. Mycol. Soc., 39:* 301–313.

Christensen, M., and M. P. Backus. 1964. Two varieties of *Monocillium humicola* in Wisconsin forest soils. *Mycologia, 56:* 498–504.

———, and W. F. Whittingham. 1965. The soil microfungi of open bogs and conifer swamps in Wisconsin. *Mycologia, 57:* 882–896.

———, ———, and R. O. Novak. 1962. The soil microfungi of wet-mesic forests in Southern Wisconsin. *Mycologia, 54:* 374–388.

Chupp, C. 1953. *A Monograph of the Genus Cercospora.* Cornell University Press, Ithaca, New York.

Ciferri, R. 1958. *Mauginiella* a synonym of *Sporendonema. Atti. Ist. Bot. Univ. Pavia, Ser. 5, 15:* 120–133.

———, and G. Caretta. 1962. Revisione del genere *Polyscytalum* Riess. *Mycopathologia, 16:* 304–314.

Clements, F. E., and C. L. Shear. 1931. *The Genera of Fungi.* H. W. Wilson, New York.

Cooke, R. C., and C. H. Dickinson. 1965. Nematode-trapping species of *Dactylella* and *Monacrosporium. Trans. Brit. Mycol. Soc., 48:* 621–629.

———, and B. E. S. Godfrey. 1964. A key to the nematode-destroying fungi. *Trans. Brit. Mycol. Soc., 47:* 61–74.

Cooke, W. B. 1959. An ecological life history of *Aureobasidium pullulans* (de Bary) Arnaud. *Mycopathologia, 12:* 1–45.

———, and D. B. Lawrence. 1959. Soil mould fungi isolated from recently glaciated soils in south-eastern Alaska. *J. Ecol., 47:* 529–549.

Cooney, D. G., and R. Emerson. 1964. *Thermophilic Fungi: An Account of their Biology, Activities and Classification.* W. H. Freeman and Company, San Francisco, California.

Corlett, M. 1965. Perithecium development in *Chaetomium trigonosporum. Can. J. Bot., 44:* 155–162.

Dale, E. 1926. Note on three new species of *Penicillium. Ann. Mycol. Berl., 24:* 137.

Dal Vesco, G. 1957. "*Geomyces vinaceus*" n. sp., forma conidica di *Pseudogymnoascus vinaceus* Raillo. *Allionia, 3:* 1–15.

———. 1960. Contributo alla conoscenza della microflora di boschi di pino laricio della sila. *Allionia, 6:* 201–226.

———. 1961. Una nuova demaziacea isolata dal suolo, *Stephanosporium atrum* n. gen. et n. sp. descrizione ed osservazioni. *Allionia, 7:* 181–193.

———. 1963. Isolamento dal terreno di due moniliali, nuova per l'Italia. *Giorn. Botan. Ital., 70:* 637–639.

Damon, S. C. 1952. Type studies in *Dictyosporium, Speira,* and *Cattanea. Lloydia, 15:* 110–124.

———. 1952a. Two noteworthy species of *Sepedonium. Mycologia, 44:* 86–96.

Daniels, J. 1961. *Chaetomium piluliferum* sp. nov. the perfect state of *Botryotrichum piluliferum. Trans. Brit. Mycol. Soc., 44:* 79–86.

Das, A. C. 1963. Ecology of soil fungi of rice fields: 1. Succession of fungi on rice roots. 2. Association of soil fungi with organic matter. *Trans. Brit. Mycol. Soc., 46:* 431–443.

Daszewska, W. 1912. Étude sur la désagrégation de la cellulose dans la terre de bruyère et la tourbe. *Bull. Soc. Bot. Geneve II, 4:* 294

Dawson, C. O., and J. C. Gentles. 1961. The perfect state of *Keratinomyces ajelloi* Vanbreuseghem, *Trichophyton terrestre* Durie and Frey and *Microsporum nanum* Fuentes. *Sabouraudia, 1:* 49–57.

Delitsch, H. 1943. *Systematik der Schimmelpilze.* Druck J. Neumann, Neudamm, Germany.

Deshpande, K. B. and K. S. Deshpande. 1966. A new genus of Hyphomycetes from India. *Mycopathologia, 28:* 206–208.

———, and ———. 1966. *Chlamydorubra,* a new genus of Dematiaceae from India. *Mycopathologia, 29:* 270–272.

———, and ———. 1966. *Multicladium,* a new genus of Dematiaceae from soil. *Mycopathologia, 30:* 184–186.

———, and ———. 1966. *Saprophragma*: A new Hyphomycete from soils of Marathwada. *Mycopathologia, 30:* 200–202.

Dickinson, C. H. 1964. The genus *Wardomyces. Trans. Brit. Mycol. Soc., 47:* 321–325.

———. 1966. *Wardomyces pulvinata* comb. nov. *Trans. Brit. Mycol. Soc., 49:* 521–522.

——. 1968. The genus *Gliomastix*. Mycol. Paper, in press. Commonwealth Mycological Institute, Kew, England.
Dingley, J. M. 1962. *Pithomyces chartarum*, its occurrence, morphology, and taxonomy. *New Zeal. J. Agric. Res.*, *5:* 49–61.
Dominik, T., and I. Majchrowicz. 1966. Some new species of fungi from the soil of Conakry, Guinea, Africa. *Mycopathologia*, *28:* 209–219.
Donk, M. A. 1956. Notes on resupinate Hymenomycetes. III. *Fungus, Wageningen*, *26:* 3–24.
Downing, M. H. 1953. *Botryotrichum* and *Coccospora*. *Mycologia*, *45:* 934–940.
Drechsler, C. 1937. Some Hyphomycetes that prey on free-living terricolous nematodes. *Mycologia*, *29:* 446–552.
——. 1943. A new nematode-capturing *Dactylella* and several related Hyphomycetes. *Mycologia*, *35:* 339–362.
——. 1950. Several species of *Dactylella* and *Dactylaria* that capture free-living nematodes. *Mycologia*, *42:* 1–79.
Duddington, C. L. 1951. *Dactylella lobata*, predaceous on nematodes. *Trans. Brit. Mycol. Soc.*, *34:* 489–491.
Durie, E. B. 1964. *Trichophyton terrestre*. *Mycologia*, *56:* 317–318.
Durrell, L. W. 1963. Notes on *Cephalosporium* species. Colorado State University Publication, Fort Collins, Colorado.
Dutta, B. G., and G. R. Ghosh. 1965. Soil fungi from Orissa (India). IV. Soil fungi of paddy fields. *Mycopathologia*, *25:* 316–322.
Ellis, J. J., and C. W. Hesseltine. 1962. A new genus of Moniliales having penicilli subtended by sterile arms. *Bull. Torrey Bot. Club*, *89:* 21–27.
Ellis, M. B. 1957. *Haplobasidion*, *Lacellinopsis*, and *Lacellina*. Mycol. Paper No. 67, Commonwealth Mycological Institute, Kew, England.
——. 1958. *Clasterosporium* and some allied Dematiaceae-Phragmosporae. I. Mycol. Paper No. 70, Commonwealth Mycological Institute, Kew, England.
——. 1959. *Clasterosporium* and some allied Dematiaceae-Phragmosporae. II. Mycol. Paper No. 72, Commonwealth Mycological Institute, Kew, England.
——. 1960. Dematiaceous Hyphomycetes. I. Mycol. Paper No. 76, Commonwealth Mycological Institute, Kew, England.
——. 1961. Dematiaceous Hyphomycetes. III. Mycol. Paper No. 82, Commonwealth Mycological Institute, Kew, England.
——. 1963. Dematiaceous Hyphomycetes. IV. Mycol. Paper No. 87, Commonwealth Mycological Institute, Kew, England.
——. 1965. Dematiaceous Hyphomycetes. VI. Mycol. Paper No. 103, Commonwealth Mycological Institute, Kew, England.
——. 1966. Dematiaceous Hyphomycetes. VII. *Curvularia*, *Brachysporium* etc. Mycol. Paper No. 106, Commonwealth Mycological Institute, Kew, England.
Embree, R. W. 1963. The status of *Gliocephalis*. *Mycologia*, *55:* 127–128.
Emmons, C. W. 1949. Isolation of *Histoplasma capsulatum* from soil. Pub. Health Rep., *64:* 892–896.
——, C. H. Binford and J. P. Utz. 1963. *Medical Mycology*. Lea and Febiger, Philadelphia.
England, C. M., and E. L. Rice. 1957. A comparison of the soil fungi of a tall-grass prairie and an abandoned field in central *Oklahoma*. *Bot. Gaz.*, *118:* 186–190.
Farrow, W. M. 1954. Tropical soil fungi. *Mycologia*, *46:* 632–646.
Fergus, C. L. 1957. *Myrothecium roridum* on gardenia. *Mycologia*, *49:* 124–127.
——. 1960. A note on the occurrence of *Peziza ostracoderma*. *Mycologia*, *52:* 959–961.
Fragoso, R. G., and R. Ciferri. 1927. Hongos parásitos y saprofitos de la República Dominicana. *Bol. Real. Soc. Espan. Host. Nat.*, *27:* 267–280.
Frey, D. 1965. Isolation of keratinophilic and other fungi from soils collected in Australia and New Guinea. *Mycologia*, *57:* 202–215.
——, and D. M. Griffin. 1961. *Ctenomyces serratus* Eidam. *Trans. Brit. Mycol. Soc.*, *44:* 449–452.
Friend, R. J. 1965. What is *Fumago vagans*? *Trans. Brit. Mycol. Soc.*, *48:* 371–375.

Gams, W. 1962. *Mycelium radicis atrovirens* in forests soils, isolation from soil microhabitats and identification. In *Soil Organisms*, North-Holland Publishing Company, Amsterdam.
Gerdemann, J. W. 1953. An undescribed fungus causing a root rot of red clover and other Leguminosae. *Mycologia, 45:* 548–554.
Ghosh, G. R. and B. G. Dutta. 1962. Soil fungi from Orissa (India). I. *Mycologia, 52:* 915–918.
Gilman, J. C. 1957. *A Manual of Soil Fungi*, Ed. 2. Iowa State College Press, Ames, Iowa.
Glenn-Bott, J. I. 1955. On *Helicodendron tubulosum* and some similar species. *Trans. Brit. Mycol. Soc., 38:* 17–30.
Goddard, H. N. 1913. Can fungi living in agricultural soil assimilate free nitrogen? *Bot. Gaz., 56:* 249–305.
Goidanich, G. 1935. Una nova specie di *Ophiostoma* vivente sul pero ed alcune osservazioni sull'estatta posizione sistematica della forma ascofora e delle forme metagenetiche del genere. *R. Staz. Pat. Veg. Bol. Rome (n.s.), 15:* 122–168.
Goos, R. D. 1956. Classification of the Fungi Imperfecti. *Proc. Iowa Acad. Sci., 63:* 311–320.
———. 1962. Soil fungi from Costa Rica and Panama. *Mycologia, 52:* 877–883.
———. 1963. Further observations on soil fungi in Honduras. *Mycologia, 55:* 142–150.
———. 1964. A new record of *Cephaliophora irregularis*. *Mycologia, 56:* 133–136.
———. 1965. Growth and survival of *Histoplasma capsulatum* in soil. *Can. J. Microbiol., 11:* 979–985.
———. 1967. Observations on *Riessia semiophora*. *Mycologia, 59:* 718–722.
———, and E. F. Morris. 1965. *Murogenella terrophila*—a new dematiaceous fungus from soil. *Mycologia, 57:* 776–781.
———, and M. I. Timonin. 1962. Fungi from the rhizosphere of banana in Honduras. *Can. J. Bot., 40:* 1371–1377.
Graniti, A. 1962. *Scolecobasidium anellii* n. sp. agente di annerimenti superficiali di Stalatiti. *Giorn. Bot. Ital., 69:* 360–365.
Gregory, P. H., and M. E. Lacey. 1964. The discovery of *Pithomyces chartarum* in Britain. *Trans. Brit. Mycol. Soc., 47:* 25–30.
Grove, W. B. 1937. *British Stem and Leaf Fungi, Vol. II.* Cambridge University Press, Cambridge, England.
Groves, J. W., and A. J. Skolko. 1946. Notes on seed-borne fungi. IV. *Acremoniella, Chlamydomyces*, and *Trichocladium*. *Can. J. Res. C., 24:* 74–80.
Guba, E. F. 1956. *Monochaetia* and *Pestalotia* vs. *Truncatella, Pestalotiopsis*, and *Pestalotia*. *Ann. Microbiol., 7:* 74–76.
———. 1961. *A Monograph of Monochaetia and Pestalotia.* Harvard University Press, Cambridge, Massachusetts.
Guillemat, J., and J. Montégut. 1956. Première contribution à l'étude de la microflore fongique des sols cultivés. *Ann. Épiphyt., 7:* 471–540.
———, and ———. 1957. Deuxième contribution à l'étude de la microflore fongique des sols cultivés. *Ann. Épiphyt., 8:* 185–207.
Haskins, R. H. 1958. Hyphomycetous fungi: *Volucrispora aurantiaca* n. gen., n. sp.; *V. ornithomorpha* (Trotter) n. comb.; and *Tricellula curvatis* n. sp., with the genus *Tricellula* emended. *Can. J. Microbiol., 4:* 273–285.
Hennebert, G. L. 1962. *Wardomyces* and *Asteromyces*. *Can. J. Bot., 40:* 1203–1216.
Hodges, C. S. 1962. Fungi isolated from southern forest tree nursery soils. *Mycologia, 54:* 221–229.
Hotson, J. W. 1917. Notes on bulbiferous fungi, with a key to described species. *Bot. Gaz., 64:* 265–284.
———. 1942. Some species of *Papulaspora* associated with rots of *Gladiolus* bulbs. *Mycologia, 34:* 391–398.
Howell, A. 1939. Studies on *Histoplasma capsulatum* and similar form species. I. Morphology and development. *Mycologia, 31:* 191–216.
Hudson, H. J. 1963. Pyrenomycetes of sugar cane and other grasses in Jamaica. II. Conidia of *Apiospora montagnei*. *Trans. Brit. Mycol. Soc., 46:* 19–23.
———. 1963. The perfect state of *Nigrospora oryzae*. *Trans. Brit. Mycol. Soc., 46:* 355–360.

Hughes, S. J. 1949. Studies on micro-fungi. I. The genus *Fusariella* Saccardo. Mycol. Paper No. 28, Commonwealth Mycological Institute, Kew, England.
———. 1951. Studies on micro-fungi. IX. *Calcarisporium*, *Verticicladium*, and *Hansfordia* (gen. nov.). Mycol. Paper No. 43, Commonwealth Mycological Institute, Kew, England.
———. 1951a. Studies on micro-fungi. X. *Zygosporium*. Mycol. Paper No. 44, Commonwealth Mycological Institute, Kew, England.
———. 1951b. Studies on micro-fungi. XII. *Triposporium*, *Tripospermum*, *Ceratosporella*, and *Tetrasporium* (gen. nov.). Mycol. Paper No. 46, Commonwealth Mycological Institute, Kew, England.
———. 1951c. *Stachylidium*, *Gonytrichum*, *Mesobotrys*, *Chaetopsis* and *Chaetopsella*. *Trans. Brit. Mycol. Soc.*, *34:* 551–576.
———. 1952. *Dactylosporium* in Britain. *Naturalist*, April-June, 63–64.
———. 1952a. Fungi from the Gold Coast. I. Mycol. Paper No. 48, Commonwealth Mycological Institute, Kew, England.
———. 1952b. Four species of *Septonema*. *Naturalist*, Jan.-Mar., 7–12.
———. 1952c. *Trichocladium* Harz. *Trans. Brit. Mycol. Soc.*, *35:* 152–157.
———. 1953. Conidiophores, conidia and classification. *Can. J. Bot.*, *31:* 577–659.
———. 1953a. Some foliicolous Hyphomycetes. *Can. J. Bot.*, *31:* 560–576.
———. 1955. Microfungi. I. *Cordana*, *Brachysporium*, *Phragmocephala*. *Can. J. Bot.*, *33:* 259–268.
———. 1957. Microfungi. III. *Mammaria* Cesati. *Sydowia Beih.*, *1:* 359–363.
———. 1958. Revisiones Hyphomycetum aliquot cum appendice de nominibus rejiciendis. *Can. J. Bot.*, *36:* 727–836.
———. 1959. Microfungi. IV. *Trichocladium canadense* n. sp. *Can. J. Bot.*, *37:* 857–859.
———. 1960. Microfungi. VI. *Piricauda* Bubák. *Can. J. Bot.*, *38:* 921–924.
———. 1965. New Zealand Fungi. 3. *Catenularia* Grove. *New Zeal. J. Bot.*, *3:* 136–150.
———, and W. B. Kendrick. 1963. Microfungi. IX. *Menispora* Persoon. *Can. J. Bot.*, *41:* 693–718.
———, and ———. 1966. New Zealand Fungi. 4. *Zanclospora* gen. nov. *New Zeal. J. Bot.*, *3:* 151–158.
Hunt, J. 1956. Taxonomy of the genus *Ceratocystis*. *Lloydia*, *19:* 1–58.
Indoh, H., and A. Oyatsu. 1965. On *Heterocephalum aurantiacum* Thaxter, newly found in Okinawa. *Trans. Mycol. Soc. Japan 5:* 79–82.
Ingold, C. T. 1942. Aquatic Hyphomycetes of decaying alder leaves. *Trans. Brit. Mycol. Soc.*, *25:* 339–417.
———. 1956. The conidial apparatus of *Trichothecium roseum*. *Trans. Brit. Mycol. Soc.*, *39:* 460–464.
———, and V. J. Cox. 1957. On *Tripospermum* and *Campylospora*. *Trans. Brit. Mycol. Soc.*, *40:* 317–321.
Ito, S. 1930. On some new ascigerous stages of the species of *Helminthosporium* parasitic on cereals. *Proc. Imp. Acad. Tokyo*, *6:* 352–355.
Jacques, J. E. 1941. Studies in the genus *Heterosporium*. *Contrib. Inst. Bot. Univ. Montreal*, *39:* 1–46.
Jensen, C. W. 1912. Fungus flora of the soil. *N. Y. (Cornell) Agric. Expt. Sta. Bull.*, *315:* 414–501.
Jensen, H. L. 1931. The fungus flora of the soil. *Soil Sci.*, *31:* 123–158.
Jensen, V. 1963. Studies on the micoflora of Danish beech forest soils. IV. Yeasts and yeast-like fungi. *Zb. Bakt.*, *117:* 41–65.
Johnson, J. 1916. Host plants of *Thielavia basicola*. *J. Agric. Res.*, *7:* 289–300.
Johnson, L. F., and T. S. Osborne. 1964. Survival of fungi in soil exposed to gamma radiation. *Can. J. Bot.*, *42:* 105–113.
Joly, P. 1964. Le genre *Alternaria*, *Recherches Physiologiques*, *Biologiques et Systematiques*. Encyclopédie Mycologique No. 33. Paul Lechevalier, Paris.
———. 1964a. Recherches sur la nature et le mode de formation des spores chez le genre *Torula*. *Bull Trimest. Soc. Mycol. Fr.*, *80:* 186–196.
Kamyschko, O. P. 1960. Fungi novi terrestres regionis Leningradensis. *Acad. Sci. U.S.S.R.*, *13:* 162–167.

———. 1961. Genera et species novae fungorum terricolarum e regione Leningradensi. *Bot. Mater.* (*Notul. Syst. Sect. Crypt. Inst. Bot. Acad. Sci. U.S.S.R.*), *14:* 221–227.

———. 1962. De Monilialibus terrestribus novis notula. *Bot. Mater.* (*Notul. Syst. Sect. Crypt. Inst. Bot. Acad. Sci. U.S.S.R.*), *15:* 137–141.

———. 1963. De fungis terrestribus novis notula. *Bot. Mater.* (*Notul. Syst. Sect. Crypt. Inst. Bot. Acad. Sci. U.S.S.R.*), *16:* 95–99.

Kaufman, D. D., and L. E. Williams. 1964. Effect of mineral fertilization and soil reaction on soil fungi. *Phytopathology*, *54:* 134–139.

Kendrick, W. B. 1958. *Sympodiella*, a new Hyphomycete genus. *Trans. Brit. Mycol. Soc.*, *41:* 519–521.

———. 1961. Hyphomycetes of conifer leaf litter. *Thysanophora* gen. nov. *Can. J. Bot.*, *39:* 817–832.

———. 1961a. The *Leptographium* complex. *Phialocephala* gen. nov. *Can. J. Bot.*, *39:* 1079 1085.

———. 1962. The *Leptographium* complex. *Verticicladiella* Hughes. *Can. J. Bot.*, *40:* 771–797.

———. 1963. The *Leptographium* complex. Two new species of *Phialocephala*. *Can. J. Bot.*, *41:* 1015–1023.

———. 1964. The *Leptographium* complex. *Hantzchia* Auerswald. *Can. J. Bot.*, *42:* 1291–1295.

———, and G. C. Bhatt. 1966. *Trichocladium opacum*. *Can. J. Bot.*, *44:* 1728–1730.

Klebahn, H. 1930. Zur Kenntnis einiger Botrytis-Formen vom Typus der *Botrytis cinerea*. *Z. Bot.*, *23:* 251–272.

Korf, R. P. 1960. Nomenclatural notes. IV. The generic name *Plicaria*. *Mycologia*, *52:* 648–651.

Krzemieniewska, H., and L. Badura. 1954. Przyczynek do znajomości mikroorganizmów ściólki i gleby lasu bukowego. *Acta Soc. Bot. Polon.*, *23:* 727–781.

———, and ———. 1954a. Z badàn nad mikoflora lasu bukowego. *Acta Soc. Bot. Polon.*, *23:* 545–587.

Lagerberg, T., G. Lunderg, and E. Melin. 1927. Biological and practical researches into blueing in pine and spruce. *Svensk. Skogsvardsf. Tidskr.*, *25:* 145–272, 561–691.

Langeron, M., and R. Vanbreuseghem. 1952. *Précis de Mycologie.* Masson et Cie, Paris.

Latch, G. C. M. 1965. *Metarrhizium anisopliae* (Metschnikoff) Sorokin strains in New Zealand and their possible use for controlling pasture inhabiting insects. *New Zeal. J. Agric. Res.*, *8:* 384–386.

Le Gal, M., and F. Mangenot. 1961. Contribution à l'étude des Mollisioidées. IV. *Rev. Mycol.*, *26:* 263–331.

Lentz, P. L. 1966. *Dactylaria* in relation to the conservation of *Dactylium*. *Mycologia*, *58:* 965–966.

Limber, D. P. 1940. A new form of the *Moniliaceae*. *Mycologia*, *32:* 23–30.

Lindau, G. 1900. Fungi Imperfecti. In *Die naturlichen Pflanzenfamilien*, Engelman, Leipzig.

Linder, D. H. 1929. A monograph of the helicosporous Fungi Imperfecti. *Ann. Missouri Bot. Gard.*, *16:* 227–388.

———. 1942. A contribution towards a monograph of the genus *Oidium*. *Lloydia*, *5:* 165–207.

Lodder, J., and N. J. W. Kreger-van Rij. 1952. *The Yeasts. A Taxonomic Study*. North-Holland Publishing Company, Amsterdam.

Loughheed, T. C. 1961. The effect of nutrition on synnemata formation in *Hirsutella gigantea* Petch. *Can. J. Bot.*, *39:* 865–873.

———. 1963. Studies on the morphology of synnemata of *Hirsutella gigantea* Petch. *Can. J. Bot.*, *41:* 947–952.

Lucas, M. T., and J. Webster. 1964. Conidia of *Pleospora scirpicola* and *P. valesiaca*. *Trans. Brit. Mycol. Soc.*, *47:* 247–256.

Luttrell, E. S. 1963. Taxonomic criteria in *Helminthosporium*. *Mycologia*, *55:* 643–674.

Luttrell, E. S. 1964. Systematics of *Helminthosporium* and related genera. *Mycologia*, *56:* 119–132.

Maciejowska, Z. 1964. Grzyby z rodzaju *Trichosporon* wyizolowane z gleby torfowej. *Pr. Nauk. Inst. Ochr. Rosl.*, *6:* 61–91.

———., and E. B. Williams. 1963. Studies on morphological forms of *Staphylotrichum coccosporum*. *Mycologia*, *55:* 221–225.

MacLeod, D. M. 1954. Investigations on the genera *Beauveria* Vuill. and *Tritirachium* Limber. *Can. J. Bot.*, *32:* 818–890.

Mains, E. B. 1948. Entomogenous fungi. *Mycologia*, *40:* 402–416.

Majchrowicz, I. 1963. Grzyby towarzyszące obumieraniu niektórych owadów w glebie. *Biul. Inst. Ochr. Rosl. Pozn.*, *24:* 229–230.

Mangenot, F. 1953. Sur quelques Hyphales dématiées lignicoles. *Rev. Mycol.*, *18:* 133–148.

Mason, E. W. 1925. Fungi received at the Imperial Bureau of Mycology. Mycol. Paper No. 1, Commonwealth Mycological Institute, Kew, England.

———. 1933. Annotated account of fungi received at the Imperial Mycological Institute. List II, Fasc. 2, Mycol. Paper No. 3, Commonwealth Mycological Institute, Kew, England.

———. 1937. Annotated account of fungi received at the Imperial Mycological Institute. List II, Fasc. 3. Mycol. Paper No. 3, Commonwealth Mycological Institute, Kew, England.

———. 1941. Annotated account of fungi received at the Imperial Mycological Institute. List II, Fasc. 3 (special part). Mycol. Paper No. 5, Commonwealth Mycological Institute, Kew, England.

———, and M. B. Ellis. 1953. British species of *Periconia*. Mycol. Paper No. 56. Commonwealth Mycological Institute, Kew, England.

Mathur, P. N., and M. J. Thirmulachar. 1960. A new *Emericellopsis* species with *Stilbella*-type of conidia. *Mycologia*, *52:* 694–697.

Matruchot, M. L. 1899. Notes mycologiques. I. *Gliocephalis hyalina*. *Bull. Soc. Mycol. Fr.*, *15:* 254–262.

Matturi, S. T., and H. Stenton. 1964. Distribution and status in the soil of *Cylindrocarpon* species. *Trans. Brit. Mycol. Soc.*, *47:* 577–587.

McLennan, E. I., and S. C. Ducker. 1954. The ecology of the soil fungi of an Australian heathland. *Aust. J. Bot.*, *2:* 220–245.

McVey, D. V., and J. W. Gerdemann. 1960. The morphology of *Leptodiscus terrestris* and the function of setae in spore dispersal. *Mycologia*, *52:* 193–200.

Mehrlich, F. P., and H. M. Fitzpatrick. 1935. *Dichotomophthora portulacae*, a pathogene of *Portulaca oleracea*. *Mycologia*, *27:* 543–550.

Mehrotra, M. D. 1963. A new species of *Beniowskia* from India. *Sydowia*, *17:* 148–150.

———. 1965. A new species of *Nodulisporium*. *Current Sci.*, *34:* 353.

Menna, M. E. 1957. The isolation of yeasts from soil. *J. Gen. Microbiol.*, *17:* 678–688.

———. 1965. Yeasts in New Zealand soils. *New Zeal. J. Bot.*, *3:* 194–203.

Melin, E., and J. A. Nannfeldt. 1934. Researches into the blueing of ground woodpulp. *Saertryck Svenska Skogsvardsforeningens Tidskr.* Haft, *3, 4:* 397–616.

Meredith, D. S. 1962. Spore discharge in *Cordana musae* (Zimm.) Höhnel and *Zygosporium oscheoides* Mont. *Ann. Bot.*, *26:* 233–241.

Meyer, J. A. 1959. Moisissures du sol et des litières de la région de Yangambi (Congo Belge). *Publ. Inst. Nat. Agron. Congo Belge, Sér. Sci.*, *75:* 1–211.

———. 1963. Écologie et sociologie des microchampignons du sol de la Cuvette centrale congolaise. *Publ. Inst. Nat. agron. Congo Belge, Sér. Sci.*, *101:* 1–137.

Miller, J. H., J. E. Giddens, and A. A. Foster. 1957. A survey of the fungi of forest and cultivated soils of Georgia. *Mycologia*, *49:* 779–808.

Misra, P. C. 1967. *Torula terrestris* n. sp. from soil. *Can. J. Bot.*, *45:* 367–369.

———, and P. H. B. Talbot. 1964. *Phialomyces*, a new genus of the Hyphomycetes. *Can. J. Bot.*, *42:* 1287–1290.

Moore, R. T. 1959. The genus *Piricauda* (Deuteromycetes). *Rhodora*, *61:* 87–120.

Moreau, C. 1963. Morphologie comparée de quelques *Phialophora* et variations du *P. cinerescens* (Wr.) van Beyma. *Rev. Mycol.*, *28:* 260–276.

Moreau, F., and F. Moreau. 1941. Première contribution à l'étude de la microflore des dunes. *Rev. Mycol.*, *6:* 49–94.

Morquer, R., G. Viala, J. Rouch, J. Fayret, and G. Bergé. 1963. Contribution à l'étude morphogenique du genre *Gliocladium*. *Bull. Soc. Mycol. Fr.*, *79:* 137–241.

Morris, E. F. 1955. A new genus of Dematiaceae. *Mycologia*, *47:* 602–605.

———. 1963. *The Synnematous Genera of the Fungi Imperfecti*. Western Illinois University, Biological Science Publication No. 3.

Morton, F. J. and G. Smith. 1963. The genera *Scopulariopsis* Bainier, *Microascus* Zukal, and *Doratomyces* Corda. Mycol. Paper No. 86, Commonwealth Mycological Institute, Kew, England.

Mosca, A. M. 1960. Sulla micoflora del terreno di un pascolo alpino in Val di Lanzo. Alpi Graie. *Allionia, 6:* 17–34.

Mosca, A. M. L. 1964. Micoflora di un terreno agrario a poirino (Torino). *Allionia, 10:* 7–16.

Nadakavukaren, M. J., and C. E. Horner. 1959. An alcohol agar medium selective for determining *Verticillium* microsclerotia in soil. *Phytopathology, 49:* 527–528.

Neergaard, P. 1945. *Danish Species of Alternaria and Stemphylium. Taxonomy, Parasitism and Economical Significance.* Einar Munksgaard, Copenhagen.

Nelson, P. E., and S. Wilhelm. 1956. An undescribed fungus causing root rot of strawberry. Mycologia, *48:* 547–551.

Nelson, R. R. 1964. The perfect stage of *Curvularia geniculata. Mycologia, 56:* 777–779.

———, and F. A. Haasis. 1964. The perfect stage of *Curvularia lunata. Mycologia, 56:* 316–317.

Nicholls, V. O. 1956. Fungi of chalk soils. *Trans. Brit. Mycol. Soc., 39:* 233–238.

Nicot, J. 1956. Une moisissure nouvelle isolée des sols africains: *Nodulisporium didymosporum* sp. nov. *Rev. Mycol., 21:* 112–118.

———. 1958. Une moisissure arénicole du Littoral Atlantique: *Dendryphiella arenaria* sp. nov. *Rev. Mycol., 23:* 87–99.

———. 1958 a. Quelques micromycetes des sables littoraux. *Bull. Soc. Mycol. Fr., 74:* 223–235.

———, and J. Meyer. 1956. Un Hyphomycete nouveau des sols tropicaux: *Staphylotrichum coccosporum* nov. gen. sp. *Bull. Soc. Mycol. Fr., 72:* 318–323.

———, and C. Olivry. 1961. Contribution à l'étude du genre *Myrothecium* Tode. I. Les espèces à spores striées. *Rev. Gen. Bot., 68:* 673–685.

———, and A. Parguey. 1963. Obtention de la forme parfaite *Hypomyces* dans des cultures de l'Hyphomycete fungicole *Didymocladium ternatum* (Bon.) Sacc. *C.R. Acad. Sci. Paris, 257:* 1331–1334.

Omvik, A. 1955. Two new species of *Chaetomium* and one new *Humicola* species. *Mycologia, 47:* 748–757.

Onions, A. H. S., and G. L. Barron. 1967. Monophialidic species of *Paecilomyces.* Mycol. Paper No. 107, Commonwealth Mycological Institute, Kew, England.

Orpurt, P. A. 1964. The microfungal flora of bat cave soils from Eleuthera Island, the Bahamas. *Can. J. Bot., 42:* 1629–1633.

Orr, G. F., H. H. Kuehn, and O. A. Plunkett. 1963. The genus *Myxotrichum. Can. J. Bot., 41:* 1457–1480.

Ôta, M. 1928. Champignons parasites de l'homme. (Études morphologiques et systematiques). VI. *Japan. J. Derm. Urol., 28:* 381–424.

Oudemans, C. A., and C. J. Koning. 1902. Prodrome d'une flore mycologique obtenu par la culture sur gelatine préparée de la terre humeuse du Spanderswoud près Bussum. *Arch. Néerl. Sci. Nat., Ser. 2, 7:* 286–298.

Papavizas, G. C., and C. B. Davey. 1961. Saprophytic behaviour of *Rhizoctonia* in soil. *Phytopathology, 51:* 693–699.

Papendorf, M. C. 1967. Two new genera of soil fungi from South Africa. *Trans. Brit. Mycol. Soc., 50:* 69–75.

Parker, A. K. 1957. *Europhium,* a new genus of the Ascomycetes with a *Leptographium* imperfect state. *Can. J. Bot., 35:* 173–179.

Parmeter, J. R., H. S. Whitney, and W. D. Platt. 1967. Affinities of some *Rhizoctonia* species that resemble mycelium of *Thanatephorus cucumeris. Phytopathology, 57:* 218–223.

Petch, T. 1930. Notes on entomogenous fungi. *Trans. Brit. Mycol. Soc., 16:* 55–75.

———. 1934. *Isaria. Trans. Brit. Mycol. Soc., 19:* 34–35.

Petersen, R. H. 1962. Spore formation in *Tricellula* van Bever. and *Volucrispora* Haskins. *Bull. Torrey Bot. Club, 89:* 287–293.

Peyronel, B. 1916. Una nuova mallattia del lupino prodotta da *Chaluropsis thielavioides* Peyr. nov. gen. et nov. sp. *Staz. Sper. Agric. Ital., 49:* 583–596.

———. 1918. *Helicodendron paradoxum. Nuovo G. Bot. Ital.* (n.s.), *25:* 460.

———, and G. Dal Vesco. 1955. Ricerche sulla micoflora di un terreno agrario presso Torino. *Allionia, 2:* 357–417.
Pirozynski, K. A. 1963. *Beltrania* and related genera. Mycol. Paper No. 90, Commonwealth Mycological Institute, Kew, England.
Preston, N. C. 1943. Observations on the genus *Myrothecium* Tode. I. The three classic species. *Trans. Brit. Mycol. Soc., 26:* 158–168.
———. 1961. Observations on the genus *Myrothecium.* III. The cylindrical-spored species of *Myrothecium* known in Britain. *Trans. Brit. Mycol. Soc., 44:* 31–41.
Pugh, G. J. F. 1962. Studies on fungi in coastal soils. I. *Cercospora salina* Sutherland. *Trans. Brit. Mycol. Soc., 45:* 255–260.
———. 1962a. Studies on fungi in coastal soils. II. Fungal ecology in a developing salt marsh. *Trans. Brit. Mycol. Soc. 45:* 560–566.
———, J. P. Blakeman, and G. Morgan-Jones. 1964. *Thermomyces verrucosus* sp. nov. and *T. lanuginosus. Trans. Brit. Mycol. Soc., 47:* 115–121.
———, ———, ———, and H. O. W. Eggins. 1963. Studies on fungi in coastal soils. IV. Cellulose decomposing species in sand dunes. *Trans. Brit. Mycol. Soc. 46:* 565–571.
———, and C. H. Dickinson. 1965. Studies on fungi in coastal soils. VI. *Gliocladium roseum* Bainier. *Trans. Brit. Mycol. Soc., 48:* 279–285.
———, and J. Nicot. 1964. Studies on fungi in coastal soils. V. *Dendryphiella salina* (Sutherland) comb. nov. *Trans. Brit. Mycol. Soc., 47:* 263–267.
Purdy, L. H. 1955. A broader concept of the species *Sclerotinia sclerotiorum* based on variability. *Phytopathology 45:* 421–427.
Rai, J. N., K. G. Mukerji, and J. P. Tewari. 1961. Two new records in soil fungi. *Current Sci., 30:* 231–232.
———, and J. P. Tewari. 1963. A new *Acremoniella* from Indian soils. *Can. J. Bot., 41:* 331–334.
Rall, G. 1965. Soil fungi from the alpine zone of the Medicine Bow Mountains, Wyoming. *Mycologia, 57:* 872–881.
Ramarao, P. 1962. *Beltraniella humicola* sp. nov. *Current Sci., 31:* 479–480.
Rao, P. N. 1966. A new species of *Dichotomophthora* on *Portulaca oleracea* from Hyderabad, India. *Mycopathologia 28:* 137–140.
Rao, P. R. 1963. A new species of *Myrothecium* from soil. *Antonie van Leeuwenhoek, 29:* 180–182.
Rambelli, A. 1956. *Chaetopsina,* nuovo genere de ifali demaziacei. *Atti. Accad. Sci. Bolog. Ser. XI, 3:* 191–196.
———. 1960. Su di una interessante helicosporea isolata da terreno ad Eucalitto: *Helicosporina veronae* n. sp., in coltura pura. *Mycopathologia, 13:* 107–112.
Raper, K. B., and D. I. Fennell. 1952. Two noteworthy fungi from Ligerian soil. *Amer. J. Bot., 39:* 79–86.
———, and ———. 1965. *The Genus Aspergillus.* The Williams & Wilkins Co., Baltimore.
———, and C. Thom. 1949. *A Manual of the Penicillia.* The Williams & Wilkins Co., Baltimore.
Rapilly, F. 1966. Limites proposées au genre *Helminthosporium,* relations avec les genres voisins. *Bull. Soc. Mycol. Fr., 82:* 221–240.
Rayss, T., and S. Borut. 1958. Contribution to the knowledge of soil fungi in Israel. *Mycopathologia, 10:* 142–174.
Redaelli, P., and R. Ciferri. 1934. Studio comparativo di ventum ceppi di *Sporendonema epizoum* (Corda) Nobis (=*Hemispora stellata* Vuill; *Oospora d'Agatae* Sacc. etc.). *Atti Inst. Bot. Univ. Pavia (Ser. 4), 5:* 145–198.
Rifai, M. A. 1964. *Stachybotrys bambusicola* sp. nov. *Trans. Brit. Mycol. Soc., 47:* 269–272.
———, and R. C. Cooke. 1966. Studies on some didymosporous genera of nematode-trapping Hyphomycetes. *Trans. Brit. Mycol. Soc., 49:* 147–168.
Robak, H. 1932. Investigations regarding fungi on Norwegian ground woodpulp and fungal infections at woodpulp mills. *Saert. Nyt Mag. Naturv., 71:* 185–330.
Routien, J. B. 1957. Fungi isolated from soils. *Mycologia, 49:* 188–196.
Roy, R. Y., R. S. Dwivedi, and R. R. Mishra. 1962. Two new species of *Scolecobasidium* from soil. *Lloydia, 25:* 164–166.

———, and S. Gujarati. 1965. A new species of *Dactylaria* from soil. *Lloydia, 28:* 53–54.
———, and ———. 1966. *Abgliophragma setosum* gen. et sp. nov., from soil. *Trans. Brit. Mycol. Soc., 49:* 363–365.
———, and K. M. Leelavathy. 1966. *Phialotubus microsporus* gen. et sp. nov., from soil. *Trans. Brit. Mycol. Soc., 49:* 495–498.
Saccardo, P. A. 1886. *Sylloge Fungosum, 4*, Pavia.
———. 1906. *Sylloge Fungorum, 18*, Pavia.
Saksena, H. K., and O. Vaartaja. 1960. Descriptions of new species of *Rhizoctonia. Can. J. Bot., 38:* 931–943.
Saksena, S. B. 1955. A new fungus, *Monocillium indicum* gen. et sp. nov., from soil. *Ind. Phytopath., 8:* 8–12.
Schanderl, H. 1936. Untersuchungen über die systematische Stellung und die Physiologie des Kellerschimmels *Rhacodium cellare* Persoon. *Zbl. Bakt. Abt., 2:* 112–127.
Schmidt, E. W. 1910. *Oedocephalum glomerulosum* Harz. Nebenfruchtform zu *Pyronema omphalodes* (Bull.). *Fuckel. Zbl. Bakt., 25:* 80–85.
Schneider, R. 1954. *Plicaria fulva* n. sp., ein bisher nicht bekannter Gewächshausbewohner. *Zentralb. Bakt. II, 108:* 147–153.
Schol-Schwarz, M. B. 1959. The genus *Epicoccum* Link. *Trans. Brit. Mycol. Soc., 42:* 149–173.
Semeniuk, G., and J. W. Carmichael. 1966. *Sporotrichum thermophile* in North America. *Can. J. Bot., 44:* 105–108.
Sewell, G. W. F. 1959. Studies of fungi in a *Calluna*-heathland soil. I. Vertical distribution in soil and on root surfaces. *Trans. Brit. Mycol. Soc., 42:* 343–353.
Shear, C. L., and B. O. Dodge. 1921. The life history, and identity of '*Patellina fragariae*', '*Leptothyrium macrothecium*', and '*Peziza oenotherae.*' *Mycologia, 13:* 135–170.
Shoemaker, R. A. 1959. Nomenclature of *Drechslera* and *Bipolaris*, grass parasites segregated from *Helminthosporium. Can. J. Bot., 37:* 879–887.
———. 1962. *Drechslera* Ito. *Can. J. Bot., 40:* 809–836.
———. 1964. Staining asci and annellophores. *Stain Tech., 39:* 120–121.
Sidorova, I. I., M. V. Gorlenko, and L. N. Nalepina. 1964. K. sistematike rodov *Trichothecium* Link i *Arthrobotrys* Corda. *Bot. Zh. S.S.S.R., 49:* 1592–1599.
Simmons, E. G. 1967. Typification of *Alternaria, Stemphylium* and *Ulocladium. Mycologia, 59:* 67–92.
Smalley, E. B., and H. N. Hansen. 1957. The perfect stage of *Gliocladium roseum. Mycologia, 49:* 529–533.
Smith, G. 1952. *Masoniella* nom. nov. *Trans. Brit. Mycol. Soc., 35:* 237.
———. 1960. *An Introduction to Industrial Mycology*, Ed. 5. Edward Arnold, London.
———. 1961. Some new and interesting species of microfungi. II. *Trans. Brit. Mycol. Soc., 44:* 42–50.
———. 1962. Some new and interesting species of microfungi. III. *Trans. Brit. Mycol. Soc., 45:* 387–394.
Srinivasan, K. V. 1958. Fungi of the rhizosphere of sugarcane and allied plants. I. *Hyalostachybotrys* gen. nov. *J. Ind. Bot. Soc., 37:* 334–342.
Stenton, H. 1953. The soil fungi of Wicken Fen. *Trans. Brit. Mycol. Soc., 36:* 304–314.
Steyaert, R. L. 1949. Contribution à l'étude monographique de *Pestalotia* de Not. et *Monochaetia* Sacc. (*Truncatella* gen. nov. et *Pestalotiopsis* gen. nov.). *Bull. Jard. Bot. Brux., 19:* 285–354.
———. 1955. *Pestalotia, Pestalotiopsis,* et *Truncatella. Bull. Jard. Bot. Brux., 25:* 191–199.
———. 1956. A reply and an appeal to Professor Guba. *Mycologia, 48:* 767–768.
Stockdale, P. M. 1963. The *Microsporum gypseum* complex (*Nannizzia incurvata* Stockd., *N. gypsea* (Nann.) comb. nov., *N. fulva* sp. nov.). *Sabouraudia, 3:* 114–126.
Stotzky, G., R. D. Goos, and M. I. Timonin. 1962. Microbial changes occurring in soil as a result of storage. *Plant Soil, 16:* 1–18.
Subramanian, C. V. 1953. Fungi Imperfecti from Madras. IV. *Proc. Ind. Acad. Sci., 37:* 96–105.
———. 1954. Fungi Imperfecti from Madras. VI. *J. Ind. Bot. Soc., 33:* 36–42.
———. 1955. Some species of *Periconia* from India. *J. Ind. Bot. Soc., 34:* 339–361.

———. 1956. *Phaeotrichonis*, a new genus of the Dematiaceae. *Proc. Ind. Acad. Sci. Sect. B, 44:* 1–2.
———. 1956a. Hyphomycetes. II. *J. Ind. Bot. Soc., 35:* 446–494.
———. 1956b. On *Spicularia terrestris* Timonin. *Proc. Ind. Acad. Sci., 43:* 276–278.
———. 1961. *Stemphylium* and *Pseudostemphylium*. *Current Sci., 30:* 423–424.
———. 1962. The classification of the Hyphomycetes. *Bull. Bot. Surv. India, 4:* 249–259.
———. 1963. *Dactylella, Monacrosporium* and *Dactylina*. *J. Ind. Bot. Soc., 42:* 291–300.
———, and B. C. Lodha. 1964. Four new coprophilous Hyphomycetes. *Antonie van Leeuwenhoek, 30:* 317–330.
———, and ———. 1964. Two interesting Hyphomycetes. *Can. J. Bot., 42:* 1057–1063.
Sukapure, R. S., and M. J. Thirmulachar. 1966. Conspectus of species of *Cephalosporium* with particular reference to Indian species. *Mycologia, 58:* 351–361.
Sutton, B. C. 1961. Coelomycetes. I. Mycol. Paper No. 80, Commonwealth Mycological Institute, Kew, England.
———. 1963. Coelomycetes. II. *Neobarclaya, Mycohypallage, Bleptosporium* and *Cryptostictis*. Mycol. Paper No. 88., Commonwealth Mycological Institute, Kew, England.
———. 1964. *Melanconium* Link ex Fr. *Persoonia, 3:* 193–198.
Swart, H. J. 1958. A new species of *Beltrania* from soil. *Antonie van Leeuwenhoek, 24:* 221–224.
———. 1959. A comparative study of the genera *Gonytrichum* and *Bisporomyces*. *Antonie van Leeuwenhoek, 25:* 439–444.
———. 1964. A study of the production of coremia in three species of the genus *Trichurus*. *Antonie van Leeuwenhoek, 30:* 257–260.
———. 1965. Conidial formation in *Haplographium fuligineum*. *Trans. Brit. Mycol. Soc., 48:* 459–461.
Swift, M. E. 1929. Contributions to a mycological flora of local soils. *Mycologia, 21:* 204–221.
Szilvinyi, A. von. 1941. Milkrobiologische Bodenuntersuchungen in Lunzer Gebiet, *Zentralbl. Bakt. Parasit. Inf. (Abt. II), 103:* 133–189.
Taber, W. A. 1961. Nutritional factors affecting the morphogenesis of the synnema. *Recent Advan. Bot., 1:* 289–293.
———, and L. C. Vining. 1959. Studies on *Isaria cretacea*. Nutritional and morphological characteristics of two strains and morphogenesis of the synnemata. *Can. J. Microbiol., 5:* 513–535.
Thaxter, R. 1903. New or peculiar North American Hyphomycetes. III. *Bot. Gaz. 35:* 153–159.
Thompson, K. V. A., and S. C. Simmens. 1962. Appendages on the spores of *Myrothecium verrucaria*. *Nature, 193:* 196–197.
Thorton, R. H. 1956. Fungi occurring in mixed oakwood and heath soil profiles. *Trans. Brit. Mycol. Soc., 39:* 485–494.
Thrower, L. B. 1954. Rhizosphere effect of heathland plants. *Aust. J. Bot., 2:* 246–267.
Timonin, M. I. 1940. The interaction of higher plants and soil microorganisms. I. Microbial population of rhizosphere of seedlings of certain cultivated plants. *Can. J. Res. C., 18:* 307–317.
———. 1961. New species of the genus *Pseudobotrytis*. *Ceiba, 9:* 27–29.
Traaen, A. E. 1914. Untersuchungen über Bodenpilze aus Norwegen. *Nyt. Magaz. Naturvidensk., 52:* 19–120.
Tsiklinsky, P. 1899. Sur les mucédinées thermophiles. *Ann. Inst. Pasteur, 13:* 500–505.
Tubaki, K. 1955. Studies on Japanese Hyphomycetes. II. Fungicolous group. *Nagao, 5:* 11–40.
———. 1958. Studies on Japanese Hyphomycetes. V. Leaf and stem group with a discussion of the classification of Hyphomycetes and their perfect stages. *J. Hattori Bot. Lab., 20:* 142–244.
———. 1960. An undescribed species of *Hypomyces* and its conidial stage. *Nagao, 9:* 29–34.
———. 1961. On some fungi isolated from the Antarctic Materials. Special Publication, No. 14, Seto Marine Biological Laboratory, pp. 3–9.

———. 1963. Taxonomic study of Hyphomycetes. *Ann. Rep. Inst. Fermentation, Osaka, 1:* 25–54.

———. 1963a. Notes on the Japanese Hyphomycetes. I. *Chloridium, Clonostachys, Isthmospora, Pseudobotrytis, Stachybotrys* and *Stephanoma. Trans. Mycol. Soc. Japan, 4:* 83–90.

———, and I. Asana. 1965. Additional species of fungi isolated from Antarctic materials. *Jare Sci. Rept. Ser. E, No. 27.* National Science Museum, Tokyo.

Upadhyay, H. P. 1966. Soil fungi from North-East Brazil. II. *Mycopathologia, 30:* 276–286.

———. 1966a. Soil fungi from North-East Brazil. V. Two fungi parasitic on Nematodes. *Inst. Micol. Univ. Recife, Brasil, No. 493.*

Vaidehi, B. K., A. V. Lakshminarasimhan, and P. Ramarao. 1967. *Murogenella terrophila* from rhizosphere of paddy. *Current Sci., 36:* 163.

Vanbreuseghem, R. 1952. Technique biologique pour l'isolément des dermatophytes du sol. *Ann. Soc. Belge Méd. Trop., 32:* 173–178.

———. 1952a. Intérêt théorique et pratique d'un nouveau dermatophyte isolé du sol: *Keratinomyces ajelloi* gen. nov., sp. nov. *Bull. Acad. Belg. Clin. Sci., 38:* 1068–1077.

De Vries, G. A. 1952. *Contribution to the Knowledge of the Genus Cladosporium Link ex Fr.* Vitgeverij and Drukkerij, Hollandia Press, Baarn.

———. 1962. *Cyphellophora laciniata* nov. gen., nov. sp. and *Dactylium fusarioides* Fragoso et Ciferri. *Mycopathologia, 16:* 47–54.

———. 1964. Keratinophylic fungi. *Ann. Soc. Belge Méd. Trop. 44:* 795–802.

Vuillemin, P. 1910. Les conidiospores. *Bull. Soc. Sci. Nancy, 11:* 129–172.

———. 1911. Les aleuriospores. *Bull. Soc. Sci. Nancy, 12:* 151–175.

Wakefield, E. M., and G. R. Bisby. 1941. List of Hyphomycetes recorded for Britain. *Trans. Brit. Mycol. Soc., 25:* 49–126.

Wang, C. J. K. 1965. Fungi of pulp and paper in New York. State Univ. Coll. Forestry Tech. Publ. No. 87.

———. 1966. Annellophores in *Torula jeanselmei. Mycologia, 58:* 614–621.

Warcup, J. H. 1957. Studies on the occurrence and activity of fungi in a wheat-field soil. *Trans. Brit. Mycol. Soc., 40:* 237–262.

———, and P. H. B. Talbot. 1962. Ecology and identity of mycelia isolated from soil. *Trans. Brit. Mycol. Soc., 45:* 495–518.

———, and ———. 1963. Ecology and identity of mycelia isolated from soil. II. *Trans. Brit. Mycol. Soc., 46:* 465–572.

Watson, P. 1955. *Calcarisporium arbuscula* living as an endophyte in apparently healthy sporophores of *Russula* and *Lactarius. Trans. Brit. Mycol. Soc., 38:* 409–414.

———. 1965. Further observations on *Calcarisporium arbuscula. Trans. Brit. Mycol. Soc., 48:* 9–17.

Webster, J. 1956. Conidia of *Acrospermum compressum* and *A. gramineum. Trans. Brit. Mycol. Soc., 39:* 361–366.

———. 1964. Culture studies on *Hypocrea* and *Trichoderma.* I. Comparison of perfect and imperfect states of *H. gelatinosa, H. rufa* and *Hypocrea* sp. 1. *Trans. Brit. Mycol. Soc., 47:* 75–96.

———, and N. Lomas. 1964. Does *Trichoderma viride* produce gliotoxin and viridin? *Trans. Brit. Mycol. Soc., 47:* 535–540.

———, M. A. Rifai, and M. Samy El-Abyad. 1964. Culture observations on some Discomycetes from burnt ground. *Trans. Brit. Mycol. Soc., 47:* 445–454.

Westcott, C. 1960. *Plant Disease Handbook,* Ed. 2. D. van Nostrand Company, New York.

Whetzel, H. H. 1945. A synopsis of the genera and species of the Sclerotinaceae, a family of stromatic inoperculate Discomycetes. *Mycologia, 37:* 648–714.

———, and F. L. Drayton. 1932. A new species of *Botrytis* on rhizomatous Iris. *Mycologia, 24:* 469–476.

White, W. L., and M. H. Downing. 1951. *Coccospora agricola* Goddard, its specific status, relationships and cellulolytic activity. *Mycologia, 43:* 645–657.

———, and ———. 1953. *Humicola grisea,* a soil inhabiting cellulolytic Hyphomycete. *Mycologia, 45:* 951–963.

Williams, L. E., and A. F. Schmitthenner. 1956. Genera of fungi in Ohio soils. University of Ohio, Res. Circular No. 39.
Wilsenach, R., and M. Kessel. 1965. Microspores in the cross-wall of *Geotrichum candidum*. *Nature*, *207:* 545–546.
Wiltshire, S. P. 1938. The original and modern concepts of *Stemphylium*. *Trans. Brit. Mycol. Soc.*, *21:* 211–239.
Windisch, S. 1951. Zur Biologie und Systematik des Milchschimmels und einiger ahnlicher Formen I. *Beitr. Biol. Pflanzen*, *28:* 69–130.
Wolf, F. A. 1955. Another *Mycotypha*. *J. Elisha Mitchell Soc.*, *71:* 213–217.
Wright, E. F., and R. F. Cain. 1961. New species of the genus *Ceratocystis*. *Can. J. Bot. 31:* 1215–1230.
Yadav, A. S. 1960. *Dactylella arnaudii* sp. nov. *Trans. Brit. Mycol. Soc.*, *43:* 603–606.
Yarwood, C. E. 1946. Isolation of *Thielaviopsis* from soil by means of carrot disks. *Mycologia*, *38:* 346–348.
Zuck, R. K. 1946. Isolates between *Stachybotrys* and *Memnoniella*. *Mycologia*, *38:* 69–76.

# Index

Page numbers set in **boldface** type indicate principal content.